EVOLUTION

EVOLUTION

THE STORY OF LIFE

DOUGLAS PALMER

ILLUSTRATED BY PETER BARRETT

MITCHELL BEAZLEY

IN ASSOCIATION WITH
THE NATURAL HISTORY MUSEUM, LONDON

EVOLUTION: THE STORY OF LIFE

First published in Great Britain in 2009 by Mitchell Beazley,
an imprint of Octopus Publishing Group Limited,
2–4 Heron Quays, London E14 4JP
www.octopusbooks.co.uk

An Hachette Livre UK Company
www.hachettelivre.co.uk

ISBN: 978 184533 3393

A CIP catalogue record for this book is available from the British Library.

Designed by Philip Gilderdale

Commissioning Editor: Peter Taylor, Jon Asbury
Art Direction: Yasia Williams-Leedham, Tim Foster
Project Manager: Giles Sparrow
Picture Researcher: Jenny Faithfull
Proofreader: Karin Fancett
Indexer: Sue Farr
Production Manager: Peter Hunt

Typeset in Kievit and Emona
Colour reproduction by Altaimage, London
Printed and bound by Toppan, China

CONTENTS

THE STORY OF EVOLUTION: DETAILED CONTENTS

FOREWORD

IN NOVEMBER 1859, AN EXTRAORDINARY BOOK WAS PUBLISHED THAT WOULD FOREVER CHANGE THE WAY IN WHICH WE SEE THE WORLD. *THE ORIGIN OF SPECIES BY MEANS OF NATURAL SELECTION*, WRITTEN BY THE 50-YEAR-OLD CHARLES DARWIN, MARSHALLED A HUGE AMOUNT OF EVIDENCE IN SUPPORT OF A NEW THEORY OF LIFE AND ITS DEVELOPMENT – EVOLUTION BY NATURAL SELECTION.

Darwin's theory, which has only been strengthened by a further 150 years of research and discovery, provided the key to understanding the story of life on Earth, but it has taken a long time to piece together the details of this tale, and it is still far from complete. Several independent lines of evidence, including the record of fossils preserved in rocks, and the biology and genetics of living organisms, have helped to fill in the gaps.

Evolution focuses on the history of life as preserved in fossils, and in particular on its development over the last 600 million years or so. But we must not forget that there is good fossil evidence to show that life originated over 3500 million years ago – it took almost three billion years to evolve into visible multicellular organisms, and for our planet to develop conditions that were hospitable enough for complex life, not only to survive but to move out of the sea and on to land.

When Darwin outlined his theory, he avoided saying much about fossils as evidence for evolution. He was all too aware that fossil record of the time was still very patchy – and indeed that many of the expert paleontologists of the day were deeply antagonistic to the very idea of evolution.

Since then, the predictive power of the theory of evolution has been verified time and time again by a succession of outstanding discoveries in modern biology, genetics, and paleontology – I think that Darwin would be pleased. *Evolution* illustrates the amazing story of life's evolutionary descent from a common ancestry, through a selection of 100 privileged windows into the past, preserved in fossil-bearing rock strata at sites around the world.

Our "snapshots" in time depict the vicissitudes of the struggle for survival and adaptation amid the Earth's ever-changing environments. Thanks to the phenomenon of evolution, life has managed to survive the most violent setbacks, and even when most species have been wiped out in catastrophic extinction events, abundant reproduction has allowed survivors to repopulate empty environments and habitats, diversifying into new forms as they do so.

We can see the same processes at work today, on the once-bare flanks of Mount St. Helens, the tsunami-swept coasts of the Indian Ocean, and more recently the wildfire-charred landscapes of Australia. Life soon regains a hold, and there are very few parts of Earth that are entirely barren. Just as moss and lichen can grow on bare rock surfaces provided there is light and some water, so all living organisms can continue to evolve and adapt.

Evolution tells the remarkable success story of life on our dynamic planet, but the story is still far from complete. Less than one percent of the fossil record has been uncovered and described, and three billion years of Precambrian history, chronicling the origins and early development of life, are still almost entirely unknown. Beyond the boundaries of our current knowledge, there is still a great terra incognita for future generations to explore.

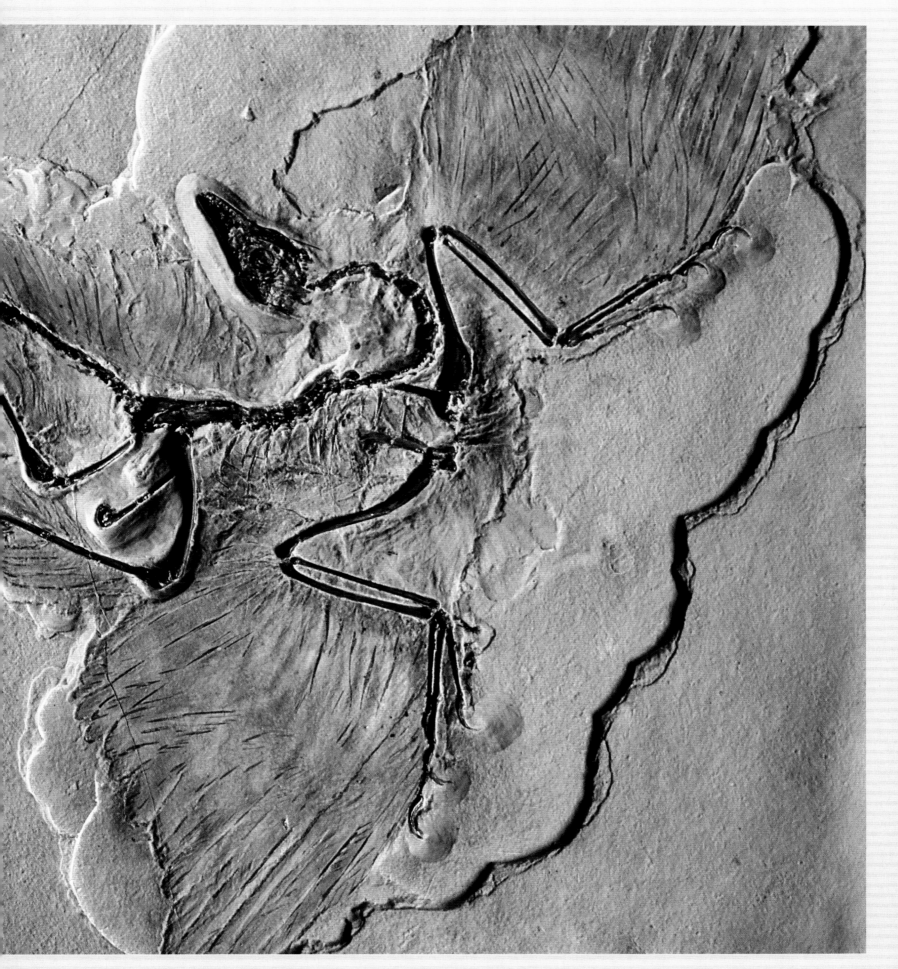

WHAT IS EVOLUTION?

PLANET EARTH HAS EXISTED FOR MORE THAN 4.5 BILLION YEARS, AND LIFE OF ONE SORT OR ANOTHER HAS FLOURISHED ON ITS SURFACE FOR MOST OF THAT TIME. THE FOSSIL RECORD SHOWS THAT ORGANISMS INHABITING THE EARTH HAVE UNDERGONE PROFOUND CHANGE THROUGH TIME. THIS IS THE PHENOMENON THAT TODAY WE CALL "EVOLUTION".

Reproduction can result in remarkable similarities between offspring, just "like peas in a pod", that prove an underlying common genetic makeup. Generally these can also be linked back directly to one or other of the parents (*above*).

While some characters tend to dominate offspring, a minority may show "recessive" traits that are less commonly expressed (*above*).

The competition between a fast predator as a cheetah and a wild pig might seem, very unequal. However the reproductive success and population size of plant-eating wild pigs greatly outnumbers that of top predators whose hunting does not always succeed (*right*).

Throughout the history of life, organisms have changed in size, number, and diversity – present-day life represents only a snapshot from an ongoing process. From its origins as single-celled microbes in the oceans, life has diversified through more than 3.5 billion years to the panoply of organisms that have today conquered land, freshwater, and airways. The catalogue of life is huge, unfinished, and perhaps unfinishable. So far, scientists have *described* two million species from a total of perhaps ten million.

Life has diversified to fill all of Earth's varied and challenging environments, from ocean depths to mountain peaks, through polar wastes and baking deserts, to boiling mud pools and arid salt lakes. Animals range from millimetre-sized insects to 30 metre long (100ft) whales that weigh over 150 tonnes. The entire range of life is even more extreme – from microscopic bacteria to towering trees that can weigh as much as 2000 tonnes.

EXPLAINING VARIETY

As the abundance and diversity of life on Earth became apparent in the 18th century, largely as a result of the age of European colonization and the ensuing scientific "Enlightenment", naturalists struggled to make sense of its profusion by grouping like with like in various classification schemes. At around the same time, a growing awareness of the fossil record made it clear that life also has a history of constant change – change that demanded an explanation. One possibility (known as "descent with modification") was that species changed over time, so that today's animals and plants were the descendants of those in the fossil record. One early advocate of this 'evolutionary' idea, around 1800, was French biologist Jean-Baptiste Lamarck.

However, it was not until the mid-19th century that a viable theory to explain such biological change over time emerged. Two British naturalists, Charles Darwin and Alfred Russel Wallace, independently came up with the same approach to the problem, first published in an 1858 paper entitled "On the tendency of species to form varieties; and on the perpetuation of varieties and species by means of natural selection".

While Wallace had stumbled on the mechanism for evolution around 1850, Darwin had in fact been developing his ideas since 1837, soon after his return from the great *Beagle* survey expedition. Therefore he was in a far better position than Wallace to expand their initial theory, and in 1859 he published the now-famous *The Origin of Species by Means of Natural Selection, or the Preservation of Favoured Races in the Struggle for Life*.

NATURAL SELECTION

For Darwin, the root cause of change in the history of life was the "natural selection" of favourable variations that

Breeding of fruit flies has shown that genetic mutations, such as the forked bristles on the face of the left hand fly, can arise spontaneously in populations (*above*).

occur spontaneously and randomly in offspring. By 1838, Darwin had realized that the interaction between these variations and the changing environment produced constant instability in nature, with "those forms slightly favoured getting the upper hand and forming species".

As Darwin was well aware, individuals produce offspring that are similar to themselves, and in greater numbers than can survive to reproduce as adults in a descendent population. Individuals within a population vary, and much of this variation is heritable – passed on to the offspring along with novel genetic features known as mutations that have arisen spontaneously during reproduction.

But, since living space on Earth is finite, there is competition both within and between populations for space and resources. Those individuals that possess favourable "characters" tend to out-compete others and pass their beneficial characters on to offspring who are, in turn, more successful than less well-endowed individuals.

The human population has exploded and diversified over the last 10,000 years from a million or so to over six billion. Separation of human populations and exposure to different physical and cultural influences has introduced differences in appearance, language, and customs, but genetically we are still the same species, *Homo sapiens*, and are capable of interbreeding (*above*).

Furthermore, since the environment varies in time and space, the heritable characters that fit a particular environment are "selected for" in that place. Thus populations in different environments diverge from one another, and each becomes increasingly well adapted to its circumstances. This process has, over geological time, produced all the diverse life found on Earth.

Since its publication 150 years ago, Darwin's theory has come to unify and direct our understanding of biology and the history of life. In 1866, Austrian priest Gregor Mendel revealed the results of groundbreaking experiments that laid the foundation for modern genetics (though they went unrecognized for several decades). In 1953, Cambridge scientists Francis Crick and James Watson produced the first accurate model of DNA, the complex molecule responsible for passing on gnentic traits and driving evolution. Today, there is a wealth of evidence from the biological sciences, and particularly genetics and paleontology, to support Darwin's great theory.

PUTTING LIFE IN ORDER

OVER TIME, EVOLUTION ACTS TO TRANSFORM ONE SPECIES INTO ANOTHER, AND CAUSES DIFFERENT POPULATIONS TO DIVERGE FROM COMMON ANCESTORS. IN ORDER TO UNDERSTAND HOW THESE PROCESSES WORK, WE MUST FIRST BE ABLE TO IDENTIFY ONE SPECIES FROM THE OTHER, AND WORK OUT WHICH ARE MOST CLOSELY RELATED. AFTER 150 YEARS OF THIS ONGOING PROCESS, MUCH WORK REMAINS TO BE DONE.

Swedish botanist Carl Linnaeus systematically ordered life into genera and species (*above*).

Although Linnaeus introduced a modern approach to classification, he still included fantastical creatures such as troglodytes and Lucifer into his classification of human relatives or "Anthropomorpha" (*above*).

With several million living organisms populating Earth today, and countless extinct fossil forms, it is often convenient to use a hierarchical scheme of classification to group like with like. Depending upon the criteria used in these classifications, they can demonstrate the true closeness of relationships between organisms or simply be a diagnostic key for identification, such as is commonly used for plants. Of course, a prerequisite for any such scheme is the ability to identify different organisms so that they can be distinguished one from another.

People have always found it useful to name plants and animals for practical reasons – identifying potential foodstuffs or poisons for example. Such nomenclature tends to use local or vernacular names, but for successful communication and consistency of identification, an internationally accepted nomenclature is needed. In fact, naturalists have been using just such a scheme for several centuries, based on binomial classification in Latin. In this system, an organism is given separate genus and species names – for example we humans are scientifically known as

Homo sapiens. Translated, this name means "wise man" – *Homo* ("man") is the genus name and *sapiens* ("wise") the species name.

This naming process acknowledges that many organisms, while identifiable as separatex species, are so similar to one another that they share a common genus. For instance, some ten species have now been identified and named within our genus *Homo*, of which some of the best known are *Homo neanderthalensis* (Neanderthal man) and *Homo habilis* ("handy man").

BINOMIAL CLASSIFICATION

In 1735, Swedish botanis Carl Linnaeus (1707–78) (better known as Linnaeus) published a systematic listing of all animals, plants, and minerals known at the time, using a hierarchical classification to place species within genera, orders, and classes. This slim volume, the *Systema Naturae*, included just a few hundred species at first, but the list grew over the following years as Linnaeus began to collect more information.

By the time of the 10th edition, published in 1758, the *Systema* incorporated more than 4200 species of animals

grouped into six classes, and 7200 species of plant. It was here that Linnaeus introduced the Latin binomial system and formally classified our species as *Homo sapiens*, placed with the monkeys and apes in the Order Anthropomorpha (later changed to the Order Primates). This edition was taken as the basis for scientific "Linnaean" classification, although it did not have any evolutionary implications.

Nevertheless, such classification schemes encouraged naturalists to consider the bases on which they were naming groups of organisms. Some categories, such as the animal classes of fish (Pisces), amphibians (Amphibia), reptiles (Reptilia), birds (Aves), and mammals (Mammalia) seemed to be very natural ones, although today most are seen as being more complicated. Only the Mammalia survive at the class level. In fact many of Linnaeus's zoological grouping harked back to Aristotle, thus he grouped fish by their fin bones, birds by their feet and beaks and mammals by their toes and teeth.

As Darwin himself spelled out in *The Origin of Species*, the "... natural system [of classification] is founded on descent with modification; that the characters ... showing true affinity between two or more species, are those ... inherited from a common parent, and, in so far, all true classification is genealogical ... and not some unknown plan of creation". This history of descent is what the German evolutionist, Ernst Haeckel (1834–1919), called a phylogeny.

Initially, branching phylogenies were inevitably based on Linnaean classification with notions of evolutionary "progress" in straight lines from primitive ancestors to more advanced descendents with humans placed at the top of the "tree". The faults in such a view are now widely recognized, and biological classification today has adopted a new approach pioneered by a German entomologist Willi Hennig (1913–76) in the 1950s. Called phylogenetic systematics or "cladistics", it attempts to produce a natural classification in which any branch or "clade" consists only of an ancestor and all of its descendents, (known as a "monophyletic" group). This new approach has revolutionized biological classification over the past few decades, and has helped to uncover many unexpected evolutionary relationships (see pp. 248–9 for more details). For instance, the birds can now be seen as part of the dinosaur clade and share a common ancestor with the crocodilians. This phylogenetic reclassification of the birds means that the dinosaurs are not entirely extinct. Some of their feathered descendants survived the end Cretaceous extinction and underwent an enormously successful explosive radiation to evolve into the modern birds – the neognaths.

Naturalists still use Linnaeus's Latin binomial scheme of genus name followed by species (*left*).

German biologist Ernst Haeckel was a great supporter of Darwinian evolution. Unlike most modern scientists, he believed that humans first arose in Asia rather than Africa (*above*).

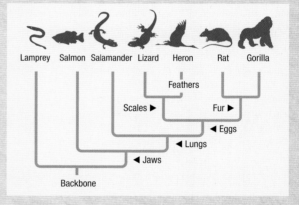

Evolutionary relationships between organisms with characters in common can be portrayed with branching diagrams known as cladograms (*above*).

Darwin was an avid collector of coleopteran beetles, whose diversity and abundance fascinated him. They are by far the most common living organisms with some 350,000 named species. The task of classifying them has been daunting, and the techniques of cladistics were developed to deal with the problem (*left*).

HOW TO USE THIS BOOK

EVOLUTION TELLS THE EXTRAORDINARY STORY OF LIFE ON EARTH – ITS DEVELOPMENT OVER SEVERAL BILLION YEARS FROM A MICROBIAL STATE IN THE OCEANS TO ITS PRESENT ABUNDANCE AND DIVERSITY IN THE SEAS, ON LAND, AND IN THE AIR. THIS REMARKABLE TALE IS REVEALED THROUGH A SEQUENCE OF PICTORIAL "WINDOWS ON THE PAST", EACH PRESENTING A SNAPSHOT OF A PARTICULAR TIME AND PLACE FROM EARTH'S DEEP HISTORY.

ARTWORK SEQUENCE

A selection of 100 of the finest fossil sites from around the world is presented as a running sequence of illustrated reconstructions. The details of the environments and their inhabitants are portrayed as accurately as permitted by current scientific knowledge.

The sites have been carefully chosen to represent the world's best known and most important fossil sites. They vary from well-preserved and accessible World Heritage Sites such as the 510-million-year-old Burgess Shale of British Columbia, to famous historic sites that have long since been exhausted and abandoned, such as Belgium's Bernissart coal mine, whose fossils can only be seen in museum collections.

The organisms portrayed in each illustration have been chosen as representatives of their time and place, and also for their importance in the evolution of life. They range from primitive microbial stromatolites and more familiar trilobites and ammonites, through the giant plants and insects of the Carboniferous coal measures and the famous

dinosaurs of Mesozoic times, to extinct human ancestors such as the australopithecines, who lived in Africa just a few million years ago.

The raw data for all these scenes consists of the rock strata and fossils found at each site, often through many decades of study. These stones, bones, and shells have mostly lost their original tissue and colour, flattened by compression and mineralized by the processes of fossilization. Interpreting and reconstructing them is an ongoing process, especially where the organisms are extinct and have no living relatives for comparison – as seen in the many radical reinterpretions given to the fossils of dinosaurs over the past 150 years,

Arranged chronologically, the artworks form a sequence of "screen grabs" or "timeframes", with an average intervals of around 6 million years. This visual storyline is designed to run like a film sequence, taking the reader seamlessly from one environment to the next, moving around the world and between land and sea as necessary.

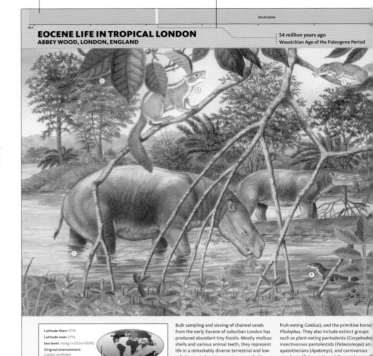

COMMONLY USED TERMS AND ABBREVIATIONS DEFINED

Adaptation A biological change that makes an organism better suited for its particular way of life and environment.

Advanced A feature of an organism that has been transformed from its primitive state.

Biota The overall flora and fauna of a particular region or site.

BP An abbreviation for "before present".

Character An attribute that can be used to determine relationships between organisms.

Deposit A concentration of material in a specific layer of rock strata.

Glaciation The spread of slow-moving ice across a landscape.

Eon The longest timescale used in the mapping of Earth's history.

Epoch A geological time division shorter than a period. Epochs are divided into ages.

Era A long timespan within Earth's history, often marked at its beginning and end by a major change in the variety of life. An era is composed of several periods.

Genus A group of species with a shared common ancestry.

MA An abbreviation for "million years ago".

Period A long timespan of Earth history, often lasting tens of millions of years.

Primitive A feature of an organism that is relatively close to the condition of the organism's ancestors.

Species A group of organisms whose genetic similarity allows them to interbreed successfully.

Stage A relatively brief geological time division distinguished by the presence of a particular fossil fauna.

SITE GAZETTEER

All the sites shown in the artwork sequence are listed in alphabetical order in the gazetteer. This offers additional information about the site, its history, visiting opportunities,

further reading of various levels, and useful websites. Each entry is also accompanied by an expanded species list that can be used to cross-refer to the main species listing section.

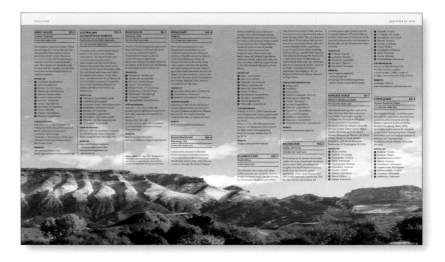

Colour bands
The colours in the upper bars of the spread match those in the international "IUGS" timescale.

Climate and biota information
Text information in the main bar provides a quick introduction to the main features of the site, including climate and common fossils.

Artwork magnifier
This icon shows that the species indicated is described in more detail at the bottom of the page.

Panoramic artwork
Each landscape reconstruction incorporates the latest information about the site and its ancient life.

Camera magnifier
This icon shows that the species indicated is illustrated by a photograph of fossil evidence below the artwork.

○ MARINAVIS (7) This long-legged procellariform or pelecaniform bird had a wingspan of around 50cm (20in), with beak bones similar to those of the modern shearwater.

● PLIOLOPHUS (12) This dog-sized primitive horse is known at Abbey Wood from jaw fragments, teeth, and other bones – the more complete skull shown here comes from sediments at nearby Harwich, where the same species has been found.

● CANTIUS (10) Fragments of teeth and skulls of Cantius represent the first adapiform primate fossils. Found in the early Eocene strata of North America and Europe, these small mammals weighed around 5kg (6.6lb), and had unspecialized small incisors with non-shearing cheek teeth. They are likely to have been fruit-eaters.

193

Photographic evidence
Where available, fossils show the evidence behind the reconstructions and provide more information about certain interesting species.

Species listing
This box identifies all the major species and some other important objects in the artwork, numbered from left to right.

A closer look
Some interesting species are shown in more detail with artwork close-ups and additional information in caption form.

○ Coryphodon
○ Ficus
○ Paramys
○ Palaeosinopa
○ Apatemys
○ Iauracean
○ Marinavis?
○ Ceriops
○ Palaeonictis
○ Cantius
○ Oxyaena
○ Pliolophus

TREE OF LIFE

A sequence of 22 detailed "family tree" diagrams, known as cladograms, tell the story of life and its ancestry in a different way, revealing the sometimes unexpected relationships between different groups of species based on fossil, anatomical, and genetic evidence.

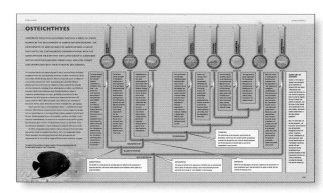

Cladistics
The Tree of Life section uses cladistics – a fairly recent method of grouping species according to their ancestry.

Introduction
The science of cladistics, and the design of the cladograms, are described in more detail on page 250.

SPECIES LISTING

This large section of the book describes the major divisions of multicellular life in detail, chronicling their evolutionary history and unique features, and providing indexes that allow the representatives of each group to be found in the context of the artwork sequence.

Groupings of life
Information at the top of each listing shows the relationships with other groups, for easy cross-referencing.

Latin names
The different groupings are listed in their internationally recognized "latinate" forms.

LANDMARK EVENTS GATEFOLDS

Two pivotal landmarks in the history of life are put into context by expanded gatefold spreads. The first is the so-called "Cambrian explosion", in which multicellular animals went through a sudden increase in number and diversity around 520 million years ago. The second is the End-Cretaceous extinction event (one of several "mass extinctions" in the history of life), which wiped out the dinosaurs and set the stage for the rise of the mammals in the modern era of life. Each gatefold describes the event in detail and explains its importance and the evidence behind it. Detailed diagrams also map out the story of life before, during, and after these pivotal events, as well as how their legacy still affects life on Earth today.

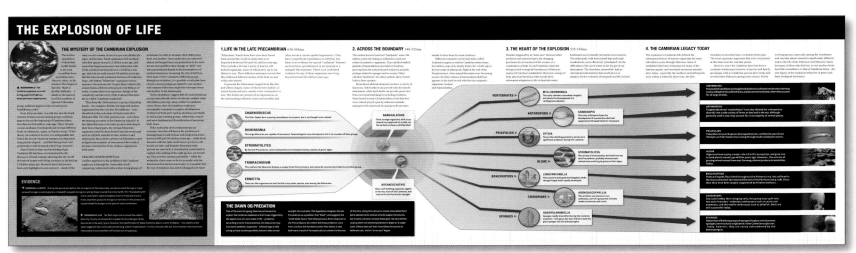

THE
SCIENCE
OF
EVOLUTION

In 1859, Charles Darwin published
The Origin of Species, unveiling to the
world his ingenious theory of evolution
by natural selection. The culmination of
two decades of work, Darwin's theory was
a response to the growing evidence for an
unimaginably ancient Earth, populated
with strange extinct creatures – a simple
but elegant model of the way in which
species are transformed over time.
In the 150 years since then, further fossil
discoveries have revealed the true variety
of life's history. Meanwhile, genetics has
unlocked the secrets of the mechanism
behind evolution, and now offers new
ways to study the relationships
between living species.

DARWIN AND THE *BEAGLE*

BORN IN SHROPSHIRE, ENGLAND, ON 12 FEBRUARY 1809, CHARLES DARWIN WAS THE FIFTH OF SIX CHILDREN IN THE FAMILY OF A WEALTHY SOCIETY DOCTOR. YET, DESPITE A PROMISING START IN LIFE AND AN ILLUSTRIOUS FAMILY, IT WAS FIVE YEARS OF TRAVEL AND EXPLORATION ABOARD HMS *BEAGLE* THAT WOULD SHAPE DARWIN'S LATER CAREER AND, WITH IT, OUR OWN UNDERSTANDING OF THE NATURAL WORLD.

Born on 12 February 1809, Charles Darwin was just 22 years old when he joined the *Beagle* (*above*).

By 1837, just a year after he had returned from the *Beagle* expedition, Darwin had produced his first sketch of an evolutionary tree – evocatively titled with a hesitant "I think" (*above*).

HMS *Beagle* was an armed survey ship just 30m (90ft) long, captained by Robert Fitzroy, who was himself just 26 years old when they left England in 1831 (*right*).

Darwin's early life was not what might be expected for an emerging "hero" of science. As a boy, he did not show exceptional interest in any particular field, did not do particularly well at school, and showed no clear ideas about what he might do in adult life. Natural history, riding, and shooting game were the main passions of his youth. His first choice of a medical training at Edinburgh University ended in failure, and he returned home to a disappointed father. However, while at Edinburgh he did meet the up-and-coming naturalist Robert Grant, a man who introduced Darwin to the ideas of Lamarckian evolution, and also to the study of the marine invertebrates that were to remain a particular interest for the rest of his life (culminating in his 1842 book on *The Structure and Distribution of Coral Reefs*).

CAMBRIDGE AND BEETLES

Darwin's next choice was a career path followed by many young gentlemen in the early part of the 19th century: Cambridge University followed by a comfortable rural life as a Church of England clergyman. At Cambridge, Darwin was, like many of his peers, renowned for having a good time and spending more than his allowance, but he was also known to disappear on long expeditions into the surrounding countryside to collect beetles. He was at least serious about natural history, and was encouraged in his studies by the Professor of Botany, the Reverend John Henslow – a connection that was later to prove invaluable. Even though the formal university course consisted primarily of classics, divinity, and mathematics, Darwin

made sure to attend Henslow's lectures and field trips, and was soon known as "the man who walked with Henslow". Meanwhile, through reading Alexander von Humboldt's account of his travels in South America, Darwin became intrigued by the richness and diversity of nature to be found in the little-explored continent, especially in the tropics.

VOYAGE OF THE *BEAGLE*

In 1831, the chance of a lifetime came Darwin's way when Henslow recommended him to one Captain Robert Fitzroy of HMS *Beagle*. Fitzroy had travelled to South America before, and was now due to make a return voyage in order to complete a coastal survey. As part of the elite scientific network of the day, Henslow had heard that Fitzroy, a young aristocratic and very able naval officer, was looking for a suitable gentleman companion with an interest in geology and natural history. Although Darwin's father objected, his uncle, Josiah Wedgwood (of the famous English pottery dynasty) intervened on his behalf. Darwin was duly interviewed by Fitzroy who, despite some initial doubts, decided to take him. Darwin had little time to prepare himself for his new role as geologist and naturalist, but his Cambridge connections provided him with a "crash course" in geology, courtesy of the Reverend Adam Sedgwick, Woodwardian Professor in the University. And so, armed with hammer, guns, abundant other collecting materials, and a small library that included the recently published first volume of Charles Lyell's

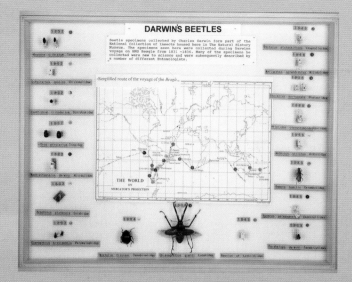

DARWIN'S BEETLES

Some of the many beetles collected by Darwin on the Beagle expedition are displayed with a map of the journey and the locations from where the beetles were collected (*above*).

Principles of Geology (a work that famously expounded the evidence for an ancient Earth and held that "the present is the key to the past"), Darwin set sail on 27 December 1831, at the beginning of a voyage that would not return until 2 October 1836.

Darwin had by now formed an ambition to make his mark as a geologist and naturalist, writing to Henslow that "geology & the invertebrate animals will be my chief object of pursuit during the whole voyage". To begin with, he found to his frustration that Humboldt and his other naturalist predecessors seemed to have already found everything. Nevertheless, he was an assiduous collector and, whenever possible, sent huge numbers of specimens back to Henslow in Cambridge. When Darwin eventually received an enthusiastic reply from Henslow about the novelty of many of the specimens, he began to feel increasingly confident that he might pursue a life devoted to natural history, rather than (as a friend and fellow student put it), "rusticating in a country Parsonage and shewing people a road I don't know – to Heaven".

Despite popular conceptions of the role played by the *Beagle* in Darwin's own intellectual voyage, much of the significance of what he was seeing and collecting only gradually dawned on the young man. For instance, the

potential importance of the varied species found on the different islands of the Galapagos Archipelago did not become apparent until Darwin was alerted to it by ornithologist John Gould after the voyage. Other sights, such as active volcanoes and earthquakes, made a more immediate impression on the young and inexperienced geologist-come-naturalist.

There was also the continuing problem of identifying the multitude of rocks, fossils, and living organisms he was collecting with the limited resources at his disposal, but by now Darwin was simply 'hooked' by the wonder of it all. Slowly he began to question the received wisdom of the day – an intellectual journey revealed in the journals and series of notebooks that he began to keep on a succession of topics from 1836 onwards. The species notebooks began in 1837 and culminated in the 1859 publication of his book *The Origin of Species*. Those on human beings began in 1838, but had a much longer gestation before publication as *The Descent of Man and Selection in Relation to Sex* in 1871.

French naturalist Jean-Baptiste Lamarck (1744–1829) developed an early theory of evolution that is now considered incorrect (*above*).

Darwin did not realize the evolutionary significance of the Galapagos finches until alerted to their uniqueness by ornithologist John Gould (*above*).

Darwin disliked the marine iguanas of the Galapagos, describing them as "most disgusting clumsy lizards", but he also noted their unique habit of feeding on particular marine algae. Iguanas also provided the model for one of the first dinosaurs to be described – *Iguanodon* (*below*).

ORIGIN **AND BEYOND**

DESPITE THE INSPIRATION AND WEALTH OF DATA HE HAD GATHERED DURING HIS YEARS ABOARD THE BEAGLE, DARWIN TOOK MANY YEARS TO FORMULATE HIS THEORY AND READY IT FOR PUBLICATION – SO LONG, IN FACT, THAT HE WAS ALMOST BEATEN TO PUBLICATION. NEVERTHELESS, WHEN IT EMERGED, DARWIN'S WORK HAD A PROFOUND EFFECT.

During a long life, Charles Darwin wrote numerous scientific papers, monographs, and some seven books. Apart from *The Origin of Species*, these included such topics as the biology of coral reefs and the ancestry of human beings (*above*).

After his five-year round the world voyage, Darwin arrived back at the family home in Shrewsbury on 5 October 1836. The following year he moved to London, dispersed his collections to appropriate experts, and, in 1838, was appointed Secretary to the Geological Society of London, one of the most dynamic scientific societies of the day. He published the first edition of his Journal of researches, married his cousin Emma Wedgwood in January 1839, and later in the year their first child was born. The young naturalist was quickly transformed into an established family man and junior member of a London elite that led the world in newly developing academic sciences such as geology, zoology, and botany. Between 1842 and 1846

Darwin saw himself largely as a geologist, and published books on coral reefs, volcanic islands, and geological observations on South America.

DEVELOPING THE IDEA

Meanwhile, however, Darwin was also developing his own ideas through a series of notebooks on species transmutation, biological evolution, and the implication of such ideas for mankind. In 1839 he wrote to Henslow, his Cambridge mentor, "I keep on steadily collecting every sort of fact, which may throw light on the origin & variation of species". He also began to receive important information about his specimens from experts such as the ornithologist John

Thomas Henry Huxley's vehement support for the Darwin–Wallace theory of evolution through his writing and popular lectures helped the idea to gain more general acceptance and earned him the nickname of "Darwin's bulldog" (*above*).

Gould, who alerted him to the fact the Galapagos finches were distinct but closely related species. Darwin investigated the breeding and artificial selection of domesticated animals, and learned about species, time, and the fossil record from the anatomist Richard Owen, who had worked on many of Darwin's vertebrate specimens and, in 1842, had "invented" dinosaurs as a separate category of reptiles.

By 1842, Darwin's evolutionary ideas were sufficiently advanced for him to produce a 35-page sketch and, by 1844, a 250-page synthesis, a copy of which he sent in 1847 to the botanist, Joseph Dalton Hooker. This trusted friend was sympathetic to his approach and was one of the first converts to 'Darwinian evolutionism'. By the 1850s, Darwin's following was extended to include the dynamic young zoologist Thomas Henry Huxley (1825–95). Meanwhile, the results of an 8-year study of barnacles, both living and fossil, were published between 1851 and 1854, establishing Darwin's credentials as a very able zoologist.

This research helped him develop the principle of divergence in speciation, which is most active with intense competition for limited resources. In other words, Darwin recognized that competition is a constant presence in nature and, as there is always some variation in populations, the result is natural selection of those adaptations that best fit the circumstances. Geographical isolation was just one of several possible conditions for speciation, with ecological pressures being equally if not more important.

UNEXPECTED COMPETITION

In 1855, Darwin read with interest a theoretical paper by Alfred Russel Wallace, a young naturalist working in south east Asia, who argued that new species tend to arise in areas already occupied by a related species. By this time Darwin's friends were encouraging him to publish his theory before someone else came up with a similar one, and by spring of 1858 he had completed ten chapters of a projected two-volume work entitled *Natural Selection*. In June 1858, however, Darwin received a bombshell with the arrival of a new manuscript from Wallace, outlining his theory that continuance of certain varieties of species might be perpetuated by processes of natural selection.

To Darwin's dismay Wallace had independently come up with a key aspect of his evolutionary theory. Luckily, Darwin's network of scientific friends arranged a compromise co-publication by the Linnean Society, ensuring that Darwin's independent and earlier formulation of the idea was recognized. Although the Linnean Society papers were largely ignored by the

ALFRED RUSSEL WALLACE (1823–1913)

Alfred Russel Wallace was a school teacher and naturalist who gave up teaching to earn his living as a professional collector of exotic plants and animals from the tropics. He collected extensively in South America, and from 1854 in the islands of the Malay archipelago. From these experiences, Wallace realized that species exist in variant forms and that changes in the environment could lead to the loss of any ill-adapted variants with the continuing success and survival of those that were adapted. In other words, he had independently come to the same conclusions as Darwin over a key aspect of the theory of evolution. Early in 1858 Wallace sent his paper to the Linnean Society in London, and it was published under the title "On the tendency of species to form varieties: and on the perpetuation of varieties and species by natural selection", alongside an extract from Darwin's manuscript on evolution and part of a letter sent by Darwin to the American botanist Asa Gray in 1857 outlining his ideas.

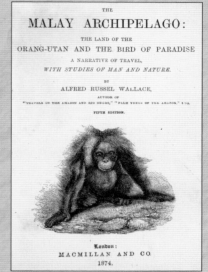

scientific community, the emergence of Wallace as a competitor in the field of evolutionary theory shocked Darwin into action. In July 1858 he set to work on a book-length 'abstract' of his ideas, in preference to the longer work that he had planned. By May 1859 he was working on proofs, and on 24 November the 500-page *The Origin of Species by Means of Natural Selection, or the Preservation of Favoured Races in the Struggle for Life* was published. All 1250 copies of the initial print run sold on the first day.

Darwin had been forewarned about the hostile reception his book would receive by the criticism that had been heaped on the evolutionary ideas contained in *Vestiges of the Natural History of Creation*, an anonymous work published in 1844. He deliberately avoided discussing the sensitive topic of human evolution, save only to predict that "Light will be thrown on the origin of man and his history". However, since he also concluded that 'all organic beings which have ever have lived on this earth have descended from some one primordial form, into which life was first breathed', his readers could easily draw their own conclusions. It was not until 1871, when the initial battles had largely been fought and won, that Darwin outlined his detailed views on human origins and the importance of sexual competition in all evolutionary stories.

Because of his ideas about the descent of man, Darwin was frequently caricatured as a monkey – as in this 19th-century French cartoon where he is seen leaping through hoops of credulity, superstition, and ignorance. Philosopher and physician Émile Littré (1801–81), a well-known French supporter of Darwin's supposedly "irreligious" ideas, holds the hoops (*opposite*).

TIME AND LIFE

William Buckland (1784–1856) drew one of the first geological sections, showing the divisions of rock strata and associated fossils (*opposite*).

IN ORDER TO MAKE SENSE OF THE STORY OF EVOLUTION, IT IS VITAL TO PLACE EXTINCT ORGANISMS IN A CHRONOLOGICAL CONTEXT – SO WE KNOW THE SPECIES THAT PRECEDED AND SUCCEEDED THEM, AND THE CREATURES THAT COEXISTED ALONGSIDE THEM. IN ORDER TO DO THIS, SCIENTISTS USE THE MANY DIVISIONS AND SUBDIVISIONS OF THE GEOLOGICAL TIMESCALE.

The 1780 discovery of giant fossil jaws from an unknown marine reptile (now named *Mosasaurus*) proved for the first time that extinctions had happened in the past, since it was clearly different from any living animal (*above*).

This timeline shows the internationally recognized divisions of geological time in a compact form (*below*).

By the early decades of the 19th century, geological mapping was progressively revealing the diverse nature and great thicknesses of successions of rock strata exposed across Earth's landscapes. National geological surveys were established as demand for materials such as coal, iron ore, and limestone increased during the Industrial Revolution.

The methodology and techniques of geological mapping were initially developed in Europe. The layered and sequential nature of sedimentary strata was discovered in Renaissance Italy; German mining from Medieval times revealed the three-dimensional geometry of folded and faulted strata; and in the latter part of the 18th century British and French geologists independently realized that fossils could be used to identify successive strata and correlate between separate outcrops.

Historically, quarrymen had named regional rock types according to their most prominent characteristics, such as the "Terrain Bituminifere" in northern France and Belgium (later recognized as the Carboniferous "coal measures"), the Muschelkalk in Germany (Middle Triassic age), and the Blue Lias in England (Early Jurassic age). As geological mapping improved it became necessary to develop new and more universal subdivisions that could be recognized across regions, countries, and continents.

Gradually, a hierarchy of more widely accepted names for strata was developed with "systems" such as the Jurassic being the major divisions of strata that are taken to represent "periods" of geological time. Systems are subdivided into lower, middle, and upper "epochs", corresponding with early, middle, and late chronological divisions, since more recent strata are deposited on top of earlier ones. For clarity, this book uses the chronological terms hereafter. Many of the divisions also have historical names, such as the Wenlock Epoch of the Silurian Period. Epochs subdivide into stages, most of which also have historical names, such as the Maastrichtian Stage of the Late Epoch of the Cretaceous Period.

DEEP TIME AND ANCIENT LIFE

By 1860 most of the period names had been established, and John Phillips, Professor of Geology at Oxford University, recognized from the distribution of particular fossils within them that the periods could usefully be grouped into Cenozoic, Mesozoic, and Paleozoic "eras", representing recent, middle, and ancient life. So the Cenozoic was characterized as the "Age of mammals", the Mesozoic as the "Age of reptiles", the late Paleozoic as the

EON	ARCHEAN				PROTEROZOIC																	
ERA	EOARCHEAN	PALEOARCHEAN	MESOARCHEAN	NEOARCHEAN	PALEOPROTEROZOIC PROTEROZOIC				MESOPROTEROZOIC PROTEROZOIC			NEOPROTEROZOIC PROTEROZOIC			PALEOZOIC							
Period					Siderian	Rhyacian	Orosirian	Statherian	Calymmian	Ectasian	Stenian	Tonian	Cryogenian	Ediacaran	Cambrian	Ordovician	Silurian	Devonian	Carboniferous (Mississippian)	Carboniferous (Pennsylvanian)		
Epoch															Early / Middle / Furongian	Early / Middle / Late	Llandovery / Wenlock / Ludlow / Pridoli	Early / Middle / Late	Early / Middle / Late	Early / Middle / Late	Cisuralian	
Stage															Paibian	Tremadocian / Darriwilian	Himantian / Rhuddanian / Aeronian / Telychian / Sheinwoodian / Homerian / Gorstian / Ludfordian	Lochkovian / Pragian / Emsian / Eifelian / Givetian / Frasnian / Famennian	Tournaisian / Visean / Serpukhovian	Bashkirian / Moscovian / Kasimovian / Gzhelian	Asselian / Sakmarian / Artinskian / Kungurian	

"Age of fishes", and the early Paleozoic the "Age of shellfish". However, while these names are still retained for the major eras of geological time, their association with fossil groups has been blurred by more recent discoveries.

But boundaries between eras are still very significant – they are now recognized as major extinction events. The end of the Paleozoic saw more than 85 per cent of life become extinct, while the end of the Mesozoic (marked by an enormous asteroid impact) saw 65 per cent of life driven to extinction, including the dinosaurs (other than birds).

Cumulatively, the successions of geological strata were found to be tens of kilometres deep, and underlain by an unknown thickness of Precambrian strata. Originally, these were referred to as "Azoic" as they were thought to be devoid of life. Today we know that life originated very early in Earth's development, and the immensity of "Precambrian" geological time has been subdivided into seven further eras, of which the most recent three are also divided into periods.

It has taken the international geological community some 200 years to carve up known strata into universally recognized units. Accompanying this has been the realization that it must have taken an immense amount of time for all these rocks to be alternately laid down, elevated, worn away, and redeposited by slow geological processes. Despite this, nobody knew just *how* old the Earth was until advances in radiometric dating during the 1950s.

HOW DO WE CALCULATE THE AGE OF ROCKS?

Radiometric dating of rocks is based on the natural process by which radioactive isotopes of certain elements decay into so-called "daughter isotopes" at steady rates. By measuring the ratio of parent to daughter isotopes of an element present in a crystal, it is possible to calculate when the parent isotope started the decay process. This normally coincides with crystallization from a molten state or, in the case of radiocarbon isotopes, the formation of the parent organic material such as wood or bone. Since the half-life of radiocarbon daughter isotopes is relatively short, this method can only be used for materials younger than around 55,000 years. Radiometric dating of geological time requires rocks that have been in a molten state, such as lavas. Most sedimentary deposits cannot be dated by the radiometric method.

PHANEROZOIC														
	MESOZOIC								CENOZOIC					
	Triassic			Jurassic			Cretaceous		Paleogene			Neogene		
Lopingian	Early	Middle	Late	Early	Middle	Late	Early	Late	Paleocene	Eocene	Oligocene	Miocene	Pliocene / Pleistocene	Holocene
Wuchiapingian, Changhsingian	Induan, Olenekian	Anisian, Ladinian	Carnian, Norian, Rhaetian	Hettangian, Sinemurian, Pliensbachian, Toarcian	Aalenian, Bajocian, Bathonian, Callovian	Oxfordian, Kimmeridgian, Tithonian	Berriasian, Valanginian, Hauterivian, Barremian, Aptian, Albian	Cenomanian, Turonian, Coniacian, Santonian, Campanian, Maastrichtian	Danian, Selandian, Thanetian	Ypresian, Lutetian, Bartonian, Priabonian	Rupelian, Chattian	Aquitanian, Burdigalian, Langhian, Serravallian, Tortonian, Messinian	Zanclean, Piacenzian / Gelasian	

THE PATTERN OF LIFE

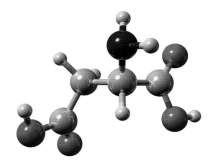

DARWIN DEVELOPED HIS IDEAS ABOUT EVOLUTION BASED ALMOST ENTIRELY ON THE STUDY OF LIVING SPECIES – AT THE TIME, THE STUDY OF LIFE'S LONG HISTORY AND THE FOSSIL RECORD WAS IN ITS INFANCY, AND IT WAS SOME TIME BEFORE THE FOSSILS ULTIMATELY PROVED HIS THEORY RIGHT.

Amino acids are the building blocks of proteins that are, in turn, the essential ingredients of organic life (*above*).

The idea that all life, no matter how complex, has descended from microscopic single-celled microbes that live in the sea was revolutionary and disturbing for many Victorian minds (*above*).

Museums built to house the wealth of new discoveries were conceived as "cathedrals", where the public could worship the newly emerging "gods" of science and technology (*right*).

By the mid-19th century, when Darwin was developing his theory of evolution, it was clear that there was some pattern to the distribution of fossils throughout the rock record. The earliest fossiliferous strata were Cambrian in age, and dominated by sea-living invertebrates, including extinct groups such as trilobites and graptolites. The first fish and land plants appeared in the Devonian, and reptiles and amphibians arrived in the Carboniferous "coal measures" – remains of the first extensive forests.

The Mesozoic Era was seen as the Age of Reptiles, following the discovery of extinct marine forms such as the ichthyosaurs and plesiosaurs, the flying pterosaurs, and finally the dinosaurs, which were not recognized as an independent fossil group until 1842. By this time, it was also known that primitive mammals had been around in Jurassic times, and by the 1860s birds were known to have first appeared in the Late Jurassic. The one group that did not seem to turn up in the fossil record were humans.

In the 1820s the eminent French anatomist Georges Cuvier had debunked earlier claims regarding the existence of human remains (thought to be victims of

Noah's Flood), although he had also discovered and named the first fossil primates from Cenozoic strata in France. Consequently, it was still possible to claim that humanity was the result of some act of special creation by a deity. However, by the mid-19th century there was mounting archaeological and fossil evidence that human-like remains occurred alongside those of the extinct animals of the Ice Age. By 1868, the first extinct human-related species – *Homo neanderthalensis* – was named, but it was not for another 20 or so years that the fossil antiquity of humans and the growing evidence for human evolution was generally accepted by the academic community.

MISSING LINKS

Darwin was well aware of the nature of the fossil record in the first half of the 19th century. His theory of evolution required that there should be fossil evidence for ancestral forms shared by descendent groups, because ultimately all life has diverged and descended from a common ancestor. But he knew only too well that such common ancestral forms for major groups had not been found, and he blamed their absence primarily on the incompleteness of the rock and fossil record.

His other major problem was the lack of fossils from Precambrian strata. Since several different invertebrate groups, such as brachiopods and trilobites, appeared in early Cambrian strata, they must have had ancestors that lived in Precambrian times. Darwin admitted that their absence was a problem for his theory, but predicted that Precambrian fossils would turn up eventually, He was right, though it was not until the 1950s that the first convincing evidence of Precambrian life was found in Russia and Canada. Since then, the record has been extended back to at least 3500 million (3.5 billion) years ago, providing indications that the sudden diversity of Cambrian life is more apparent than real.

Among the most ancient remains of life are chemical fossils, so called because all that remains of the original cellular material are complex organic molecules that can be distinguished from inorganic molecules. These are preserved as particles of graphitic carbon within metamorphosed shales from Greenland, over 3.7 billion years old .

BRIDGING THE GAPS

Modern genetic analyses of the simplest known living organisms show that life can be separated into three major domains. The Bacteria are the most ancient and primitive, followed by the Archea from which the Eukaryota evolved less than 2 billion years ago. The eukaryotes include all the more familiar organisms, from single-celled amoebae to multicellular plants, animals, and fungi.

Today, Darwin's problem with the incompleteness of the fossil record has been largely resolved with the discovery of many extinct fossil groups. These amply demonstrate many major evolutionary innovations such as the development of the tetrapod limb with the transition from aquatic fish to land-living tetrapods, the appearance of feathers and wings in dinosaurs and their flying descendants the birds, and the evolution of upright bipedal walking in our own primate ancestors.

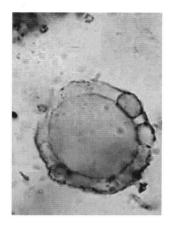

Darwin's evolutionary expectations that fossils of primitive organisms would eventually be found in Precambrian strata were not confirmed until the 1950s, when microbial fossils were found in 2-billion-year-old chert from Canada's Lake Superior (*above*).

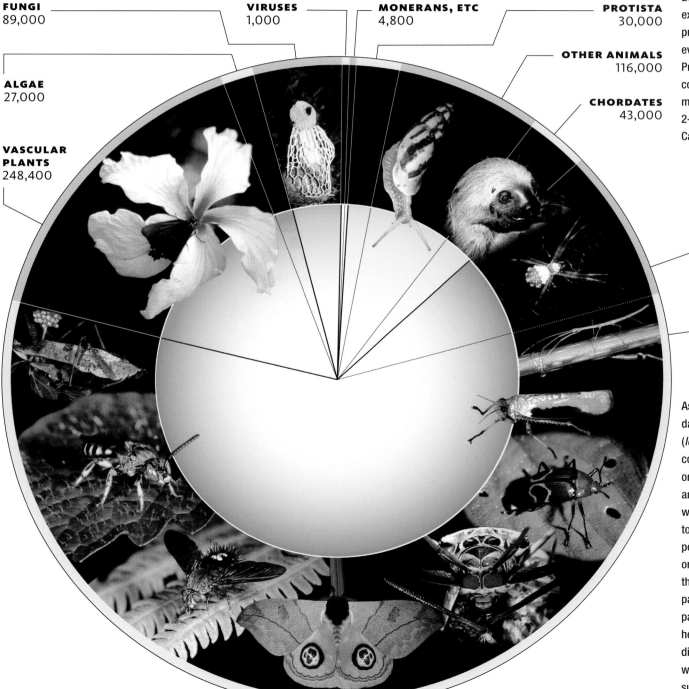

FUNGI
89,000

VIRUSES
1,000

MONERANS, ETC
4,800

PROTISTA
30,000

OTHER ANIMALS
116,000

ALGAE
27,000

CHORDATES
43,000

VASCULAR PLANTS
248,400

ARTHROPODS
NON HEXAPOD ARTHROPODS
123,000

ARTHROPODS
HEXAPOD (INSECTS)
925,000

As this chart of known present-day species demonstrates (*left*), much of life's variety is concentrated in a few groups of organisms — notably arthropods and vascular plants. However, when more is known about the total diversity of life today, it is possible that microbial organisms will outnumber even these groups. While life in the past probably followed a similar pattern, the fossil record is heavily skewed in other directions — towards organisms with preservable hard parts, such as shelled molluscs.

WHAT ARE FOSSILS?

FOSSILS ARE A RECORD OF PAST LIFE PRESERVED IN THE ROCKS BENEATH OUR FEET. THEY TELL US THAT LIFE BEGAN IN EARTH'S OCEANS MORE THAN 3.5 BILLION YEARS AGO, AND SINCE THEN HAS EVOLVED AND DIVERSIFIED FROM AQUATIC MICROBES TO AN ESTIMATED TOTAL OF 10 MILLION DIFFERENT KINDS OF LIVING ORGANISMS. MORE THAN 99 PER CENT OF ALL LIFE THAT HAS EXISTED IS NOW EXTINCT, SO TELLING THE FULL STORY OF EVOLUTION RELIES ON OUR STUDY AND INTERPRETATION OF THESE SCARCE REMNANTS OF PAST LIFE.

The preservation of the entire exoskeleton of this extinct trilobite is a relatively uncommon occurrence, since trilobites were arthropods, and most of their remains consist of the separate parts that were regularly moulted to allow for growth (*above*).

These bones are all that remains of a whale after a few weeks on the seabed. Soft tissues are rarely preserved even underwater, since they provide a good food source for a variety of microbes, not to mention larger scavengers ranging from sharks to crabs. Yet there is still a chance these remnants will be buried, preserved, and one day revealed again like those opposite (*below*).

Most fossils are the physically and chemically robust parts of extinct organisms. They are found buried within layers of ancient sedimentary deposits that have accumulated over hundreds of million years to form what is known as the stratigraphic rock record. The way that organic remains become part of the rock record is so complex and highly selective that only a tiny percentage of past life is preserved. Of all the hundreds of millions of species that have ever existed only a few hundred thousand fossil species have been found and described so far. And, the process of fossilization often transforms organic remains beyond recognition.

As a visit to any museum of paleontology will show, fossils are typically stone-coloured and rock-like in appearance, often completely flattened on the rock surface so that they look like old sepia photos. Yet they were once breathing, moving, vital organisms. Three-dimensional remains that preserve the original form of shells and bones can be found, but even then fossilization makes organic remains more rock-like, heavier, and harder than the original bone or shell, and removes any original colour.

In a curious way, the processes that reduce living creatures to fossils with the loss of so much information may also contribute to their huge popularity. It seems that people can be just as excited by strange-looking fossils as they are by stuffed animals and even some live ones. Fossil museums might be mausoleums and memorials full of the monochrome petrified remains of past life, but they also preserve the only publicly displayed record of the wonderful and sometimes weird life of past eons.

Over the centuries, paleontologists have struggled to understand the nature of the fossil record and read the record as a history of evolving life on Earth. Surely Darwin himself would celebrate the extent to which modern paleontology has confirmed so many of the predictions of his theory of evolution.

THE TROUBLE WITH FOSSILS

Ever since the times of ancient Greece and China when scholars such as Xenophanes of Colophon (c.570–490BC) first began to write about fossils, their true nature has been a cause of prolonged and intense debate. While many fossils look like the remains of past life, they are often preserved as inorganic materials that differ significantly from living matter. For instance, a fossil might have the outward appearance of a sea creature, such as a clam or sea urchin, but be composed of a mineral such as siliceous flint, that is quite unlike the calcium carbonate from which the shells of living clams and sea urchins are constructed.

Today, we understand that fossilization can have a radical effect on the chemical composition of fossil material, but early naturalists saw only that the minerals from which fossils are made were more akin to the minerals that form rocks than to the shells and skeletons of living organisms. Some

Animals caught in catastrophic events such as these wildebeest swept away in a flash flood may be rapidly buried in sediment (*left*).

Chondrichthyan sharks and rays have cartilaginous skeletons that are rarely fossilized, so all that normally remains are their mineralized teeth – just one example of the way in which the fossil record is biased to favour certain organisms (*above*).

argued that fossils had perhaps grown within rock strata, and only bore a superficial and misleading resemblance to living organisms. To support this line of reasoning, they noted that many fossils were found in very different environments from their living counterparts . For instance, fossils that looked like sea urchins could be found in rock strata on mountainsides far removed from the waters where such creatures live today. It is only recently that we have come to understand how earth movements can displace rocks far from where they were originally formed.

It was not until Renaissance times that many of these questions began to be resolved. By using the newly emerging methods of scientific investigation, a number of naturalists produced carefully argued examples of the truly organic nature of fossils. For instance, in the mid-17th century, Niels Stensen (also known as Steno), a Danish physician working in Italy, dissected a modern shark and carefully compared its teeth with fossils known as "tongue stones" (glossopetrae), commonly found around the Mediterranean. Stensen argued that "tongue stones", even those found inland, are most reasonably interpreted as ancient shark teeth. Furthermore, they also provided evidence that, in the past, the sea had reached far inland.

Now even young school children are familiar with the reality of fossils as the remains of once-living organisms. But despite this, there are still plenty of pitfalls for the unwary and even expert professional paleontologists can have trouble deciding whether some remains are genuine fossils or not – witness, for example, the arguments over so-called microbes preserved in early Australian chert (see p.38), and alleged microbes in meteorites from Mars.

This virtually complete skull of an iguanodont dinosaur was one of many complete skeletons recovered from a coal mine in Belgium, providing detailed information about this plant-eating ornithopod dinosaur (*above*).

Geological uplift, erosion, and weathering have resurrected an ancient seabed and revealed the fossil bones of an ancient whale, stretched out just as it came to rest millions of years ago (*left*).

27

THE VARIETY OF FOSSILS

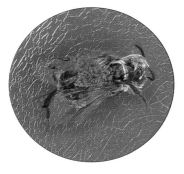

WHILE THE TRADITIONAL IMAGE OF A FOSSIL MAY BE THAT OF AN ANCIENT BONE OR TOOTH TURNED TO STONE, THE TRUE DIVERSITY OF FOSSILS IS FAR GREATER, SINCE A FOSSIL CAN BE ANY TRACE LEFT BY FORMER LIFE AND SOMEHOW PRESERVED – THIS INCLUDES A VARIETY OF DIFFERENT MEANS OF PRESERVATION, AND ALSO EXTENDS TO TRACES OF THE WAY ANIMALS INTERACTED WITH THEIR ENVIRONMENTS.

The long history of the interdependence between insects and plants is revealed by this bee, preserved complete with orchid pollen in 15- to 20-million-year-old Miocene amber from the Dominican Republic (*above*).

The cadaver of "Otzi", a Neolithic hunter, found freeze-dried in a glacier in the Tyrolean Alps, preserves not only soft tissues, but also DNA. However, the long-term survival of such "protofossils" depends on the persistence of permafrost, which is much more ephemeral than rock (*below*).

Fossils provide the main evidence for the history and evolution of life on Earth. But this simple statement hides a more complex reality that has taken centuries to resolve. Essentially, fossils are the toughest and least destructible parts of organisms, although occasionally more delicate structures and tissues are preserved, such as when an entire body is freeze-dried in frozen ground.

Fossils vary in composition from mineralized bones and shells, to organic molecules preserved as blobs of bitumen, and the compressed, carbonized plant remains that we know as coal. However, they can generally be separated into just a few different kinds.

Chemical fossils are residual organic chemicals, such as the bituminous biomolecules recovered from Archean strata more than three billion years old. Generally, their true organic nature and chemical composition can only be resolved by sophisticated analytical equipment.

Body fossils, the remains of original tissues, are the most common fossils, represented by countless shells and bones, and their impressions left in the rock. Some have been chemically altered, and a rare few preserve soft tissues. This fossilized skin and muscle is usually preserved in ancient sediment when it has been replaced by inorganic minerals, such as the apatite (a phosphate) or pyrite (an iron sulphide).

Trace fossils are the marks left by a living organism on or within the sediment substrate – for example footprints, burrows and tooth or cutmarks. Most common are the burrows, and root traces of organisms that live or grow within sediment. Rarely is the maker of the trace preserved, but trace fossils provide very important evidence of certain environments such as tidal sandflats, and can prove the existence of ancient behaviours such as herding among certain plant-eating dinosaurs and the meat preparation techniques of Neanderthal hunters.

As we have seen, fossils preserved in the rock record are usually the most robust parts of an organism, especially mineralized skeletal materials such as shell, bones, and the tough woody tissues of plants, which can survive long after death of an organism. As a result, the fossil record is heavily biased towards organisms that have such tissues, and does not fully represent the diversity of life, especially

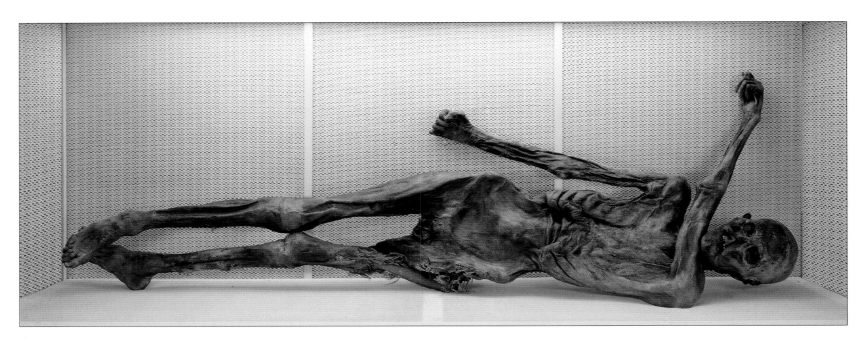

These trilobite arthropods, perhaps buried alive while breeding in shallow water, were preserved by internal sediment moulds even as their mineralized exoskeletons dissolved away (*above*).

among soft-bodied organisms that range from viruses and bacteria to giant squid.

The bulk of body tissue in most organisms is composed of water and organic compounds that degrade rapidly following death. This organic matter also represents potential food for other organisms. In most natural environments any dead body is scavenged, consumed, and biologically degraded to some degree, leaving just the hard parts as potential fossils. Occasionally, however, a body may come to rest in a naturally preservative medium or location before it has deteriorated to any great extent. Fortunately for paleontologists there are many excellent media and circumstances under which soft tissues can and have been preserved. These range from freeze-drying in subzero temperatures to soaking in oil, salt, or resin.

HOW ARE FOSSILS PRESERVED?

While it is true that an entire mammoth or human can be freeze-dried in glacial sediments with their soft tissues and even some DNA preserved, such events are exceedingly rare. Frozen mammoths and humans such as the 5200-year-old Tyrolean Ice-man "Otzi" are not, strictly speaking, true fossils, since their enclosing icy sediments are themselves ephemeral on a geological timescale. Fossils preserved in amber can survive for much longer, but while organisms trapped in this ancient tree resin may seem perfectly preserved, appearances are deceptive. During the 1990s, attempts were made to recover DNA from amber insects, but claims of success were unfounded and did not pass the critical scientific test of consistently reproducible results. However, other proteins have been recovered from fossil remains, most recently collagen from the 68-million-year-old bones of a *Tyrannosaurus rex* dinosaur.

Our everyday experience of terrestrial environments shows why it is difficult to preserve any remains of land-living animals or plants. Our landscapes are full of animal and plant life, but what happens when they die? How often do we come across the bones of a bird, or even leaves buried in soil? Deciduous plant leaves may cover the ground in autumn, but over weeks and months they are degraded by fungi or bacterial decay and a variety of animals from snails to myriapods and earthworms. The bones may survive for a year or so, but acids in the soil and oxidation of the organic matrix soon weaken them – only teeth, with their tough dentine and enamel, last longer.

Some biological structures do survive well in soil – tough coated spores, seeds, and pollen that are adapted for survival in such conditions. Indeed, such structures may last long enough to be fossilized. Preservation of the soil and other surface sediments requires special conditions such as rapid burial instead of the normal processes of surface weathering, and erosion.

FLUKES OF PRESERVATION

So how is it that any fossil record of terrestrial life survives? And how come there are significant global reserves of terrestrial deposits such as coal, the compressed remains of ancient tropical forests, swamps, and bogs? The answer to both questions is that a combination of geological conditions has made it possible. Deposition of sediments and their organic remains in a generally subsiding landscape has buried them to such a depth that they survive subsequent erosion.

However, the bulk of the fossil record consists of the shells and bones of marine organisms that lived in shallow seas on the continental margins, and within their waters. Yet even ocean-floor sediments and their organic remains are ultimately destroyed by another inexorable geological process – subduction as a result of tectonic movements.

More than 99.9 per cent of all the life that has existed is extinct, but fossils can give us some idea of what life was like in the past, and how it evolved.

A MOUSE CADAVER

Decomposition after 3 days

Decomposition after 5 days

Decomposition after 7 days

Decomposition after 9 days

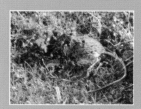

Decomposition after 15 days

Decomposition after 23 days

This sequence of photos shows how processes of decomposition and scavenging rapidly destroy organic remains in most conditions.

RECONSTRUCTING THE PAST

IN ORDER TO UNDERSTAND THE PROCESSES OF EVOLUTION, WE MUST LOOK AT PLANTS AND ANIMALS WITHIN THEIR WIDER CONTEXT – THE ENVIRONMENT IN WHICH THEY EXISTED, THE FOOD SOURCES THAT WERE AVAILABLE TO THEM, AND THE COMPETITORS AND THREATS THEY FACED. PIECING TOGETHER A COMPLETE PICTURE OF SUCH ANCIENT ECOSYSTEMS CALLS FOR A VARIETY OF TECHNIQUES.

Millimetre-sized shells of single-celled organisms, such as this foraminiferan, are abundant in the fossil record. They can be used as proxy measures of past climates, since the shells record ocean water chemistry at the time they were built (*above*).

Today we take for granted scenes of the deep past populated with dinosaurs and other extinct fossil animals and plants. And, with the use of computer graphics, such images are getting more and more superficially realistic. However, they often use modern land and seascapes for the background and even modern plants that are not usually appropriate. In this book we rely on the more traditional art techniques that give a greater degree of flexibility, detail, and accuracy.

Even so, all such reconstructions are to some degree imaginative fictions, since there are no visual records beyond the 30,000-year-old artworks made by early modern humans who saw and depicted extinct animals such as the woolly mammoth, woolly rhinoceros, and giant deer. Beyond this in time, we know little of the coloration of extinct animals and plants apart from some indications of camouflage patterning. Detailed body shapes can be equally problematic, except through indirect inference and reconstruction of soft tissues, using our

understanding of comparative anatomy. However, a few spectacular new finds of soft tissue preservation, dating back as far as Cambrian times, do provide some accurate information about body shapes.

Over the last 200 years, enormous progress has been made in our understanding of the life of the geological past. Most, but by no means all, fossil organisms can now be reconstructed with some degree of certainty as to their general appearance. The main exceptions are the plants and some of the larger vertebrates, because their entire body form is so rarely preserved in the sedimentary rock record. Plants are especially difficult because their numerous anatomical parts, such as pollen, leaves, woody tissues, and roots, tend to be separated one from another both during life and following death, and may be deposited widely in different sedimentary environments.

Fortunately, interpretation of the rock record and the associations of fossils from specific sites and stratigraphic levels is now sophisticated enough to allow reconstruction

William Buckland (1784–1856), (*above*), was one of the first people to attempt the reconstruction of life in the past – in this case an Ice Age hyena den discovered in Yorkshire in 1821. His conclusions inspired his friend William Conybeare's cartoon (*right*) showing Buckland himself entering the cave.

Catastrophic events such as the eruption of Vesuvius in AD79 preserve some of the most complete records of past life – even if they show the victims' dying moments (*far left*).

Delicate tissues such as those of angiosperm flowers are normally lost to the fossil record. Rare preservations, such as this 34-million-year-old *Florissantia*, from Eocene deposits in Colorado, USA, require special conditions in fine-grained sediments (*left*).

of some past environments and the inter-relationships of the various organisms that lived in them. Most fossil environments are waterlain, although in certain circumstances low-lying terrestrial environments may be preserved, but uplands are exceedingly rare. And there are still many problems of temporal resolution that make it impossible to determine whether the organisms actually lived together, died together, or simply had their remains jumbled together long after their death.

PIECING TOGETHER THE EVIDENCE

To understand all these factors, scientists have made detailed studies of the ecological relationships between living organisms and their environments. Over the 170 years since the first reconstructions of ancient scenes were attempted, we have learned a great deal about how death occurs in the natural world, what happens to the remains of plants and animals following death, how remains may be lost or recruited to the rock record, and what happens to them after burial, during the often complicated and destructive processes of fossilization.

There are a few situations that tend to preserve organic remains particularly well. For body form and soft tissue preservation rapid entombment in a mummifying medium, such as cold dry air, ice, amber resin, or salt is necessary, but these are relatively rare in the geological record. Catastrophic and near-instantaneous natural burial processes can entomb a whole range of organisms that lived and died together. Such circumstances are known as "Pompeii" scenarios, after the pyroclastic erupion that

engulfed the Roman town in AD79, preserving much of the structure, artefacts, and some of the inhabitants.

Volcanic eruptions, avalanches of sediment, dust-storms, and floods are common catastrophic events in the natural world, both on land and in water. They can overwhelm living communities and potentially preserve much of their life. And, if the post-mortem environment lacks oxygen – for example the fine sediments of a lake bottom, they may even preserve some soft tissues.

As will be seen from the reconstructions in this book, most marine fossil locations were in shallow waters and lagoons, and most of the terrestrial locations were in lake and river deposits, so the actual land environments are largely reconstructed from indirect information preserved in waterlain sediments and surrounding rocks.

Rock art, such as these engravings from North Africa, records an abundance and diversity of life forms that are not now present because of climate change and extinction aided by human hunting (*below*).

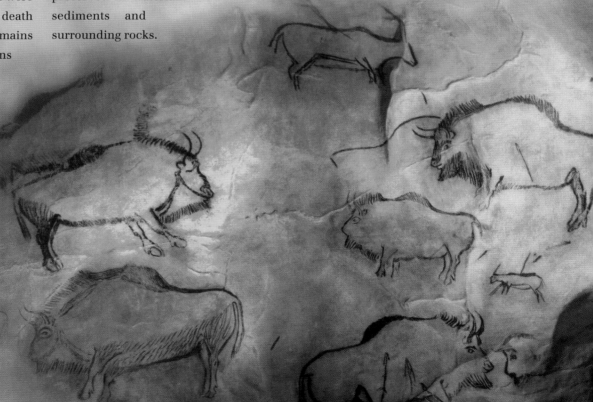

EVOLUTION TODAY

TODAY, THE STORY OF EVOLUTION IS BEING RE-EXAMINED USING THE TECHNIQUES OF GENETIC ANALYSIS. NEW TOOLS ALLOW SCIENTISTS TO ESTIMATE THE RELATEDNESS OF ANY TWO ORGANISMS, LEADING TO NEW DISCOVERIES AND MANY SURPRISES. BUT WHEN IT COMES TO THE PAST, THERE IS STILL A PLACE FOR THE TRADITIONAL METHODS.

DNA's ability to replicate itself precisely and to combine the characteristics of two different parents is the very thing that allows reproduction, but replication is not perfect, and often it is the random mutations that have the greatest impact (*above*).

Sophsiticated machines such as the scanning electron microscope allow researchers to look at rocks in unprecedented detail, discovering for the first time the minute traces of the most ancient life forms, including the fossilized bacteria shown here (*below*).

Our understanding of evolution has been revolutionized by modern genetics, the fundamental breakthrough discovery of the structure of DNA, and progress in molecular biology. These advances have provided new analytical methods and access to sequences of macromolecular characteristics that allow comparisons to be made between organisms that have no obvious anatomical similarities, such as bacteria, plants, fungi, and animals.

According to evolutionary theory, all life has descended from a common ancestor. And yet, traditional methods of classification, based on anatomy, have failed to clarify the evolutionary inter-relationships between major groups such as bacteria, plants etc, and between some groups at a much lower taxonomic level. For instance, there has been considerable debate about the inter-relationships of some mammal families. Now, thanks to the new genetic information, it is possible for the first time in the history of

biology to overcome the difficulties facing Darwin and his contemporaries in reconstructing life's history of evolutionary descent from that common ancestor.

GROUPINGS OF LIFE

Three major groups of organisms are recognized today – Eubacteria, Archea, and Eukaryota. The Eubacteria and Archea, all single-celled microbes, are the most primitive, since they are prokaryotes, lacking a cellular nucleus.

The living Archea are not very diverse or numerous, but can survive in extreme environments such as hydrothermal vents. They were once thought to include the most primitive forms of life, but this is now questioned. The group is not represented by any known fossils.

By comparison, the Eubacteria are very diverse, with 10,500 and more microscopic species living today and a fossil record, albeit a limited one. The Eubacteria can also survive in hostile environments without air and light. Despite restrictions in size, they have an extraordinary diversity of metabolic pathways, and their ability to obtain energy from minerals probably played an important role in the early evolution of life. They are of great importance in the formation of soils, without which plants would never have been able to invade the land. Others can photosynthesize energy from sunlight with or without the production of oxygen. It was the stromatolitic cyanobacteria of early Precambrian times that increased the oxygen level in Earth's atmosphere from 1 per cent to the present 20 per cent.

All other organisms belong to the vast, diverse grouping known as the Eukaryota, with over 1,738,000 described species, including single-celled forms, plants, fungi, algae, and animals from worms to whales. Although the eukaryotes vary enormously in size, their individual cells only vary between 10 and 100 microns. The DNA of the eukaryotic cell lies within a membrane-bounded nucleus and structured cytoplasm, which also contains the mitochondria that assure cellular respiration.

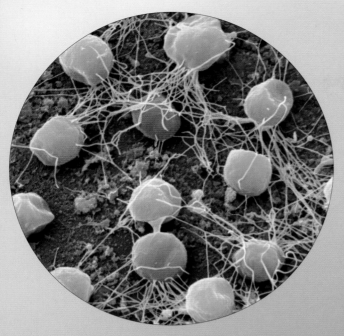

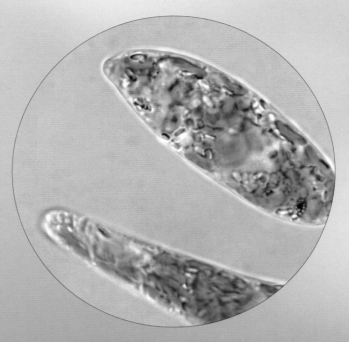

The discovery of primitive life forms (prokaryotic archaeans, *far left*) that can survive in extreme environments such as boiling brines and mud, freezing seawater, and acidic, oxygen-free waters, gives some idea of what the earliest organisms might have looked like. Evolution of a distinct membrane-surrounded nucleus, typical of eukaryotic organisms such as the flagellate protozoan *Euglena* (*left*), represented a considerable advance.

Fundamentally aerobic, eukaryotes require the presence of oxygen, and photosynthetic species always produce oxygen as a by-product of their metabolism. There are true sexes, with each sex contributing an equal portion of the genetic material to the next generation.

Analysis of molecular relationships supports the evolutionary hypothesis that all eukaryotes have diverged from a single common ancestor. That ancestor possessed mitochondria, derived from a bacterium, that had been integrated into the cell cytoplasm through a single original 'endosymbiotic' event in the remote past.

MOLECULAR CLOCKS AND EVOLUTION

The construction of these new classifications, using molecular data from living organisms, has caused problems for the organisms where such molecular data cannot be extracted and taxonomists must still rely on anatomical characters. This typically includes most fossil species, although fossil DNA fragments can sometimes be recovered from subfossil bones tens of thousands of years old. Furthermore, the molecular clock timings for the divergence of groups invariably provide dates that precede their fossil record by a considerable length of time. The molecular clock measures genetic 'distance' between living groups, and by using known rates of mutation estimates the date of divergence. For instance, the 98.4 per cent genetic similarity between chimps and humans indicates a divergence from a common ancestor some 6–8 million years ago, and the 5 per cent difference between the gibbons and them, suggest an ancestor some 18 million years ago. Yet, until the recent discovery of the 7-million-year-old fossil ape *Sahelanthropus tchadensis*, the oldest known fossil that was anywhere near the common ancestor of chimps and humans was less than 4 million years old. At the other end of the scale, the molecular clock puts the first divergence of invertebrates around 800 million years ago, but the macroscopic fossil evidence only goes back for 580 million years.

Molecular clock analysis compares the genetic distance between two living organisms and, given a known rate of evolutionary change (mutation) can estimate the date of their original divergence. Recently, the method has estimated a Late Jurassic to Early Cretaceous divergence of ants from wasps, but it was another 80 million years before the ants themselves diversified (alongside the angiosperm plants) in the early Paleogene. Today, there are some 11,880 known ant species (*below*).

THE STORY OF LIFE

The story of life on Earth is written in fossils –
a record that nevertheless extends back
through some 3.4 billion years of our planet's
history, and becomes more tangible with the
rise of multicellular life in the last 600 million
years. But fossilization requires special
circumstances, and the exquisite preservation
of soft tissues that can reveal life's most
intimate secrets is even more demanding.

   Despite these problems, there are some
localities that stand out as jewels in the
intermittent fossil record – often because of
their exceptional quality, but sometimes just
because they offer our only brief glimpse of
a major development in life's history.
The panorama spread across this section of
the book offers 100 different snapshots of
evolution in progress, based on the latest
research and discoveries, and ranging from
the first single-celled life to the present day.

IN THE BEGINNING
STRELLEY POOL, WESTERN AUSTRALIA

3460 million years ago
Paleoarchean Era of the Archean Eon

Latitude then: equatorial

Latitude now: 22°S

Sea level: generally rising but with minor fluctuations

Original environment: rocky shore to subtidal

Deposits: laminated siliceous cherts and carbonate dolomites

Status: first discovered 30 years ago, but the best examples have only recently been found

Preservation: for their great age these are amongst the best preserved stromatolites known

Tectonic reconstructions of the Archean are highly speculative

⊕ Earth c.3400 MA

• Strelley Pool

⊕ Fossil site today

For more than four billion years, Earth's landscapes were devoid of visible life. But slowly and surely life's long, slow fuse was burning away. The rock record preserves biochemical evidence of primitive life as far back as 3.5 billion years ago, while the first visible traces of life are strange, laminated sedimentary structures called stromatolites, which grew in the warm, shallow tropical seas of Paleoarchean times, 3.4 billion years ago.

Stromatolites were produced by the interaction of microbial life and sedimentary deposition – a process that continues today in the coastal waters of Western Australia and the Caribbean. Photosynthesizing "microbial mats" growing on the seabed are covered in sediment as the current washes over them. To reach the light, the algae and cyanobacteria grow up through the sediment and form new surface mats. As the process repeats, it produces laminated structures with varying shapes modified by water currents and the shape of the seabed.

Some claim that stromatolites can be produced by inorganic processes, but the stromatolites of Western Australia's 3.4-billion-year-old Strelley Pool Chert provide convincing evidence for their organic origin.

Common organisms: cyanobacteria

Volcanism: geysers and hot pools

Climate: tropical–subtropical

Biota: prokaryotic microbes

① egg carton
② cuspate swale
③ encrusting/domical

↑ EGG CARTON (1) The common coniform or "egg carton" stromatolites are thought to have developed from small clumps that deflected the growth of new microbial filaments upwards. Forms such as this are thought to depend on the wave patterns that once drove currents through the ancient, shallow waters.

↰ ↱ ENCRUSTING/DOMICAL (3) The "encrusting/domical" form varies from laminated drapes spread across beach pebbles to domes that are identical in form to those of today's stromatolites. These living examples are from Shark Bay in Western Australia.

FIRST FOSSILS?
APEX CHERT, NEAR STRELLEY POOL, WESTERN AUSTRALIA

3460 million years ago
Paleoarchean Era of the Archean Eon

Latitude then: equatorial

Latitude now: 22°S

Sea level: generally rising but with minor fluctuations

Original environment: volcanic hot springs

Deposits: siliceous cherts

Status: remote outcrops in the northwest of Western Australia

Preservation: three-dimensional "microbes" found within chert, but organic origin disputed

Darwin himself expected Precambrian fossils to exist, but the first convincing specimens, microscopic organisms around two billion years old, from Russia and Canada, were not found until the 1950s.

The fossilization of cellular material requires unusually rapid mineralization, such as occurs in sedimentary chert rock. Formation of this silica mineral is particularly associated with hot springs, such as those found in the USA's Yellowstone National Park, where many primitive microbes also flourish.

The search for ancient chert associated with hydrothermal activity targeted the Archean sedimentary rocks of Western Australia, and in the early 1990s US paleontologist William Schopf claimed to have identified the oldest known fossils in the Apex Chert. Scanning with high-power microscopes revealed tiny bacterium-like filaments that Schopf described as new prokaryotic cyanobacteria-like genera *Primaevifilum* and *Archaeoscillatoriopsis*.

Subsequently, the status of these "fossils" has been questioned, with suggestions that they are no more than mineral chains that mimic cellular filaments, and that the cherts may have formed some time after the original sedimentary rocks.

➲ PRIMAEVIFILUM (1)
Six species of this genus from the Apex Chert were named in the early 1990s, and are among a total of 11 species belonging to four genera. They were described as between 2 and 7 microns in width and belonging to cyanobacteria-like microbes with differentiated cells – cylinder-shaped in the body and rounded at each end. At the time they were hailed in the international media as indisputably the most ancient fossils. However, more recently their organic origin has been disputed.

Common organisms: cyanobacteria

Volcanism: geysers and hot pools

Climate: tropical–subtropical

Biota: prokaryotic microbes

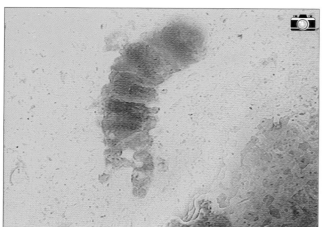

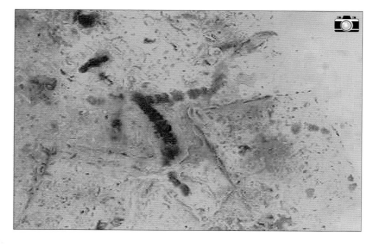

❶ *Primaevifilum*
❷ *Archaeoscillatoriopsis*

☞ ARCHAEOSCILLATORIOPSIS (2)
Between 7 and 18 microns in width,
three species of this long, sinuous
genus were originally named from
the Apex Chert in the early 1990s,
though critics say that they are
nothing more than stacks of mineral
grains formed by hydrothermal
activity. The organism, if indeed it
is an organism, had a tapered,
filamentary shape composed of
disc-shaped cells.

SNOWBALL EARTH
FLINDERS RANGE, SOUTH AUSTRALIA

Climate: widespread glaciation
Biota: low-diversity marine micro-organisms (eg acritarchs)

Latitude then: *c.*8°N

Latitude now: 31°S

Sea level: falling until end of glaciation, then rising rapidly

Original environment: global terrestrial ice sheets and sea-ice perhaps as far as the equator

Deposits: "Elatina formation" glacial deposits

Status: numerous accessible exposures of Elatina strata occur around Adelaide, South Australia

Preservation: the glacial deposits are devoid of fossils

⊕ Earth *c.*640 MA

⊕ Fossil site today

One of the most extraordinary phenomena in Earth's early history is that of repeated and extensive glaciation. According to the "Snowball Earth" theory, at least two, and perhaps four or more, glacial phases encompassed the Precambrian world from pole to pole. Theoretically, growing polar ice sheets reflected so much solar energy back into space that even the tropics were cooled sufficiently to ice over, with temperatures plummeting to -50°C, freezing the world's oceans.

The first glacial event occurred around 2.3 billion years ago, but the best known events are all late Proterozoic, dating to around 710, 640, and perhaps 580 MA – known as the Sturtian, Marinoan, and Varangian glaciations.

Supposedly, the impact on environment and life was such that, in this frozen "icehouse" state, ocean productivity and the weathering of the land were shut down. Unable to photosynthesize, marine phytoplankton died off and the oceans became anoxic. Biological activity only revived in brief, hot interglacials. Such stop–start, "freeze–fry" processes were precursors to the subsequent explosion of Ediacaran life, and may be responsible in some way for that event. At least, that is the theory, but it is still far from proven.

647–635 million years ago
Cryogenian Period of the Neoproterozoic Era

Common organisms: cold-tolerant micro-organisms

❶ dropstone

⟲⟳ DROPSTONE (1) This metre-wide boulder – one of many found 'floating', surrounded by finer grained sediment – is a dropstone, formed when a melting iceberg released it to fall into soft sediment on the sea floor. No other natural mechanism can carry such large and heavy rocks out to sea, so the presence of dropstones is a useful indicator of glacial activity. This spectacular specimen occurs within the Late-Proterozoic (Sturtian) glacial strata of Namibia's Skeleton Coast, close to Narachaamspos. Carbonate rocks in the overlying strata lack dropstones: they show that, following the glacial event, the environment quickly returned to its normal hot subtropical phase.

MYSTERIOUS EDIACARANS
MISTAKEN POINT, NEWFOUNDLAND, AND EDIACARA, AUSTRALIA

Climate: southern hemisphere polar
Biota: soft-bodied marine "Ediacarans"

Latitude then: c.80°S

Latitude now: 25°N

Sea level: rising (+200m/650ft)

Original environment: shallow sea

Deposits: fine-grained seabed muds and sands

Status: numerous sites over a wide region are actively being excavated for their fossils

Preservation: highly flattened but often with soft tissues preserved as imprints

Mistaken Point

↻ Earth c.570 MA

Mistaken Point

↻ Fossil site today

Since the late 1940s it has become increasingly clear that a diverse and widespread marine biota of strange organisms evolved towards the end of the Neoproterozoic Era. Known as the Ediacarans, they were soft-bodied and often curiously quilted, with forms ranging from flat discs through jellyfish-like "blobs" and domes, to "fronds" up to a metre (40in) or more in length. Their biological affinities are unknown, despite early attempts to shoehorn them into familiar invertebrate groups such as scyphozoan jellyfish, annelid worms, and molluscs. Although a few do bear a superficial resemblance to living organisms, detailed analysis has not supported these affinities. For instance *Charnia*, one of the longest-lived and most widespread Ediacarans, has been seen as a seapen, but recently its mode of growth has been shown to be quite different.

The oldest known Ediacaran biota comes from Mistaken Point, on Newfoundland's Avalon Peninsula. Thirty or so different Ediacarans have been found scattered through some 3km (2 miles) of strata deposited over 10 million years from about 575 million years ago. Here, the Ediacarans lived in relatively deep water (c.50m/165ft), and died as successive falls of volcanic ash overwhelmed and suffocated them.

Common organisms: "spindles" and fronds

575–565 million years ago
Ediacaran Period of the Neoproterozoic Era

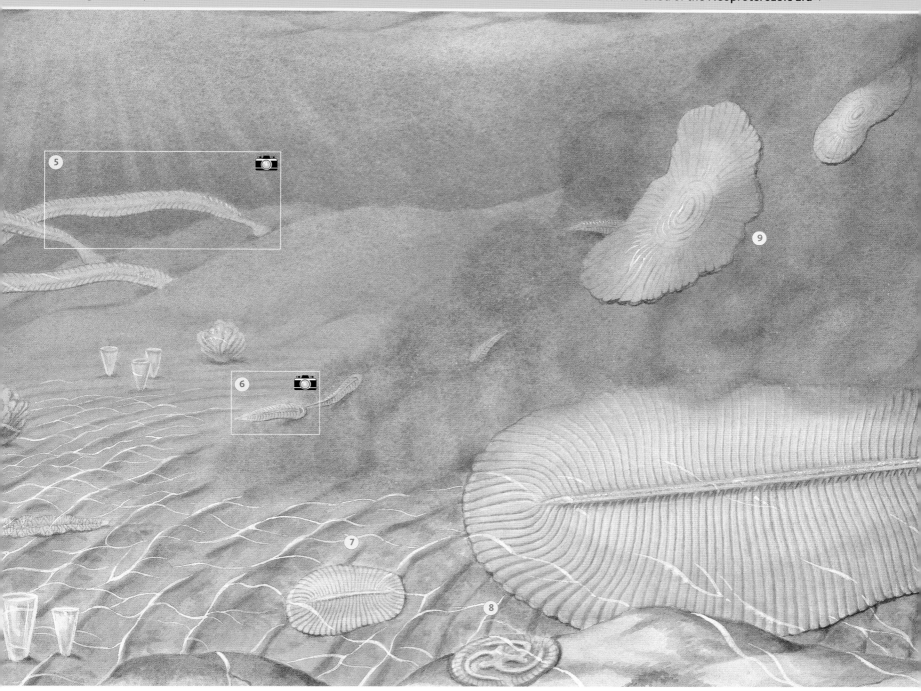

Mistaken Point:
1. *Charniodiscus*
2. *Thectardis*
3. "spindles"
4. *Bradgatia*
5. *Charnia wardi*

Ediacara:
6. *Spriggina*
7. *Dickinsonia*
8. *Tribrachidium*
9. *Cyclomedusa*

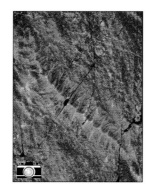

⊙ "SPINDLES" (3) Up to 15cm (6in) long, these seabed organisms are probably rangeomorphs.

⊙ CHARNIA WARDI (5) This is the largest known Ediacaran, with fronds up to 2m (6.6ft) long, a stiff central shaft, and a holdfast disc. Fronds are often found buried in parallel alignments.

⊙ SPRIGGINA (6) One of the first fossils found in the Ediacaran Hills, *Spriggina* grows up to 3cm (1.2in) long, and seems to have a symmetrical, segmented body with some sort of "head" structure.

43

EDIACARANS DIVERSIFY
EDIACARA, AUSTRALIA (CONT.) AND NAMA GROUP, NAMIBIA

Climate: tropical
Biota: soft-bodied marine "Ediacarans"

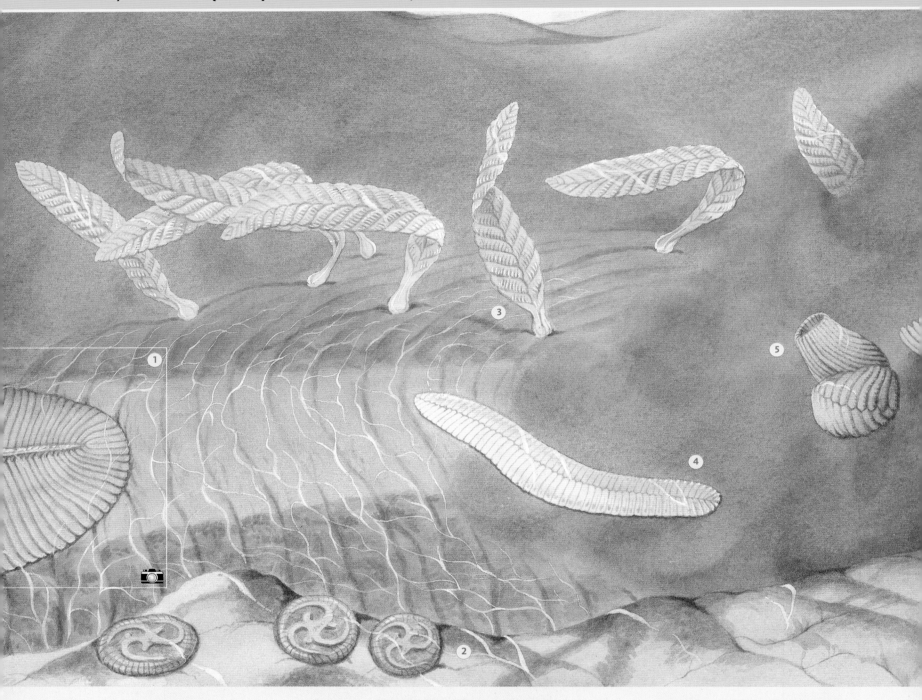

Latitude then: 33°N
Latitude now: 31°S
Sea level: rising (+75m/245ft)
Original environment: shallow sea
Deposits: seabed sands (quartzites) and muds
Status: several sites within the Ediacara Hills, Flinders Range
Preservation: slightly flattened moulds and casts of soft tissue

• Ediacara Hills

⊙ Earth *c.*545MA

Ediacara Hills •

⊙ Fossil site today

Fossils from the Ediacara Hills in Australia's Flinders Range began the modern interest in the Ediacaran biota around 1946. Their discoverer, geologist Reg Sprigg, described them as soft-bodied jellyfish-like organisms of early Cambrian age. The discovery of similar fossils in English Precambrian strata led to a reassessment of the Australian fossils' age, and they gained wide notice in the late 1950s when Austrian-born Martin Glaessner published illustrations and descriptions of these remarkable fossils.

The Australian Ediacarans occur in shallow sea sands, and questions about how "jellyfish" could be preserved in such deposits led to a questioning of their biological affinities. These organisms now help define the final, "Ediacaran" Period of the Neoproterozoic Era.

The youngest known Ediacarans (550–543MA) are found in Namibia, where the biota includes some particularly interesting and unusual examples. There are a number of sac-shaped forms, such as *Namalia* and *Ernietta*, which evidently lived in the seabed sands. The youngest strata extend to the early Cambrian, and some of the last Ediacarans such as *Pteridinium* were contemporaries of early shelled organisms such as *Cloudina* (see over).

545 million years ago
Ediacaran Period of the Neoproterozoic Era

Common organisms:

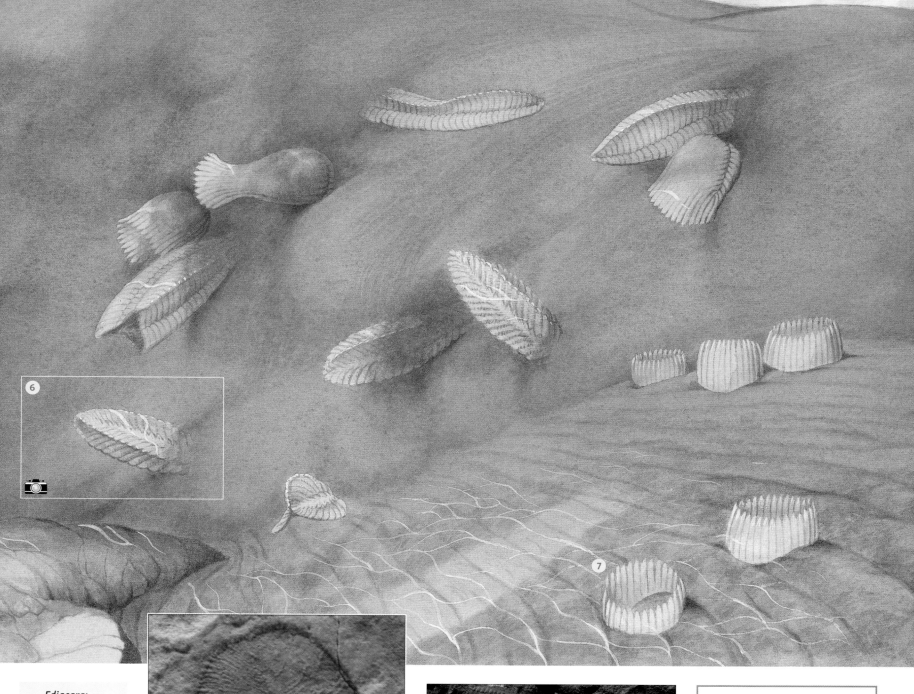

Ediacara:
1 *Dickinsonia*
2 *Tribrachidium*
3 *Charniodiscus*
4 *Phyllozoon*

Namibia:
5 *Namalia*
6 *Pteridinium*
7 *Ernietta*

Latitude then: 40°S

Latitude now: 27°S

Sea level: rising

Original environment: shallow sea

Deposits: seabed sands and carbonates

Status: scattered desert sites in central and southern Namibia

Preservation: 3D shells and moulds

• Nama Group

DICKINSONIA (1) Found in Australia, Russia, and China, this Ediacaran has a flat, oval, ribbed body, up to 80cm (31.5in) long, which may be segmented and is divided by a median line.

PTERIDINIUM (6) This widespread genus is known from Russia, Australia, India, North America, and Namibia. Up to 20cm (8in) long, it has a three-vaned frond-like shape, and seems to have lived partially buried in the seabed.

FIRST SHELLS AND REEFS
NAMA GROUP, NAMIBIA, AND SIBERIA, RUSSIA

548–535 million years ago
Late Ediacaran to early Cambrian periods

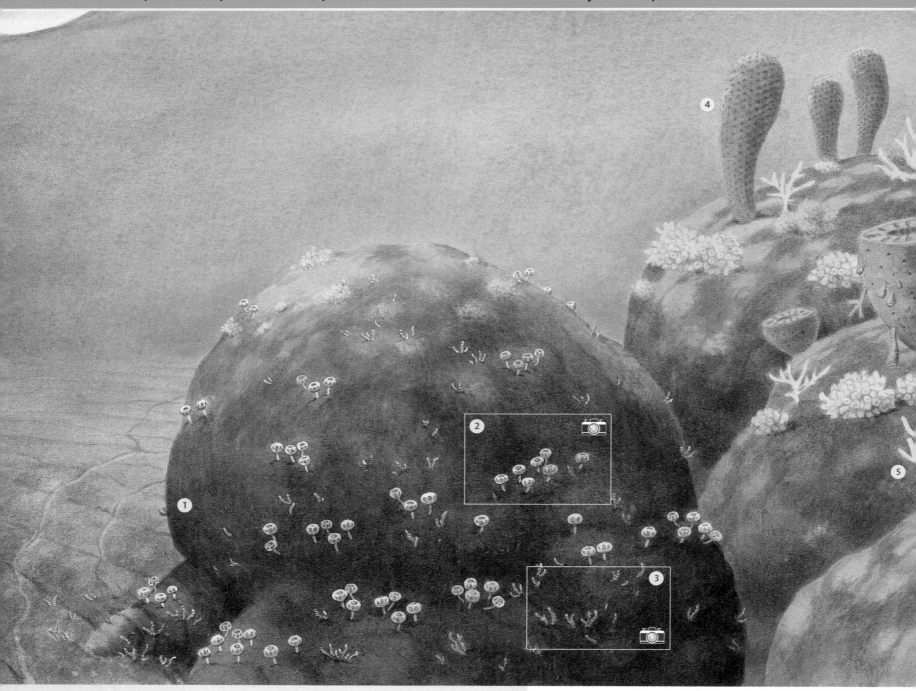

Building a shell around a body suggests the need for protection or support, and requires some "clever" biomechanisms and chemistry. One of the many momentous steps in evolution is the appearance of the first hard shells in the fossil record around 548 million years ago, during the Ediacaran Period of late Proterozoic times.

The first shells, belonging to *Cloudina* and *Namacalathus*, were tiny tubes just a few millimetres in size, made with calcium carbonate extracted from the seawater. They may have been a simple adaptation to support soft-bodied organisms and lift the feeding apparatus above the seabed, but they were also a protection against predators – and an "arms race" soon began.

Building hard, mineralized shells and skeletons allowed organisms to increase in size and colonize tough environments otherwise dominated by stromatolite-like microbial outcrops known as thrombolites. They were soon joined by the newly evolving molluscs, arthropods, and sponge-related archeocyathids with carbonate skeletons. These new associations led to the formation of the first reefs that grew in warm, shallow seas.

Namibia (cont.)
1. thrombolite
2. *Namacalathus*
3. *Cloudina*

Siberia
4. chancellorid
5. *Cambrocyathellus*
6. radiocyathid
7. coralomorphs
8. *Okulitchicyathus*

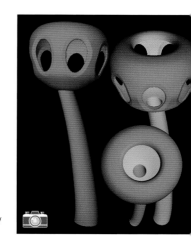

➔ NAMACALATHUS (2)
A reconstruction shows this organism's tiny (2mm-wide) calcium carbonate "goblets".

Common organisms: thrombolites, tubular shelled organisms, archeocyathids, coralomorphs

Climate: equatorial

Biota: diverse marine invertebrates

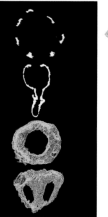

📷 ☉ **CLOUDINA (3)** This *Cloudina*'s millimetre-sized calcareous tube has been breached by an unknown predator. Despite *Cloudina*'s tough mineral shell, some predators were already ahead in the arms race.

↪ **RADIOCYATHID (6)** A cross-section cuts through the branched centimetre-sized skeleton of a primitive, and now extinct, calcareous sponge-related organism called a radiocyathid.

Latitude then: 22°S

Latitude now: 70°N

Sea level: rising

Original environment: shallow seas

Deposits: limestones

Status: the remoteness of these strata makes them virtually inaccessible

Preservation: 3D within limestone

• Kotuikan River

THE ARMS RACE GATHERS PACE
CHENGJIANG, YUNNAN PROVINCE, CHINA

520 million years ago
Stage 3 of the Cambrian Period

Latitude then: 8°N

Latitude now: 25°N

Sea level: rising (+200m/650ft)

Original environment: shallow sea

Deposits: fine-grained seabed muds

Status: numerous sites over a wide region are actively being excavated for their fossils

Preservation: highly flattened, but often with soft tissues preserved

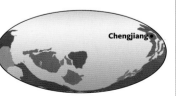
Chengjiang

⊕ Earth c.520 MA

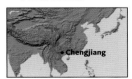
• Chengjiang

⊕ Fossil site today

Right from the earliest Cambrian times, around 542 million years ago, an extraordinary explosion of life in the seas prompted an arms race between predators and prey. Within 20 million years or so, most major groups of animals had evolved, including our remotest vertebrate ancestors. A first insight into the conflict between well-equipped predators and their increasingly armour-plated prey, is provided by the 520-million-year-old marine fossils of Chengjiang.

Top predators were the free-swimming anomalocarids, over 50 cm (20in) long, and armed with grasping appendages. Most animals living on or above the seabed were protected either with sharp spines like *Hallucigenia* and the sponges or with tough exoskeletons like the arthropods. Unarmoured animals such as *Myllokunmingia* were presumably fast swimmers, while the soft-bodied priapulid worms lived in burrows from which they ambushed their prey.

So far around 150 species, ranging from sponges and worms to vertebrates, have been found. Of these, 60 are arthropods, whose diversity supports the idea that arthropod ancestry must have extended back into Precambrian times.

Common organisms: sponges, worms, brachiopods, and certain arthropods

Climate: equatorial

Biota: diverse marine invertebrates

1. *Retifacies*
2. *Kumaia*
3. *Quadrolaminiella*
4. *Eldonia*
5. *Anomalocaris*
6. *Longtancunella*
7. *Archisaccophyllia*
8. *Canadaspis*
9. *Haikoucaris*
10. *Paraleptomitella*
11. *Myllokunmingia*
12. *Paucipodia*
13. *Paraselkirkia*
14. *Hallucigenia*

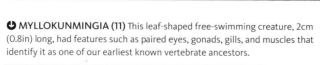

MYLLOKUNMINGIA (11) This leaf-shaped free-swimming creature, 2cm (0.8in) long, had features such as paired eyes, gonads, gills, and muscles that identify it as one of our earliest known vertebrate ancestors.

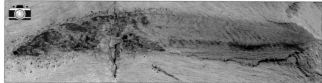

CANADASPIS (8) This little shrimp-like arthropod (up to 3cm/1.2in long) was a common swimming animal both at Chengjiang and in the Burgess Shale. Its body was covered with a bivalved carapace for protection.

PARASELKIRKIA (13) This soft-bodied carnivore, about 2cm (0.8in) long, is related to today's priapulid worms, which can introvert their hooked proboscis to pull prey into their bodies.

HALLUCIGENIA (14) This strange seabed-dwelling animal, up to 3cm (1.2in) long, was first discovered in Canada's Burgess Shale. The first reconstructions are now known to have been upside down, with what were probably defensive spines interpreted as legs.

1. LIFE IN THE LATE PRECAMBRIAN 630-543ma

"Ediacaran" fossils from have now been found from around the world in strata that were deposited between 630 and 542 million year ago. They include a diverse variety of marine soft-bodied organisms, some of which grew up to 2m (80in) in size. Their different anatomies reveal that they followed different modes of life both on and within the seabed.

In general the Ediacarans ranged from flat disc and ribbon shapes, some of which were mobile, to rooted fronds and sacs, mostly a few centimetres in size. The bodies are preserved as impressions on the surrounding sediment (casts and moulds) and often reveal a curious quilted appearance. They have a superficial resemblance to jellyfish, but there is no evidence for typical "cnidarian" features such as tissue specialization or an opening to a stomach-like structure. There is no confirmed evidence for any of these organisms surviving beyond around 543 million years ago.

2. ACROSS THE BOUNDARY 548-535ma

The earliest known fossilized "hardparts", some 548 million years old, belong to millimetre-sized and relatively primitive organisms. Tiny calcified stalked globules (Namacalathus) and tubes (Cloudina) growing from the seabed housed unknown animals, perhaps related to sponges and to worms. Other calcified "skeletons", the radiocyathids, show clearer links to later sponges.

Mineralized skeletal elements can have a variety of functions. Solid surfaces can provide sites for muscle attachment, while hard shells can provide protection from environmental dangers including predators. Holes bored in some Cloudina tubes reveal that they were indeed preyed upon by unknown animals equipped with some kind of rasping tooth structure similar to those found in some molluscs.

CHARNIODISCUS
This filter-feeder bore a passing resemblance to sea pens, but is not thought to be related.

DICKINSONIA
This large Ediacaran was capable of movement. Paleontologists have attempted to link it to a number of later groups.

STROMATOLITES
By the late Precambrian, most stromatolites were being formed by colonies of green algae.

TRIBRACHIDIUM
This cushion-like Ediacaran displays a unique three-fold symmetry, and cannot be conclusively linked to any other group.

ERNIETTA
These sac-like organisms are hard to link to any other species, even among the Ediacarans.

NAMACALATHUS
These strange organisms, with a cup-shaped top supported on a stalk, are the earliest to show a calcified shell.

ARCHAEOCYATHID
These reef-building organisms appear at the very start of the Cambrian, and may be the earliest known sponges.

THE DAWN OF PREDATION

One of the most intriguing theories put forward to explain the Cambrian explosion is that it was triggered by the appearance of a new mode of life – predation. According to some interpretations, the Ediacarans may have been symbiotic organisms – inflated bags of jelly acting as hosts to photosynthetic bacteria that turned sunlight into nutrients. This hypothesis imagines the late Precambrian as a predator-free "Eden", and suggests the "small shelly fauna" that followed was a direct response to the first predators. But where did these predators come from, and how did the shells evolve? One theory is that both were a result of increased calcium content in the seas of the time. Using this calcium to create mineralized hard parts allowed some animals to build support structures, but in some unknown animals these gave rise to primitive rasping teeth and allowed predation to begin on a larges scale. Others soon put their mineralized structures to defensive use, and an "arms race" began.

THE EXPLOSION OF LIFE

THE MYSTERY OF THE CAMBRIAN EXPLOSION

▲ **WATERWORLD** The Cambrian explosion occurred during a period of high sea levels and warm temperatures.

The sudden appearance of abundant visible fossils in the rock record has been a problem since Darwin's days. As he wrote in The Origin of Species, "there is another difficulty... I allude to the manner in which numbers of species of the same group, suddenly appear in the lowest known fossiliferous rocks".

Even 150 years later, it is still true that the fossil remains of many marine animal groups suddenly appear around the beginning of Cambrian times, now dated at 542 million years ago. These include various molluscs, brachiopods and several different kinds of arthropods. Again, as Darwin wrote, "if the theory (of evolution) be true, it is indisputable that before the lowest Cambrian stratum was deposited, long periods elapsed... and that during these vast periods the world swarmed with living creatures".

Since Darwin's time our knowledge of pre-Cambrian life has been revolutionized by the discovery of fossil remains showing that the world did indeed swarm with living creatures as far back as 3.5 billion years ago. However these discoveries have only highlighted a new mystery – much of the early record consists of microscopic unicellular life – algae and bacteria. Small organisms with multiple cells first appear around 2.5 billion years ago, and somewhat larger metazoans with eukaryotic cells (with a distinct nucleus) around 1.6 billion years ago. But it is not until around 550 million years ago that the trace fossils (sediment burrows) of relatively large, soft-bodied "bilaterian" organisms start to appear in the rock record. Despite firm evidence for simple forms of life stretching back over billions of years, it seems there was a genuine change in the complexity and diversity of life at around this time – the so-called "Cambrian explosion".

Then from the 1950s onward, a group of puzzling fossils – the remains of relatively large soft-bodied organisms from the very late Precambrian – were identified in the rock strata of South Australia's Ediacara Hills. The vital question was – were these the missing ancestors of the Cambrian animals? At first there did seem to be some promise that despite their lack of hard parts, the various Ediacaran organisms could be shoe-horned into known groups such as jellyfish, annelid worms, molluscs and arthropods. But as better-preserved Ediacarans came to light from a number of sites around the world, it became clear that few if any of these suggestions held water.

TRACING GENETIC HERITAGE

Another approach to the problem of the Cambrian explosion is through the "molecular clock", comparing certain molecules within living groups of metazoans in order to measure their differences from one another. These molecules are inherited almost unchanged from one generation to the next, but are susceptible to slow change or "drift" over many generations thanks to the accumulation of random mutations. Assuming the rate of drift has been more or less constant in different groups throughout evolution, it is possible to calculate how closely various groups are related to one another, and estimate when they might have diverged from one another in the distant past.

Such calculations suggest that the main metazoan groups were already distinct from one another some 800 million years ago, deep within Precambrian times. If true, then the Cambrian explosion presumably represents a roughly simultaneous evolution of hard parts such as skeletons and shells in various pre-existing groups, rather than a rapid and more fundamental diversification of metazoan body plans.

If the molecular clock measurements are accurate, then this still leaves the problem of a missing fossil record of those soft-bodied ancestors between 800 and 540 million years ago – while fossil deposits with the right conditions to preserve soft tissues are rare, and deposits from such early periods are rarer still, it nevertheless seems hard to explain why nothing of the right age has yet turned up. This points to another possibility – while the molecular clock seems to be tie in neatly with the fossil record on shorter timescales, is it possible that the rate of mutation, has indeed changed over time?

EVIDENCE

▼ **SNOWBALL EARTH** During the period just before the emergence of the Ediacarans, our planet went through a major series of ice ages in which glaciers extended to equatorial regions and perhaps covered the entire Earth. This "Snowball Earth" event, and Earth's rapid emergence from it in the late Precambrian, may have produced changes on land and in the oceans that may be linked to changes in the pace of marine evolution.

▶ **WONDERFUL LIFE** The first major site to reveal the sudden diversity of early Cambrian life is Canada's famous Burgess Shale, discovered in the early 1900s. However, the identification of these fossils has been a matter of debate – new studies in the 1970s suggested that many represented long-extinct "experiments" in early metazoan life, but more recently most have been interpreted as early members of familiar surviving groups.

THE
EXPLOSION

Ever since the first fossil remains from the Cambrian Period were described, it has been clear that many marine animals seem to appear suddenly in Early Cambrian strata. Yet these organisms already include recognizable and distinct forms such as sponges, worms, molluscs, brachiopods, and arthropods. In contrast, soft-bodied Ediacaran fossils are found in the latest Precambrian strata around the world, but few can be related biologically to the early Cambrian organisms. If the fossils tell the whole story, then the various major animal groups must have arisen in an evolutionary "explosion" during the earliest Cambrian times. But the record may be deceptive – another possibility is that these groups were already evolving independently in the Precambrian, and that several of these groups acquired preservable hard parts more or less simultaneously at the beginning of Cambrian times.

OF LIFE

Charles Walcott's 1909 discovery of abundant and beautifully preserved Cambrian fossils in British Columbia's Burgess Shale prompted an intense investigation of their evolution. These mid-Cambrian rocks record a well established and diverse fauna with signs that a complex ecosystem had already developed.While tiers of filter-feeding organisms of varying heights exploited the drift of food particles across the seabed surfaces, other organisms burrowed in the sediment beneath. There were mobile seabed dwellers and active swimmers, and the numerous arthropods shows advanced specialization for different lifestyles.

This early development of ecological complexity either implies a significant and, as yet unknown ancestry stretching back into late Precambrian times or, a very rapid diversification of life with the acquisition of mineralized shells and skeletal parts in the Early Cambrian. The more recent discovery of the even older but equally complex Chengjiang fauna from China has sharpened the debate. The argument in favour of a real Cambrian explosion is strengthened by an apparent lack of arthropod diversity in the very earliest Cambrian or latest Precambrian strata – but since the fossil record from this boundary period has no real equivalent to Burgess or Chengjiang, we cannot be sure.

4. THE CAMBRIAN LEGACY TODAY

The explosion in Cambrian life defined the subsequent history of marine organisms for some 300 million years through Paleozoic times. It established the basic bodyplans for many of the familiar aquatic and terrestrial metazoan animals alive today - especially the molluscs and arthropods. And, within a relatively short time, the first chordates evolved the basic vertebrate body plan. The most important organisms that were not present at this time were the vascular plants.

However, since the Cambrian there has been a constant evolutionary turnover of species and family groupings, with no Cambrian species alive today and several entire Paleozoic groups now extinct. Newly evolving groups, especially among the vertebrates and vascular plants ensured a significantly different look to the life of late Paleozoic and Mesozoic times but many of these also died out, so our modern biota has little resemblance to that of Cambrian times – the true legacy of the explosion today lies in genes and basic biological structures.

VERTEBRATES

Possession of a backbone and toughened skeleton has allowed vertebrates to develop in astounding diversity within the sea, and also to move onto land, where they account for all large animals.

ARTHROPODS

Toughened external "exoskeletons" have also allowed the arthropods to diversify into a wide variety of forms on land and in the sea. Although generally small in size, they account for a vast majority of animal species.

PRIAPULIDS

Today there are just 16 species of priapulid worms, confined to specialized lifestyles as filter feeders burrowing through mud in cold polar waters.

ALGAE

Algae continue to play a major role in Earth's ecosystem, and gave rise to land plants around 450 million years ago. However, the success of grazing animals means fewer mat-formong colonies produce stromatolites today.

BRACHIOPODS

The brachiopods flourished throughout the Paleozoic era, but suffered in the mass extinction that marked the end of the Permian (see p.108). Since then they have been largely supplanted by bivalve molluscs.

CNIDARIANS

Characterized by their stinging cells, this group soon split into two main branches –sedentary anthozoans such as corals and anemones, and the mobile medusozoa such as jellyfish. Both are still successful today.

SPONGES

At least two of the three groups of sponges (the glass and calcareous sponges) seem to have originated in the Cambrian explosion. Today, however, they are vastly outnumbered by the demospongiae.

3. THE HEART OF THE EXPLOSION 535-510ma

Different centimetre-sized fossil tubes called
barites appear in earliest Cambrian strata (some
million years old), shortly before the is little sign a
rse range of arthropods. Right at the end of the
proterozoic when manyEdiacarans were becoming
nct, the first evidence formineralised skeletons
ears in the fossil record with the tiny enigmatic
anisms, loudina and Namacalathus.

Whether triggered by an "arms race" between other
predators and armoured prey, the changing
geochemical environment of the oceans, or a
combination of both, by mid-Cambrian times the
biological and ecological foundations of Paleozoic
marine life had been established. Moreover, a range of
body plans had developed that would allow
subsequent adaptation to life in brackish-water,

freshwater and eventually terrestrial environments.
The arthropods, with their jointed legs and tough
exoskeletons, were effectively "preadapted" for the
difficulties of life out of water in the dry gas of our
atmosphere. The chordates, meanwhile, had an
internal skeletal anatomy that would prove so
adaptive for the evolution of tetrapods and life on land.

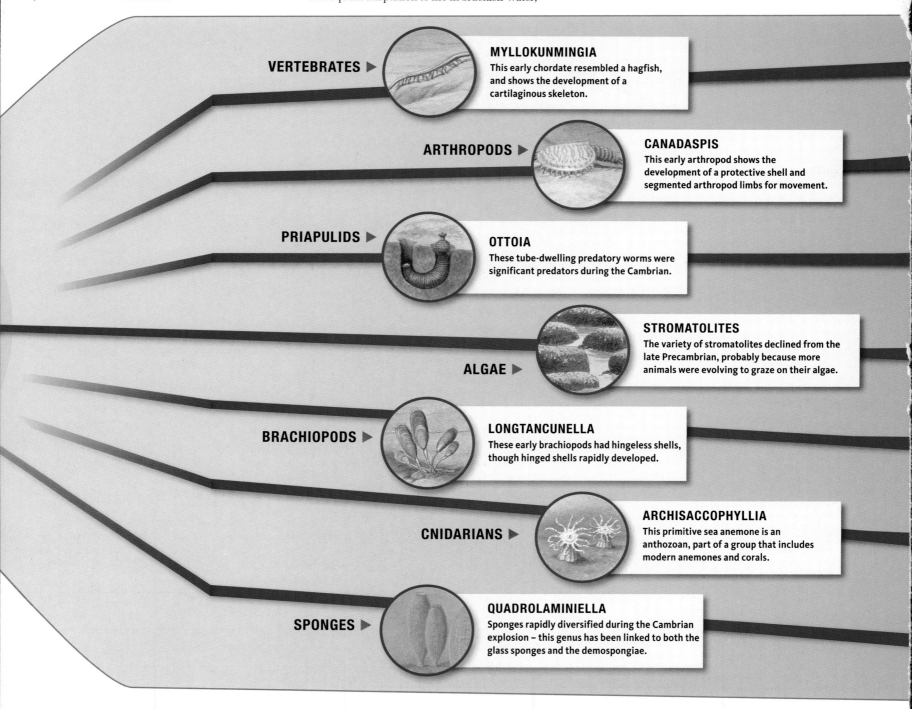

VERTEBRATES ▶

MYLLOKUNMINGIA
This early chordate resembled a hagfish,
and shows the development of a
cartilaginous skeleton.

ARTHROPODS ▶

CANADASPIS
This early arthropod shows the
development of a protective shell and
segmented arthropod limbs for movement.

PRIAPULIDS ▶

OTTOIA
These tube-dwelling predatory worms were
significant predators during the Cambrian.

ALGAE ▶

STROMATOLITES
The variety of stromatolites declined from the
late Precambrian, probably because more
animals were evolving to graze on their algae.

BRACHIOPODS ▶

LONGTANCUNELLA
These early brachiopods had hingeless shells,
though hinged shells rapidly developed.

CNIDARIANS ▶

ARCHISACCOPHYLLIA
This primitive sea anemone is an
anthozoan, part of a group that includes
modern anemones and corals.

SPONGES ▶

QUADROLAMINIELLA
Sponges rapidly diversified during the Cambrian
explosion – this genus has been linked to both the
glass sponges and the demospongiae.

A CAMBRIAN WORLD HERITAGE SITE
BURGESS SHALE, BRITISH COLUMBIA, CANADA

510 million years ago
Stages 4–5 of the Cambrian Period

Latitude then: 15°N

Latitude now: 51.5°N

Sea level: high

Original environment:
tropical marine continental shelf
slope to basin

Deposits: seabed muds

Status: World Heritage Site,
designated in 1981, between Mount
Wapta and Mount Field

Preservation: flattened and slightly
metamorphosed but with soft parts
preserved.

⊕ Earth c.510MA

⊕ Fossil site today

As with the older Chengjiang biota, the remarkable feature of the Burgess Shale is its diversity and abundance, especially among the arthropods. They are already adapted to such a huge array of ecological niches that some considerable time must have passed since they first evolved.

Did the boom in diversity happen at the beginning of Cambrian times when the arms race began, or does it stretch back into the late Precambrian? The problem with the latter scenario is that there is no fossil record of groups such as the arthropods appearing before the Cambrian.

The Burgess fossils are nearly always found in sediment at the base of a near-vertical submarine cliff. Periodic avalanches of seabed muds carried away any organisms unlucky enough to be living in the vicinity at the time. When the avalanche eventually halted downslope, the creatures entombed in the mud died from lack of oxygen.

Luckily for paleontologists, this almost instantaneous burial deterred scavengers, while the anoxic conditions helped preserve the biota's soft parts, such as gills, antennae, and internal organs, as well as the normally preserved hard parts.

Common organisms: sponges, shrimp-like arthropods, eldoniids, and priapulid worms

Climate: seasonal – tropical waters
Biota: diverse marine organisms, especially arthropods

⊙ **ANOMALOCARIS (13)** First thought to be a jellyfish, this fossil actually preserves the toothed mouth structure of the metre-long (40in) *Anomalocaris*, the Burgess Shale's top predator.

⊙ **WIWAXIA (14)** This scaly mud-crawler, up to 5cm (2in), had a snail-like soft underside and radula, and was related to polychaete worms.

1. *Pikaia*
2. *Opabinia*
3. *Thaumaptilon*
4. *Yohoia*
5. *Odontogriphus*
6. *Dinomischus*
7. *Marrella*
8. *Aysheaia*
9. *Vauxia*
10. *Sidneyia*
11. *Olenoides*
12. *Canadaspis*
13. *Anomalocaris*
14. *Wiwaxia*
15. *Canadia*
16. *Pirania*
17. *Hallucigenia*
18. *Ottoia*
19. *Eldonia*
20. *Ctenorhabdotus*

⊙ **CANADIA (15)** This 3cm-long (1.2in) segmented worm had a protective covering of long bristles that helped it move over the seabed, protected it from predators and may have given it an iridescent colouring.

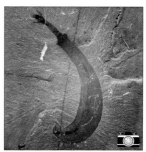

⊙ **OTTOIA (18)** This 8cm-long (3in) priapulid was a common predator of the seabed that lived in a burrow and ambushed unwary animals that passed too close.

REEF LIFE FLOURISHES
TRENTON, NEW JERSEY, USA

c.461 million years ago
Middle to Late Ordovician epochs

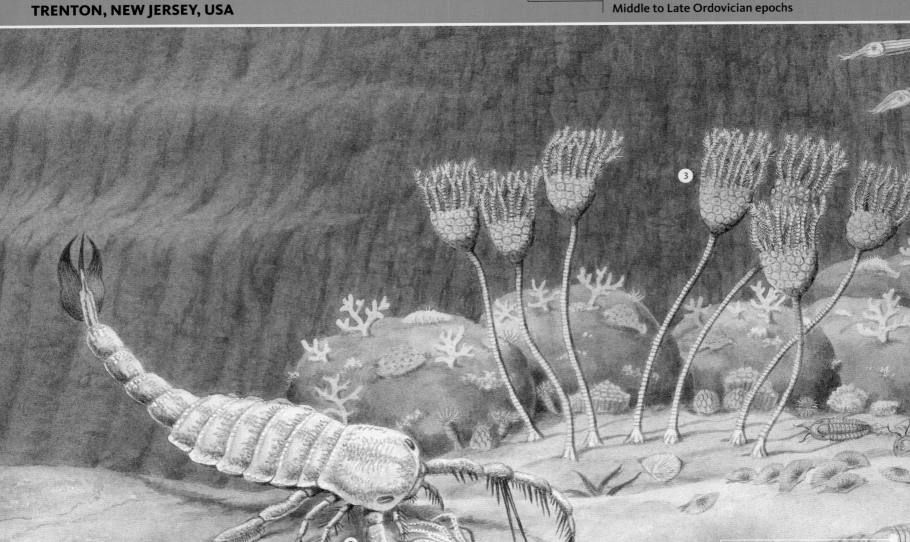

Latitude then: 13°S

Latitude now: 39°N

Sea level: high

Original environment: shallow sea

Deposits: mud carbonates

Status: many historic sites scattered over the extensive Ordovician outcrop of central and eastern North America

Preservation: three-dimensional fossils in limestone; some soft tissue preservation in pyritous shales

⊕ Earth c.461MA

⊕ Fossil site today

By Ordovician times, life in shallow seas was bursting with evolutionary innovation, and marine diversity had risen to 500 families. Most important was the rapid development of reefs, especially in the tropical waters that flooded Laurentia (North America). Increasing numbers of organisms with calcified skeletons contributed to the reef fabric.

Continually growing thrombolites formed extensive mound-shaped banks rising above the seabed in subtidal waters. This firm substrate provided anchorage for calcareous sponges, bryozoans, sea lilies (crinoids), and the newly evolving corals. Their growth –

especially the meadow-like swathes of sea lilies with strong, flexible stems – helped baffle and subdue water currents.

In these sheltered hollows, brachiopods and some clams filtered organic particles from the water, while trilobites wandered the seabed scavenging for organic debris, and snails grazed on algae. The first eurypterids lurked in the shadows waiting to grab unsuspecting prey with their pincers.

In the waters above, predatory nautiloids cruised around awaiting any opportunity to pounce. Some of these grew into the giants of the early Paleozoic seas, up to 10m (33ft) long.

Common organisms: sponges, sea lilies, trilobites, brachiopods, nautiloid cephalopods

Climate: tropical marine, high global summer temperatures
Biota: diverse reef dwellers

❶ Sinuites
❷ Megalograptus
❸ Balanacrinus
❹ Homotelus
❺ Orthoceras
❻ Salteraster
❼ Sowerbyella

➲ HOMOTELUS (4)
The discovery of many complete *Homotelus* suggests that this 8cm (3in) trilobite may have gathered together in large seasonal swarms, probably for the purposes of reproduction.

↻ SINUITES (1)
Symmetrical muscles show the shell of this early snail, 4cm (1.6in) long, opened to the rear. A slit around the shell allowed separate expulsion of deoxygenated water from respiration.

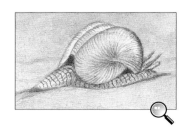

⬆ SOWERBYELLA (7) An entire shell of this common early Paleozoic brachiopod, about 1.5cm (0.6in) wide, lies on a fossilized seabed among broken fragments of other brachiopods.

59

RISE OF THE JAWLESS FISH
SACABAMBILLA, COCHABAMBA, BOLIVIA

c.455 million years ago
Katian Stage of the Late Ordovician Epoch

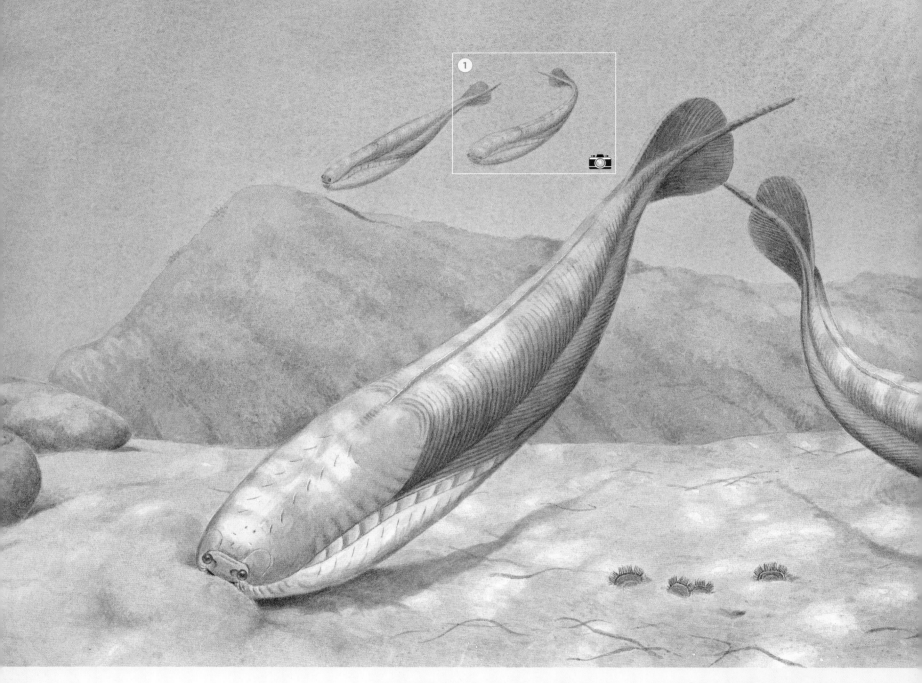

Latitude then: 50°S

Latitude now: 17°S

Sea level: falling

Original environment: shallow sea

Deposits: sandstones

Status: locality near a remote village high in the Andes

Preservation: original bone sometimes preserved, and virtually complete internal casts preserve good detail

⊕ Earth *c*.455ᴍᴀ

⊕ Fossil site today

The first fish to evolve in early Paleozoic seas were very different from modern fish. They were strange-looking creatures – jawless and toothless, and consequently known as agnathans. The most primitive lacked paired fins, and are distantly related to the surviving agnathan hagfish. The head and trunk of many fossil forms were covered with thick protective plates of an easily repairable porous bone-like material, suffused with blood.

To be used in swimming, the tail had to be flexible, and so it was less well protected, with either naked skin or thin scales. Without teeth, the agnathans were mostly mud grubbers, sucking up soft algal and microbial mats from the seafloor, along with organic, rich sediment.

The first early agnathan fossils (*Eryptichius* and *Astraspis*) were found within Colorado's Late Ordovician Harding Sandstones during the 1890s by American paleontologist Charles Walcott. Even older, Middle Ordovician remains (*Arandaspis*) were found in Central Australia in 1977. But the most complete fossils are those of *Sacabambaspis* from Bolivia, where many complete bodies were found stacked amid lingulid shells, suggesting they all died suddenly from the same cause – perhaps a rapid change in salinity.

Common organisms: brachiopods, agnathans, algae

Climate: cooling into an ice age
Biota: low-diversity shallow marine

❶ *Sacabambaspis*
❷ lingulid brachiopod

⊕ "NO JAW" (1) The scoop-shaped mouth of *Sacabambaspis* was filled with thin rows of bony platelets for scraping algae and mud from the seabed.

↻ SACABAMBASPIS (1)
Although the shape of this 24cm-long (9.5in) agnathan is captured by a well-preserved 3D sandstone mould of its interior, no details of the bony body armour have survived.

⊕ LINGULID BRACHIOPOD (2)
Only a small cluster of hair-like spikes projecting above the seabed give away the presence of this primitive burrowing brachiopod.

COLD WATERS, GIANT CONODONTS
SOOM SHALE, SOUTH AFRICA

450 million years ago
Katian Stage of the Late Ordovician Epoch

Latitude then: 60°S

Latitude now: 32°S

Sea level: falling

Original environment: shallow marine waters

Deposits: laminated marine muds with rare dropstones

Status: the Soom Shale sites are on private land and are legally protected

Preservation: the fossils are flattened but importantly preserve some soft tissues

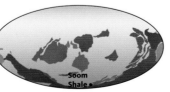

⊕ Earth c.450MA

⊕ Fossil site today

The discovery of a giant conodont fossil in 1993 first brought international attention to the Soom Shale and its cold-water sea creatures. Conodonts were eel-shaped animals – active predators with a mouth full of tiny, hard teeth, quite different from most vertebrate teeth.

For more than 100 years conodont fossils, which are common in Paleozoic seabed deposits, were only represented as isolated tiny teeth, about a millimetre in size. The parent animal was entirely unknown.

Not until 1983 was the first fossil conodont animal found in Scottish Carboniferous strata. No more than 4cm (1.6in) long, it was evidently a chordate with well-developed eyes and a complex feeding apparatus with several pairs of conodont tooth bars. By contrast the Soom Shale conodont is enormous – around 40cm (16in) long with a tooth apparatus some 2cm (0.8in) long. It is thought that the animals could project the apparatus to grab prey that was then chopped into pieces when the apparatus retracted into the mouth.

Most of the other Soom Shale animals, such as the arthropods and cephalopods, were also free swimmers, as seabed conditions were low in oxygen, although some sessile brachiopods apparently managed to survive.

Common organisms: conodonts, eurypterids, naraoiids, nautiloids, brachiopods

Climate: shallow sea cold water with some sea-ice

Biota: low-diversity, cold water marine animals

❶ *Onychopterella*
❷ orthoconic nautiloid
❸ orbiculoid brachiopod
❹ *Promissum*
❺ *Soomaspis*

↪ **HITCHING A LIFT (3)**
Filter-feeding orbiculoid brachiopods needed a firm surface for attachment, such as the shell of this cone-shaped nautiloid. The brachiopod larvae may have used their fleshy pedicles to anchor themselves to the nautiloid while it was still alive.

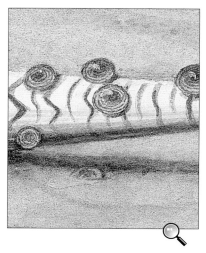

🔍 **PROMISSUM (4)** Conodont teeth consisted of several pairs of bilaterally symmetrical spiky bars, which functioned together as a feeding apparatus, capable of drawing prey into the mouth. The dark, rounded patches are fossilized eye capsules.

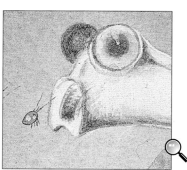

🔍 **PROMISSUM (4)** This 40cm-long (16in) conodont was a large-eyed predator that could extend its tooth apparatus to enmesh its prey.

SEA SCORPIONS AND JAWLESS FISH
LESMAHAGOW, SCOTLAND

430 million years ago
Telychian Stage of the Llandovery Epoch

Latitude then: 17°S

Latitude now: 56°N

Sea level: high

Original environment: shallow marine to brackish waters

Deposits: mud carbonates

Status: several protected historic sites throughout the Midland Valley of Scotland

Preservation: fossils flattened but mostly well preserved with some fossilized soft tissues

● Lesmahagow

⊕ Earth c.430MA

● Lesmahagow

⊕ Fossil site today

By Silurian times the agnathan fish were flourishing in the oceans of the world and extending their range into shallow waters with fluctuating salinities. At least 15 different kinds of agnathans are known from Silurian strata in Scotland's Midland Valley, where they range from sluggish bottom-dwelling and heavily armoured forms to more active non-armoured swimmers (such as *Loganellia*, *Jamoytius* and *Birkenia*) that probably fed on a variety of microbial and planktonic food. Some of the best specimens of these later swimmers have been found in the fine-grained sediments of localities such as Lesmahagow.

However, the agnathans were, by this time, facing fierce competition from both the newly evolving jawed fish, and large and active arthropod predators – the eurypterids, some of which grew up to 2m (6.6ft) long. Also known as the sea scorpions, these extinct arthropods were armed with large and powerful pincers for grabbing and pulling apart their prey.

Numerous smaller arthropods are also found, including *Ainiktozoon*, once thought to be a peculiar chordate animal. Turned upside down, it was more easily recognized as a free-swimming, predatory thylacocephalan arthropod, with a protective carapace.

Common organisms: agnathans, eurypterids, phyllocarids, brachiopods, sponges, sea lilies, trilobites, nautiloids

Climate: warm-water tropical marine

Biota: euryhaline shallow sea dwellers

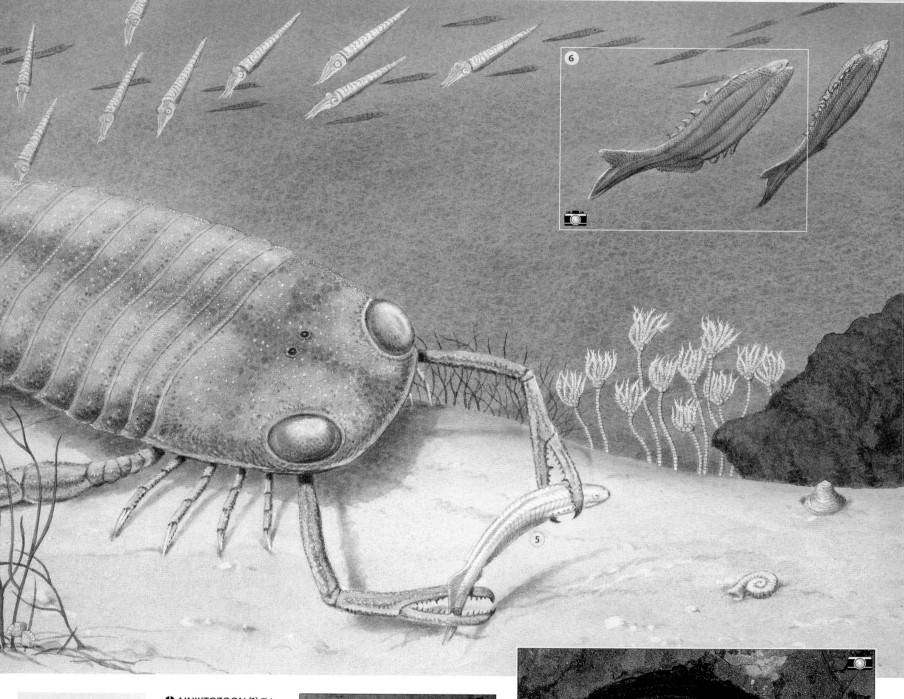

- ❶ *Ainiktozoon*
- ❷ *Loganellia*
- ❸ *Ceratiocaris*
- ❹ *Pterygotus*
- ❺ *Jamoytius*
- ❻ *Birkenia*

AINIKTOZOON (1) This strange thylacocephalan arthropod, about 15cm (6in) long, had large compound eyes for spotting its prey, which it then captured with its grasping clawed appendages.

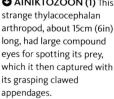

LOGANELLIA (2) This thelodont agnathan, 15cm (6in) long, has small paired fins and a body covered in naked skin with some very small scales. They are widely found across the Baltic and Germany.

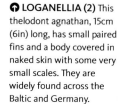

CERATIOCARIS (3) This free-swimming shrimp-like arthropod, up to 12cm (5in) long, had a bivalve carapace for protection. The muscular tail could produce a rapid flick to escape predators.

BIRKENIA (6) Here flattened on it side, the 10cm (4in) *Birkenia* shows numerous thin body scales. The mouth is to the left with a large eye above, and the tail to the right.

REEF LIFE RECOVERS
WREN'S NEST, DUDLEY, ENGLAND

*c.*425 million years ago
Homerian Stage of the Wenlock Epoch

Latitude then: 18°S

Latitude now: 52.5°N

Sea level: high

Original environment: shallow marine

Deposits: carbonates

Status: many protected historic sites scattered over the Midlands and Welsh Borders

Preservation: commonly three-dimensonal, with shell carbonate preserved

⊕ Earth *c.*425 MA

⊕ Fossil site today

Following a glaciation in the Late Ordovician, melting ice raised water levels, flooding the continents with shallow seas. Tropical reefs flourished anew with an abundance of corals, bryozoans, stromatoporoids, and calcareous algae as the main reef builders.

Silurian fossil reefs are well preserved in North America's Great Lakes region, the Baltic, and the Russian Urals. Historically important reefs are preserved in the Welsh Borders and at Dudley in the English Midlands, where their limestone fueled the Industrial Revolution.

The Dudley reefs developed mainly as small knolls, a metre (3ft) or so high, along with lower patch reefs and shell pavements. Meadows of sea lilies and algae grew on the hard reef substrates along with brachiopods and clams, while trilobites and snails searched for food in the spaces between. Shoals of predatory squid-like orthoconic nautiloids swam overhead and the waters teemed with tiny conodonts and arthropods such as the minute bivalved ostracods. Occasional colonies of graptolites drifted in from deeper waters. When they died, their strangely geometric organic skeletons, looking like serrated fretsaw blades, sank to the seabed and were buried in the soft carbonate muds surrounding the reef.

Common organisms: corals, bryozoans, stromatoporoids, sea lilies, trilobites, brachiopods, nautiloids

Climate: tropical marine, high global summer temperatures, volcanic ashfalls
Biota: high-diversity reef dwellers (some 600 species recorded)

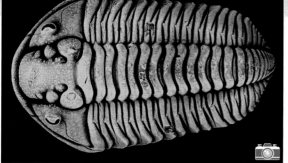

↻ CALYMENE (4)
Colloquially known as the Dudley "bug" or "locust", this trilobite grew up to 10cm (4in) long, and could roll up its body, bringing the tail under the head for protection from predators or a hostile environment.

↻ CYRTOGRAPTUS (9)
Despite their simple shapes, graptolites like this helically coiled one, around 12cm (5in) across, were complex hemichordate animals.

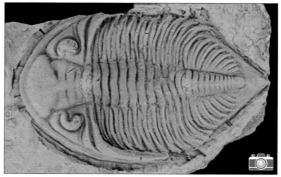

↥ DALMANITES (11) The typical trilobite body can be clearly seen in a complete specimen of this common and widespread Silurian arthropod, which grew up to 8cm (3in) long.

1. *Protochonetes*
2. *Halysites*
3. *Monograptus*
4. *Calymene*
5. *Gissocrinus*
6. *Favosites*
7. *Heliolites*
8. *Atrypa*
9. *Cyrtograptus*
10. *Ketophyllum*
11. *Dalmanites*

GREENING THE LAND
LUDFORD LANE, SHROPSHIRE, ENGLAND

419 million years ago
Ludfordian Stage of the Ludlow Epoch

The first land plants were primitive mosses and lichens (bryophytes) that evolved in the Late Ordovician but were very restricted in their growth. Life on land in a relatively dry atmosphere is difficult for all organisms. In colonizing the land, plants had the advantage of securing their energy from sunlight through photosynthesis, while animals were dependent on either plants or other animals for food. However, plant tissue still needed support to grow against gravity, and for protection from oxidation, hydration, and the damaging effect of ultraviolet light. Fossils of early vascular upright land plants, such as *Cooksonia* and

Steganotheca come from Late Silurian strata in the Welsh Borders. They had short (4cm/1.6in) forked stems ending in spore-bearing capsules. With no leaves, the stomata (cell openings for "breathing") were on stems and branches.

The first animals on land were arthropods, pre-adapted for life in the inhospitable conditions by their tough, waterproof exoskeletal armour. They included myriapod "detritivores" that ate plant material already degraded by soil bacteria, and carnivorous trigonotarbid arachnids. All these early terrestrial fossils are found as river-borne fragments stranded along shorelines.

Latitude then: 15°S

Latitude now: 52°N

Sea level: high

Original environment: coastal mudflats

Deposits: mud and sand

Status: many protected historic sites scattered over the Welsh Borders

Preservation: fossils are generally flattened but often well preserved in fine-grained strata

⊕ Earth *c.*419 MA

⊕ Fossil site today

Common organisms: *Cooksonia*, *Steganotheca*, trigonotarbids, myriapods

Climate: tropical, with global 'greenhouse' temperatures
Biota: first land plants and animals

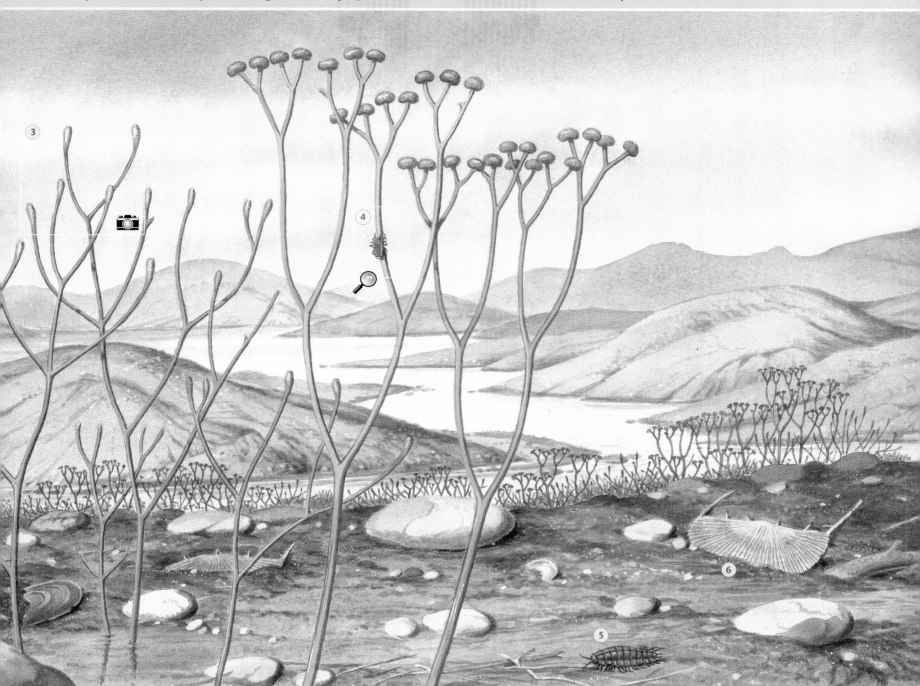

- ❶ *Lingula*
- ❷ *Cooksonia*
- ❸ *Steganotheca*
- ❹ *Palaeotarbus*
- ❺ *Eoarthropleura*
- ❻ *Strophochonetes*

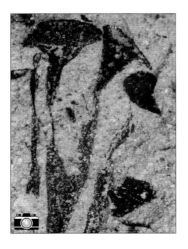

↪ COOKSONIA (2) A highly flattened and carbonized fossil shows the simple forked branches and terminal reproductive structures of this primitive land plant.

↪ COOKSONIA (2)
Chemically isolated from the rock, this specimen of *Cooksonia* shows the cap-shaped reproductive structures that bore the plant's spores. Some of the spores are still preserved within the cup, and these proved for the first time the connection between the parent plant and a particular (trilete) spore structure.

↥ STEGANOTHECA (3)
This primitive land plant offers an example of more complex, first- and second-order branching. *Steganotheca* grew to around 4.5cm (1.8in) high.

↥ PALAEOTARBUS (4)
These tiny (2–3mm/1in long), air-breathing and land-living arthropod predators had paired fangs with which to stab their prey.

THE FIRST PEAT BOG
RHYNIE CHERT, ABERDEENSHIRE, SCOTLAND

408 million years ago
Pragian Stage of the Early Devonian Epoch

Latitude then: 13°S

Latitude now: 57°N

Sea level: high

Original environment: terrestrial hot spring

Deposits: silica and organic debris

Status: an internationally important and legally protected site

Preservation: three-dimensional plant and animal tissues exceptionally well preserved in chert

⊕ Earth *c.*408 MA

• Rhynie Chert

⊕ Fossil site today

The world's oldest known bog community is preserved at Rhynie in Scotland. Around 408 million years ago, hot mineral springs created silica-rich ponds in which a number of primitive land plants and animals survived. Rapid deposition of silica mineralized their organic remains in chert, leaving an astonishingly detailed view of life at the time.

Seven different land plants have been identified, five of which are true vascular plants with strengthened water-conducting cells. Growing to 40cm (16in) in height, they resemble the living, leafless plant *Psilotum*. All these plants have two reproductive generations – an asexual "sporophyte" and a sexual "gametophyte" stage requiring germination in wet conditions. Other plants here include extinct nematophytes, the oldest known lichen (*Winfrenatia*), stoneworts (charophytes), chlorophytes, and cyanophytes.

Tiny arthropods were also well established in the community, with crustaceans, chelicerates, the world's oldest mites, euthycarcinoids, springtails, and myriapods. Many of these were predators, such as *Crussolum*, a fast-moving centipede with poison jaws. Others were detritivores that ate decaying plant matter, while wounds found on some plants show that sap-sucking arthropods were also present.

Common organisms: early vascular plants, small arthropods
Volcanism: hot springs venting from underground heat source

Climate: tropical, high global summer temperatures
Biota: a primitive bog community of plants and invertebrates

1. *Asteroxylon*
2. *Palaeocharinus*
3. *Protacarus*
4. *Leverhulmia*
5. *Aglaophyton*
6. *Horneophyton*
7. *Lepidocarus*
8. *Rhyniella*
9. *Rhynia*
10. *Nothia*

PALAEOCHARINUS (2) This 3mm-long (0.12in) spider-like trigonotarbid was a carnivore that used fangs to inject its prey and suck out liquefied flesh.

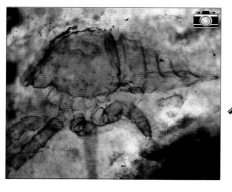

LEPIDOCARUS (7)
The commonest arthropod at Rhynie is this 4mm-long (0.15in), shrimp-like animal with a dozen pairs of limbs and long branched antennae. It lived in ephemeral hot-spring pools.

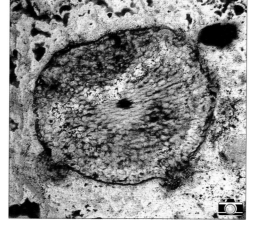

RHYNIA (9)
Just 2mm wide (0.08in), this fossil stem preserves detailed cellular structure including the central water-conducting vessels that allowed it to grow upright. The outer surface also shows two lesions (lower part of photo) where the plant has been attacked by sap-sucking mites.

71

542 MILLIONS OF YEARS   488.3   443.7   416

RISE OF THE JAWED FISH
HUNSRUCK SLATE, GERMANY

407

Pragian to Emsian stages of the Early Devonian Epoch

Latitude then: 10°S

Latitude now: 50°N

Sea level: high

Original environment: marine waters over 100m (330ft) deep

Deposits: seabed muds

Status: most fossils were found in slate quarries that are no longer worked, but some spoil tips are still accessible

Preservation: fossils flattened in slates, with some pyritized soft tissues, and best seen in X-rays

⊕ Earth *c.*407ᴍᴀ

⊕ Fossil site today

Continuing evolution of the fish jaw accelerated the marine arms race. The first jawed "spiny sharks" (acanthodians) arose in the early Silurian, flourished in the Devonian, and died out in the Early Permian. The fishes and other marine life of the Early Devonian are remarkably well preserved in the Hunsrück Slate of Germany's Rhenish Massif.

Compressed into slate, these seabed muds have been quarried and mined since Roman times. Splitting the slates exposes their superb fossils, and since the 1960s X-rays have been able to reveal fossils still hidden within the slates.

The diversity of marine life here ranged from immobile animals such as sponges, corals, brachiopods, molluscs, and many echinoderms to more mobile arthropods and swimming cephalopods, along with some of the oldest sea spiders.

Interestingly, the fish are mainly flattened bottom-dwelling agnathans, such as *Drepanaspis*, and ray-like placoderms, such as *Gemuendina*. Acanthodian "spiny sharks", some 2–3m (7–10ft) long, were more active swimmers, but all that remains of them are 40cm-long (16in) fin spines. A rarity is a single specimen of the earliest fossil sarcopterygian lungfish.

Common organisms: echinoderms, arthropods, agnathan fish

Climate: tropical, high global summer temperatures
Biota: diverse marine life from worms to fish

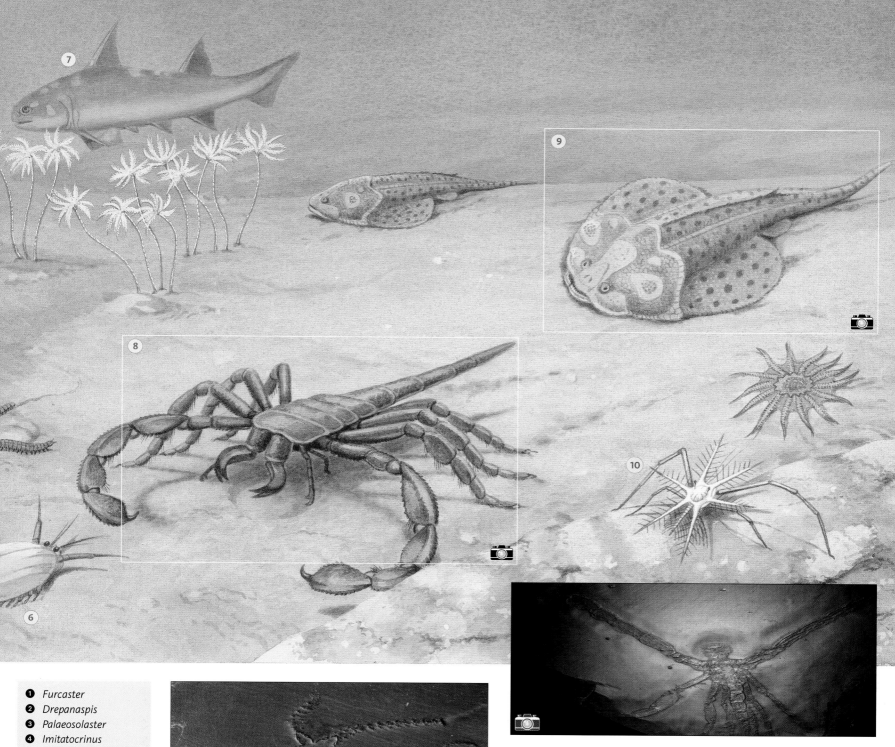

❶ *Furcaster*
❷ *Drepanaspis*
❸ *Palaeosolaster*
❹ *Imitatocrinus*
❺ *Bundenbachochaeta*
❻ *Nahecaris*
❼ acanthodian
❽ *Palaeoisopus*
❾ *Gemuendina*
❿ *Mimetaster*

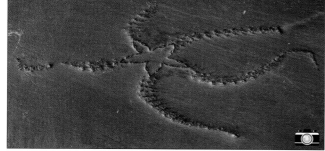

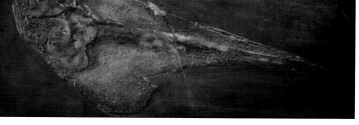

➲ **FURCASTER (1)** Starfish and brittlestars such as this slender-armed example are among Hunsrück's most common fossils.

➲ **PALAEOISOPUS (8)** With 18cm-long (7in) arms and strong pincers, this sea spider (pycnogonid) was an active seabed hunter.

➲ **GEMUENDINA (9)** Up to 1m long (40in), this large, ray-like fish was a placoderm that had bony plated jaws and a flattened body.

DEVONIAN REEF DWELLERS
GOGO, CANNING BASIN, WESTERN AUSTRALIA

Climate: tropical, high global summer temperatures
Biota: diverse reef-dwelling fish fauna (c.44 species)

Latitude then: 15°S	
Latitude now: 17°S	
Sea level: high	
Original environment: inter-reef basin around 100m (330ft) deep	
	⊕ Earth c.384 MA
Deposits: mud and sand	
Status: well-exposed reef limestones in the remote Canning Basin in the north of Western Australia	
Preservation: three-dimensional fossils preserved in carbonate nodules, requiring prolonged chemical preparation	
	⊕ Fossil site today

Around 384 million years ago, the tropical waters of Western Australia were home to over 44 different kinds of fish. Their fossil remains are preserved in the Gogo strata of the Canning Basin, and this detailed record reveals the early evolution of two fish groups that became of great importance – the bony fish (actinopterygians) that dominate today, and the lobefins (sarcopterygians) that gave rise to our tetrapod ancestors. The most common fish were the extinct armoured placoderms, but these were in decline as jawed fish took over.

Landscape erosion has resurrected an extensive seascape of stromatoporoid reefs and their abundant life forms. Between the reefs, at a depth of around 100m (330ft), the seabed muds entombed entire fish in carbonate nodules. Careful chemical etching releases the bones from their stony tombs.

Gogo's placoderms include arthrodires, antiarchs, ptyctodonts, and the shark-like camuropiscids, while the sarcopterygians include *Gogonasus*. The structure of this lobefin's head and muscular pectoral fins show significant developments towards the tetrapod condition. As we shall see, this was a prerequisite for backboned animals to move onto land.

384 million years ago
Frasnian Stage of the Late Devonian Epoch

Common organisms: stromatoporoids, shellfish, placoderm fish

❶ *Bothriolepis*
❷ *Gogonasus*
❸ *Moythomasla*
❹ *Eastmanosteus*
❺ *Campbellodus*
❻ *Holodipterus*
❼ *Griphognathus*
❽ *Onychodus*
❾ stromatoporoid

◗ GOGONASUS (2) This predatory tetrapod-like, lobe-finned osteolepiform grew up to 30cm (12in) long. It is similar to *Osteolepis*, and characteristically has a pair of external nostrils (clearly shown on this reconstructed skull) and two pairs of muscular fins.

◗ GRIPHOGNATHUS (7) Careful chemical isolation of this Gogo nodule reveals a strange-looking long-snouted, so-called "duck-billed" lungfish (technically a "rhynchodipterid dipnoan"), which grew up to 60cm (24in) long and used its snout to nose in the seafloor mud seeking worms and arthropods that lived there. It was common at Gogo, and widespread across North America, Europe, and Australia.

◗ ONYCHODUS (8) This lobe-finned predator grew up to 3m (10ft) long, and had a shark-like whorl of teeth that rotated outward to capture prey, and retracted when the mouth was closed.

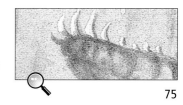

RISE OF THE LOBEFINS
MIGUASHA, ESCUMINAC BAY, QUEBEC, CANADA

Climate: tropical, high global summer temperature

Biota: diverse marine fish and arthropods

Latitude then: 28°S

Latitude now: 47°N

Sea level: high

Original environment: coastal mudflats

Deposits: mud and sand

Status: Miguasha National Park is a protected World Heritage Site

Preservation: flattened but often well-preserved remains of entire organisms

⊕ Earth *c.*380MA

Miguasha •

⊕ Fossil site today

• Miguasha

The Devonian fossil fishes of Miguasha were discovered in 1843, but it was another century before their full importance was recognized. Among the thousands of fossils collected, the lobe-finned fish *Eusthenopteron* earned international fame when Erik Jarvik showed its close link to our tetrapod ancestors.

The sands and muds of the strata at Escuminac bay were deposited in a muddy estuary that opened into a shallow gulf. Fossils buried in the sediments come from both the land and sea. Abundant land plant remains are mixed with arthropods and some 20 species of marine and freshwater fish belonging to

several major groups. There are agnathans, acanthodians (some of which grew to considerable size), and one of the earliest bony, ray-finned fish (the actinopterygian *Cheirolepis*). Six sarcopterygians include lungfish (such as *Scaumenacia*), heavily built porolepiforms (*Holoptychius*), a coelacanth (*Miguashaia*), and a tetrapod-like osteolepiform, *Eusthenopteron*.

But this is not the only "missing link" from this part of the world. *Archaeopteris* is a botanical parallel – with fern-like fronds attached to a gymnosperm-type woody trunk that grew some 18m (60ft) high, it was one of the first tree-sized plants to evolve.

384–374 million years ago
Frasnian Stage of the Late Devonian Epoch

Common organisms: fish, eurypterids

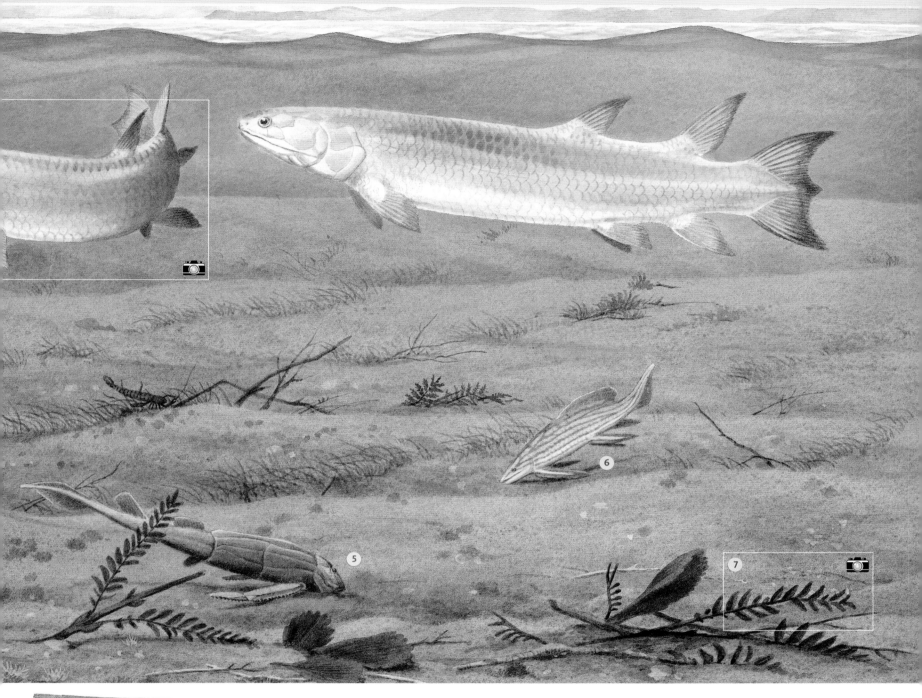

EUSTHENOPTERON (4) This large lobe-finned sarcopterygian grew up to a metre (40in) long, and had tetrapod-like paired muscular fins attached to its pelvic and pectoral girdles, a pair of internal nostrils, and an armoured bony skull – all characteristics of a tetrapod ancestor. However, *Eusthenopteron* also retained primitive fish features. It is grouped with the osteolepiforms, and was an active predator in fresh and brackish waters.

1. *Cheirolepis*
2. *Endeiolepis*
3. *Escuminaspis*
4. *Eusthenopteron*
5. *Bothriolepis*
6. *Scaumenacia*
7. *Archaeopteris*

ARCHAEOPTERIS (7) These fronds were thought to belong to a fern until 1960, when gymnospermous stems up to 10m (35ft) long were discovered.

77

A FISH WITH ARMS
ELLESMERE ISLAND, CANADIAN ARCTIC

Climate: equatorial, high global summer temperatures and seasonal rainfall
Biota: freshwater river plants and animals

Latitude then: 5°N

Latitude now: 77°N

Sea level: very high (+250m/820ft)

Original environment: swampy river channels and floodplains

Deposits: mud and sand

Status: a remote and inhospitable location on Ellesmere Island

Preservation: original bone and partially articulated skeletons

Ellesmere Island •

⊕ Earth c.380ma

Ellesmere Island •

⊕ Fossil site today

Nicknamed the "fishapod", the fossil remains of *Tiktaalik* emerged from the frozen landscapes of Arctic Canada in 2006. They combine features that link the evolution of the earliest four-limbed vertebrates (tetrapods such as *Acanthostega*) to the lobe-finned fish (sarcopterygians such as *Eusthenopteron*). Although the broad evolutionary connection between the two groups was well known, the intermediate steps of major tetrapod features such as the limbs and wrists, and lower jaw and ear, were not well represented by fossils.

Importantly, *Tiktaalik* preserves a connected skull, neck, "shoulders" (pectoral region), and "arms", plus the articulations of the "elbow" and "wrist" joints.

Detailed study shows that *Tiktaalik* had novel mechanisms for lifting its head and body on muscular front fins. After lying in wait on a river bed, this ambush predator could lift its head and launch itself towards unsuspecting prey. The "arms" could also drag the heavy animal from one body of water into another – a useful adaptation in the vegetation-choked channels of equatorial rivers. *Tiktaalik*'s respiratory system is also intermediate between fish and tetrapods – it may well have had gills and been able to breathe air.

380 million years ago
Frasnian Stage of the Late Devonian Epoch

Common organisms: *Titktaalik, Asterolepis*

① *Tiktaalik*
② *Laccognathus*
③ *Asterolepis*

↪ **TIKTAALIK (1)** With a flattened bony skull some 20cm (8in) long, a raised snout, and dorsally positioned eyes, the head of this large predatory lobefin fish resembles that of a crocodile. The overlapping ribs were also strongly built – another evolutionary link to early tetrapods.

↥ **TIKTAALIK LIMB (1)** *Tiktaalik* had strong, paired, muscular fins – especially the pectoral fins and girdle, which may have been able to lift and prop up the head and shoulders. The lack of restrictive bony plates around the neck would also have made its head more mobile.

A GIANT OF THE DEVONIAN DEEP
ROCKY RIVER VALLEY, OHIO, USA

Climate: equatorial, high global summer temperatures and seasonal rains
Biota: marine fish

Latitude then: c. 40°S

Latitude now: 42°N

Sea level: high

Original environment: marine open seas

Deposits: oxygen-poor seabed muds

Status: the Cleveland Shale occurs at a number of sites in New York State and Ohio

Preservation: mineralized bony plates within carbonate concretions

⊕ Earth *c.*370 MA

⊕ Fossil site today

Dunkleosteus is the most spectacular animal of the early Paleozoic, an armoured placoderm fish that grew to 8m (26ft) long, and was the top ocean-going predator of its time. Not only was its head enclosed in heavily armoured bony plates up to 5cm (2in) thick, but so was the shoulder region, to which it was joined by a unique bony hinge that allowed the mouth to open wide.

The primitive jaws were armed with long, sharp-edged, bony plates instead of teeth. Calculations suggest that its powerful jaw muscles produced a very rapid snapping action, capable of opening the jaws in one fiftieth of a second. The sudden intake of water would have sucked unwitting prey into the enormous mouth before the jaw closed to deliver a bone-shattering bite.

Neither the backbone nor tail of *Dunkleosteus* have been found as fossils, suggesting that the posterior skeleton was cartilaginous. Despite this, the entire animal is thought to have weighed around 3.6 tonnes. Because of the weight of its armour, the fish was a vigorous if relatively slow swimmer, and may have been an ambush predator. It is likely that it would attack anything it could find, including other *Dunkleosteus*.

370 million years ago
Famennian Stage of the Late Devonian Epoch

Common organisms: sharks such as *Cladoselache*

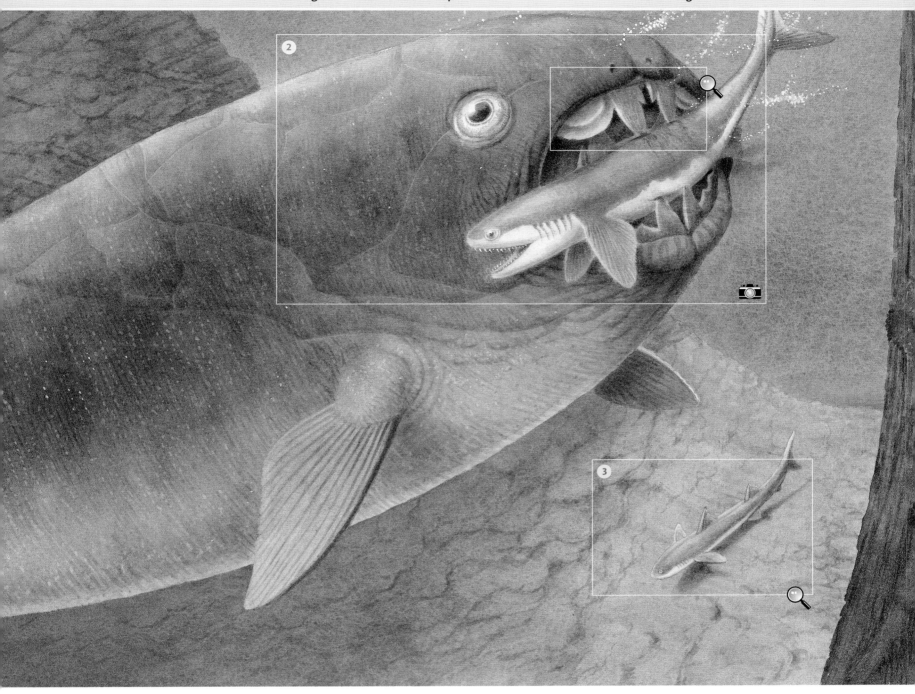

❶ *Cladoselache*
❷ *Dunkleosteus*
❸ *Ctenacanthus*

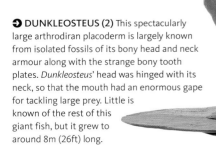

➲ DUNKLEOSTEUS (2) This spectacularly large arthrodiran placoderm is largely known from isolated fossils of its bony head and neck armour along with the strange bony tooth plates. *Dunkleosteus*' head was hinged with its neck, so that the mouth had an enormous gape for tackling large prey. Little is known of the rest of this giant fish, but it grew to around 8m (26ft) long.

➲ DUNKLEOSTEUS MOUTH (2)
Calculations show that the jaws of *Dunkleosteus* could deliver the most powerful bite of any fish in the known history of life, with a chopping force of up to 107 megapascals (16,000 pounds per square inch). Much greater than the bite force of any modern sharks, this is more in the Tyrannosaur league.

➲ CTENACANTHUS (3) Ctenacanth, meaning "comb spine", sharks had fin spines covered with rows of tubercles that give them a "comb-like" appearance. They also had a powerful jaw like that of modern sharks, though tooth replacement was less frequent.

TETRAPODS SET FOOT ON LAND
KEYSER FRANZ JOSEPH FJORD, GREENLAND

Climate: tropical, high global summer temperatures
Biota: freshwater and amphibious

Latitude then: 10°S

Latitude now: 73°N

Sea level: high (+200m/650ft)

Original environment: river channels

Deposits: sands and muds

Status: remote and inhospitable mountainside sites in East Greenland

Preservation: partially articulated skeletons in hard sedimentary rock

Keyser Franz Joseph Fjord •

⊕ Earth *c.*366 MA

Keyser Franz Joseph Fjord •

⊕ Fossil site today

Tetrapod fossils were first found high on a remote East Greenland mountainside in the late 1890s, but it took more than 50 years for the full significance of the finds to be revealed. *Ichthyostega*, a metre-long (40in), salamander-like animal, was then portrayed as an amphibious five-toed tetrapod that could emerge from water onto dry land dragging its long fish-like tail behind it. New fossils, including a more primitive genus, *Acanthostega*, have substantially revised this picture. These fossils make it quite clear that tetrapod limbs first evolved as an adaptation for survival in water – "coadaptation" for life on land came later.

Acanthostega retained its gills, and its backward-pointing hind legs indicate that it was primarily aquatic, while *Ichthyostega*'s limbs and strong barrel-shaped ribcage were able to support its body out of water (though it still retained a laterally flattened tail for swimming). Both had multi-toed feet rather than the basic five-toed (pentadactyl) arrangement of living tetrapods.

Both animals lived in large, fast-flowing rivers and lakes on the tropical continent of Laurentia (North America, Greenland, and northwestern Europe), preying upon any invertebrates and young fish they could catch.

366 million years ago
Famennian Stage of the Late Devonian Epoch

Common organisms: invertebrates, fish, tetrapods

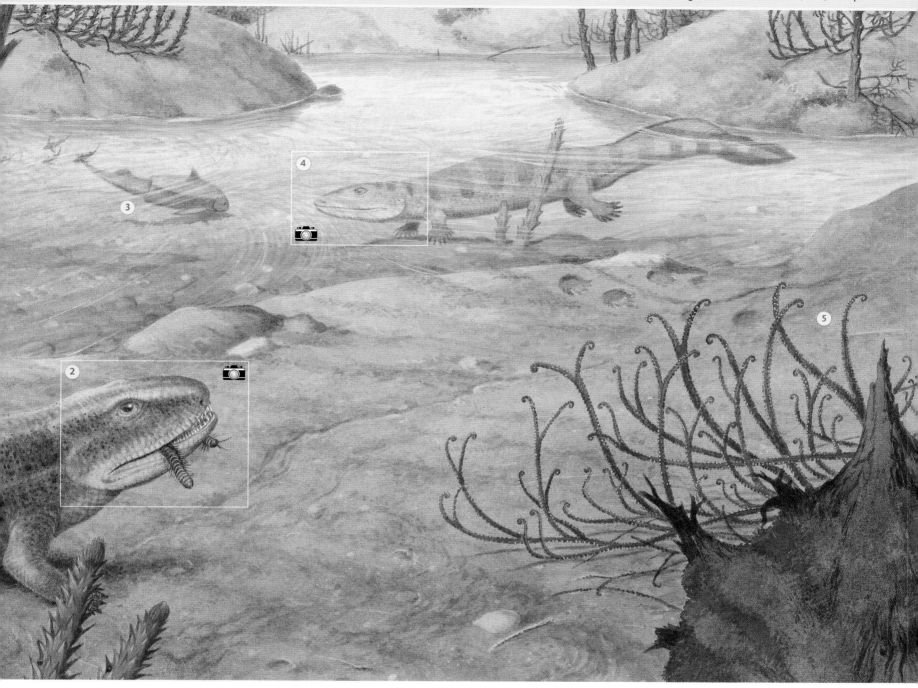

❶ *Drepanophycus*
❷ *Ichthyostega*
❸ *Groenlandaspis*
❹ *Acanthostega*
❺ *Serracaulis*

❷ **ICHTHYOSTEGA LIMB (2)** Paleontologists were surprised when detailed preparation of the hind leg revealed four well-developed toes with another three clustered together, plus a small cartilaginous supporting splint – evidence that the five-digit plan found in modern tetrapods was a later development.

ICHTHYOSTEGA (2) Although obliquely flattened, this fossil skull shows how the position of the animal's eyes high on its head gave it the field of vision of a bottom-dwelling ambush predator.

ACANTHOSTEGA (4) The hard, calcareous Greenland sandstone made the preparation of *Acanthostega*'s 60cm-long (24in) skeleton a very lengthy process. Great care was taken in order to avoid damaging the bone, which was actually softer than the surrounding rock.

LIVING ON LAND
EAST KIRKTON, SCOTLAND

Climate: warm and fairly dry
Biota: freshwater and terrestrial

Latitude then: 10°N

Latitude now: 56°N

Sea level: rising (+ 300m/980ft)

Original environment:
lake waters with hot springs
surrounded by woodland

Deposits: limestones, shales and
volcanic ash

Status: legally protected

Preservation: Mostly flattened
but with some soft tissue
preservation

● East Kirkton

↻ Earth c.328 MA

● East Kirkton

↻ Fossil site today

One of the oldest known communities of land-living vertebrates has been recovered in recent decades from early-Carboniferous strata at East Kirkton, near Bathgate, Scotland. The fossils record life in and around a freshwater lake.

Of greatest evolutionary interest are the tetrapods, mostly small lizard-sized animals such as *Westlothiana* and *Balanerpeton*, but also a metre-long legless amphibian, *Ophiderpeton*. All were evidently at home on land for many of their needs – they breathed air and *Westlothiana* had scales to protect its skin from drying out or being torn by rough surfaces. Their legs stuck out sideways, so they would

have scuttled in a similar way to modern lizards, as they looked for food that probably included small arthropods and worms.

When *Westlothiana* was first described, it was thought to be the oldest known egg-laying (amniote) reptile, but it is now regarded as a more primitive reptile-like tetrapod, still dependent on water for reproduction (like the other East Kirkton amphibians, whose shell-less, unprotected eggs were laid, fertilised, and developed in water). The hot springs and lake waters teemed with fish such as *Acanthodes*) which would have feasted upon the spawn of the tetrapods during the breeding season.

Common organisms: fish, tetrapods especially *Balanerpeton*

328 million years ago
Visean Stage of the Mississippian Epoch

1. *Sphenopteridium*
2. *Westlothiana*
3. *Ophiderpeton*
4. *Archaeocalamites*
5. opilionid
6. *Eldeceeon*
7. *Balanerpeton*
8. *Acanthodes*
9. *Pulmonoscorpius*
10. millipede

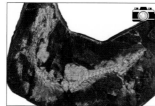

↑ WESTLOTHIANA (2) This little tetrapod (up to 20cm/8in long) had a lizard-like appearance, but its skull combines reptilian (amniote) and more primitive features. It is classed as a reptiliomorph, and probably still needed water to lay unprotected (shell-less) eggs.

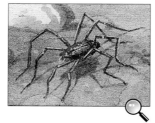

↑ HARVESTMAN (5) This tiny arthropod (around 1cm/0.4in long) with its small oval body and long legs is the earliest known representative of the opilionids (harvestmen), a surviving group of arachnids. The East Kirkton species has yet to be named.

↑ BALANERPETON (7) is one of the earliest temnospondyl tetrapods, 50cm (20in) long. Like living salamanders, it could pull its eyeballs down into its mouth to aid in swallowing food. Lidded eyes and a tympanic ear for detecting vibrations in air both show that it spent time on land.

↻ PULMONO-SCORPIUS (9) Fossils of this large scorpion (up to 80cm/32in long) from East Kirkton are remarkably well preserved. They retain details of structures such as a "booklung" that proves this predatory arthropod breathed air and could live on land.

DIVERSE CARBONIFEROUS FISH
BEAR GULCH, MONTANA, USA

Climate: tropical, high global summer temperatures

Biota: high-diversity mixed terrestrial and aquatic life

Latitude then: 12°N

Latitude now: 47°N

Sea level: high

Original environment: marine lagoonal or estuarine

Deposits: carbonate muds and silt

Status: scattered outcrops on the Potter Dome have been worked for their fossils over several decades

Preservation: soft tissue preservation in phosphate with moulds of carbonate fossils

⊙ Earth c.320 MA

• Bear Gulch

⊙ Fossil site today

More than 108 species of fish (4800 specimens) have been found in Bear Gulch's Carboniferous strata. It is one of the most diverse fossil fish faunas in the world, and particularly important because it preserves five species of cartilaginous fish – animals such as sharks and rays whose remains are not normally fossilized. Some of these fossils even preserve distinct differences between the sexes (sexual dimorphism). Additionally, many soft-bodied creatures such as polychaete and nematode worms are preserved.

In Carboniferous times these were equatorial seas, with shallow embayments where muddy, oxygen-poor sediments were deposited in water no more than 40m (130ft) deep. This depositional environment led to rapid burial and preservation of dead organisms.

The base of the food chain seems to have been abundant kelp-like algae and cyanobacteria that were grazed by snails and other invertebrates. Acanthodian sharks, tolerant of brackish waters, fed on conodonts and small crustaceans such as shrimps and ostracods that lived among the plants. The latter also attracted opportunist predators and scavengers, especially small cephalopods, and these in turn were preyed upon by bigger fish.

- ❶ *Allenypterus*
- ❷ *Caridosuctor*
- ❸ *Belantsea*
- ❹ *Falcatus*
- ❺ *Echinochimaera*
- ❻ *Stethacanthus*
- ❼ *Paratarrasius*
- ❽ *Crangopsis*
- ❾ *Harpagofututor*

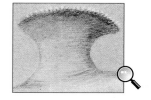

↻ FALCATUS (4) and **↺ STETHACANTHUS (6)** Several Bear Gulch sharks display unusual adaptations such as brush-like dorsal structures and long rods. These were probably used for display or during copulation.

↪ ALLENYPTERUS (1) Coelacanths were once a major group of lobefin fish, and some early forms resemble today's species. The 15cm (6in) *Allenypterus*, however, was deep-bodied like a modern reef fish.

↪ CARIDOSUCTOR (2) This coelacanth's name means "shrimp eater", a reference to its possible diet. Quite like modern coelacanths, it had a slender body up to 25cm (10in) long, and an extended central lobe.

LIFE IN THE COAL MEASURES
MAZON CREEK, ILLINOIS, USA

Climate: tropical, high global summer temperatures

Biota: high-diversity mixed terrestrial and aquatic life

Latitude then: 8°N

Latitude now: 25°N

Sea level: rising (+200m/650ft)

Original environment: swamps around the mouth of an estuary

Deposits: muds

Status: mine dumps

Preservation: inside ironstone nodules, with some soft tissue preservation

Mazon Creek •

↻ Earth c.314 MA

Mazon Creek •

↻ Fossil site today

Rocks from the coal mines of northeastern Illinois are the source of one of the most spectacular fossil biotas known, a glimpse of what life was like in the estuaries of the "Coal Measure" swamps. Over several decades, an amazing diversity of plants and animals – more than 300 species – have been found by professional and amateur fossil hunters.

Swampy forests were home to tree-sized clubmosses that grew to 40m (130ft) high, alongside seedferns and 10m (33ft) horsetails. They were populated by over 150 arthropod species including giant dragonflies, millipedes, athropleurids, and cockroaches, hunted by predatory scorpions and amphibians. Periodically, floodwaters swept plants and animals downriver into estuarine waters where their remains were deposited alongside brackish water and marine animals such as jellyfish, clams, many kinds of fish (30 species), and *Tullimonstrum*, the State fossil of Illinois – a bizarre creature of unknown affinities.

The fossils tend to occur in clay ironstone (siderite) nodules that originally formed within the estuarine muds. They preserve some soft tissues, but since the nodules are less than a metre (40in) in diameter, no large organisms are preserved in their entirety.

314 million years ago
Bashkirian Stage of the Pennsylvanian Epoch

Common organisms: seed ferns, fish, arthropods

↻ TULLIMONSTRUM (2)
This bizarre soft-bodied invertebrate was apparently a free swimmer and has no known affinities. It is named "Tully Monster" after the amateur collector who discovered it.

↻ STRANGE INVERTEBRATE (2)
The Tully Monster had a segmented body up to 19cm (7.5in) long, with a long proboscis ending in a curious claw. It may have been related to a group of shell-less gastropods called heteropodids.

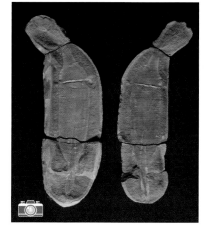

↑ SPHENOPHYLLUM (8)
This Carboniferous horsetail had a vine-like growing habit. It also had a ribbed main stem that branched at enlarged nodes, each of which had a whorl of 6 wedge-shaped leaves.

89

REPTILE BEGINNINGS
JOGGINS, NOVA SCOTIA, CANADA

Climate: humid tropical rainforest

Biota: 148 species of plants and animals belonging to 96 genera

Latitude then: 12°N

Latitude now: 46°N

Sea level: rising (+250m/820ft)

Original environment: rainforest swamps beside an inland sea

⊕ Earth c.314 MA

Deposits: alternating coal seams, limestones, sandstones, and shales

Status: accepted as a World Heritage Site in 2008

Preservation: generally flattened in the rock, except where protected within tree stumps

⊕ Fossil site today

Some of the oldest reptiles in the world have been found in rotten and burnt-out tree stumps buried within the late Carboniferous coal seams of Joggins in Nova Scotia. Fossil trees up to 6m (20ft) high were first described from here in the 1820s, but it was the 1852 discovery, by William Dawson and Charles Lyell, of the first small lizard-like tetrapod that secured the site's international fame. In 1871, Lyell declared Joggins to be the world's finest Pennsylvanian outcrop – today, it is a World Heritage Site.

Some 313 million years ago, Joggins lay close to the equator. Its poorly drained swamps and marshes, surrounding an inland sea, were home to rainforest vegetation and better drained river plains covered with dry coniferous scrub. Growth of the swamp vegetation kept pace with rising sea levels and subsiding landscapes. These environmental conditions were repeated 14 times over a period of a million years, depositing more than 900m (nearly 3000ft) of strata, including 45 coal seams.

The diverse plant habitats hosted evolving land animals, with many tetrapods (12 species, all under a metre/40in long), arthropods large and small, and the first land snails, especially around the waterholes. A total of 148 species have been found.

① *Baphetes*
② scorpion
③ *Arthropleura*
④ *Cordaites*
⑤ *Neuropteris*
⑥ *Protodiscus*
⑦ *Xyloiulus*
⑧ *Hylonomus*
⑨ *Sigillaria*

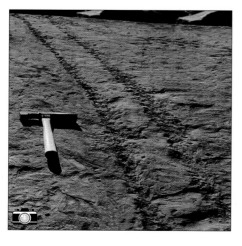

↪ ARTHROPLEURA TRAIL (3) The millipede-like giant arthropods of Joggins frequently left tracks up to 20cm (8in) across.

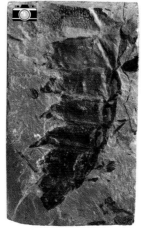

↻ ARTHROPLEURA (3) The arthropleurids were the most common and successful plant-eaters of the Carboniferous terrestrial community. They were myriapods, with segmented bodies and up to 30 pairs of legs, and evolved to produce some giant forms, including the 2m (6.6ft) *Arthropleura*. This fossil shows just one of the animal's numerous jointed legs.

↑ HYLONOMUS (8) Placed in a group known as the captorhinomorphs, *Hylonomus* has skeletal features that distinguish it as a small (20cm/8in long), slender and agile primitive amniote and anapsid reptile.

WILDFIRE IN THE CARBONIFEROUS
JOGGINS, NOVA SCOTIA, CANADA (CONT.)

Climate: humid tropical rainforest

Biota: 148 species of plants and animals belonging to 96 genera

None of the Joggins tetrapods was more than a metre (40in) long, but finds have included some examples that were becoming increasingly independent of water, such as the lizard-like *Hylonomus* and *Paleothyris*. These slender-bodied, insect-eating animals, up to 20cm (8in) long, are part of a group known as captorhinomorphs, and are regarded as being among the earliest true reptiles known. Although Joggins shows another major step forward in the conquest of the land, other tetrapods, such as *Baphetes*, were still largely aquatic and fed on fish.

Tinder-dry coniferous scrub, which grew on better drained river banks and plains, was repeatedly swept by wildfires through the site's history, and so far the remains of 24 hollow, burnt-out tree stumps (mostly the *Sigillaria*

genus) have been excavated. Detailed examination has revealed remains of some 200 small tetrapods along with millipedes, arachnids, charcoal from the wildfires, and snail shells. It is not clear whether the animals retreated into the hollow stumps to escape the fire and then became trapped, or whether they normally lived there.

The arthropod fauna was diverse and spectacular. Predators such as scorpions and the giant eurypterid *Hastimima* patrolled the peat mire. By contrast the giant arthropleurids fed in part on the abundant rotting trunk wood of lycopsid trees. There were also huge dragonfly-like megasecopterans, and small arachnids such as *Graeophonus*, an amblypigid arachnid (whip spider) also known from equatorial coal measures in Europe.

314–313 million years ago
Bashkirian Stage of the Pennsylvanian Epoch

Common organisms: plants, fish, tetrapods

❶ *Hylerpeton*
❷ *Coryphomartus*
❸ *Amynilyspes*
❹ *Sphenophyllum*
❺ *Dendrerpeton*
❻ *Sigillaria*
❼ megasecopteran
❽ *Graeophonus*

◉ HYLERPETON (1) This microsaur was a part of a group of small, mostly terrestrial, lizard-shaped lepospondyl tetrapods from the Carboniferous and early Permian.

◉ DENDRERPETON (5) Despite a superficial similarity to *Hylerpeton*, this 40cm (16in) animal was a temnospondyl, part of a lineage that survived to the Cretaceous.

◉ JOGGINS FOSSIL CLIFFS The gently tilted Carboniferous strata exposed in the seacliffs bordering the Bay of Fundy have been internationally renowned for their fossils since the early 19th century. Most famous of all are the sediment casts of coal swamp trees that still stand where they grew, surrounded by younger sediments.

◉ SIGILLARIA (6) Giant tree-sized clubmosses (lycopsids) such as *Sigillaria* (with a trunk up to 9.2m/about 30ft long) and *Lepidodendron* dominated the coal measure marshes.

TETRAPODS DIVERSIFY
NYRANY, CZECH REPUBLIC

Climate: tropical, high global summer temperatures
Biota: freshwater and terrestrial coal measure ecosystem

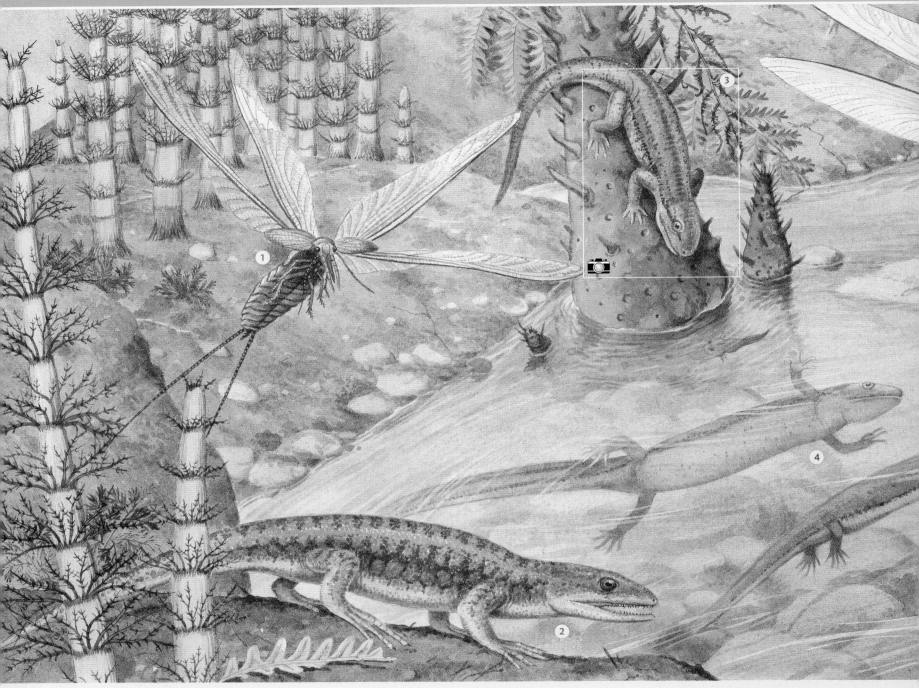

Latitude then: 3°S

Latitude now: 49°N

Sea level: high (+200m/650ft)

Original environment: lake with surrounding swamp

Deposits: lake muds and coals

Status: worked-out coal mines no longer accessible

Preservation: flattened in shale with some soft tissue preservation

⊕ Earth *c.*310 MA

⬆ Fossil site today

At the end of the 19th century, more than 700 fossil tetrapods were dug out of late Carboniferous coal deposits around the small mining town of Nyrany in the Czech Republic. They were among thousands of extremely well-preserved fossils of plants and giant insects recovered from a tropical "coal measure" lake that was over 8km (5 miles) across. Among the finds were some 20 species of basal tetrapods, including representatives of most major tetrapod groups of the time.

Land-living forms from around the lakeshore include primitive anthracosaurs, temnospondyls, microsaurs, a snake-like aistopod, and three primitive amniotes. Among the latter is *Archaeothyris*, one of the earliest synapsid reptiles, whose later Triassic descendants produced the mammals. Nyrany's shallow water and swamp dwellers include nectrideans, a branchiosaur, and a microsaur, all of which were partially aquatic predators, feeding on fish and small arthropods. The branchiosaurs are especially interesting, as several adult forms have been found to retain their tadpole-like juvenile body, with external gills and fleshy tail fins. There are also some rare open-water predators such as baphetids and an eogyrinid anthracosaur.

310 million years ago
Moscovian Stage of the Pennsylvanian Epoch

Common organisms: fish, tetrapods

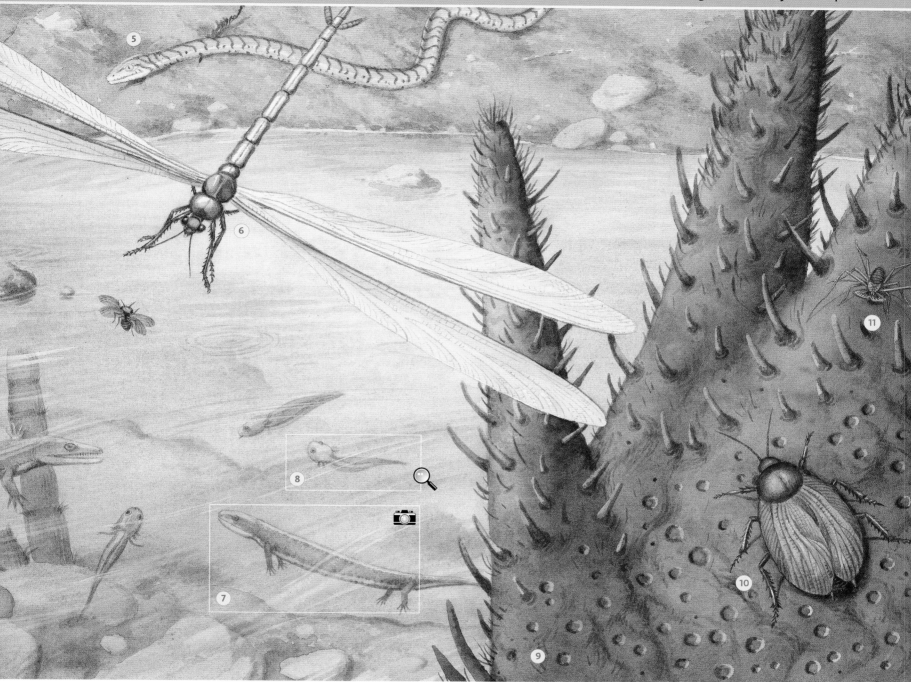

- ❶ palaeoodictyopteridan
- ❷ *Gephyrostegus*
- ❸ *Archaeothyris*
- ❹ *Sauropleura*
- ❺ *Aornerpeton*
- ❻ *Namurotypus*
- ❼ *Microbrachis*
- ❽ branchiosaur
- ❾ *Stigmaria*
- ❿ blattodean
- ⓫ amblypygid

🔼 **ARCHAEOTHYRIS (3)** This fossil jaw and teeth belong to the lizard-shaped, land-living *Archaeothyris*, which grew up to 50cm (20in) long and was one of the earliest known synapsids.

🔼 🔎 **MICROBRACHIS (7)** This small, four-toed tetrapod grew up to 30cm (12in) long. With reduced limbs, an elongated spine with more than 40 vertebrae, and a fish-like tail, it was secondarily aquatic (in other words it had returned to the water despite land-living ancestors). *Microbrachis* fed on small floating organisms in lakes such as those found at Nyrany.

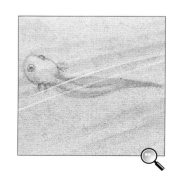

🔼 **BRANCHIOSAUR (8)** With their characteristic retention of tadpole-like "pedomorphic" features, these tetrapods may be ancestral to modern amphibians.

A LATE CARBONIFEROUS ICE AGE
KAROO BASIN, SOUTH AFRICA

Climate: ice age alternations of cold and warm
Biota: no known biota in glacial deposits

Latitude then: 65°S

Latitude now: 31°S

Sea level: high (+200m/650ft)

Original environment: terrestrial and coastal glaciation

Deposits: Dwyka tillites with dropstones

Status: numerous glacial features seen over southern Africa

Preservation: among the best fossil glacial features in the world

⊕ Earth *c.*300 MA

⊕ Fossil site today

Around 320 million years ago, Earth descended into an ice age as global climate cooled from its mid-Carboniferous temperatures, allowing mountain glaciers to form, along with small ice sheets in polar regions. At this time the continents of South America, Africa, Antarctica, and Australia were clustered in far southern latitudes. Ice centres that formed on each landmass grew into a single vast ice sheet with an area of some 70 million square kilometres (27 million square miles). The main centre of spreading ice straddled southern Africa and Antarctica.

Some of the best evidence for this glaciation, one of the most dramatic in Earth's history, comes from South Africa, where the inland Karoo Sea separated the ice sheet of the Cargonian Highlands to the north from that of the Cape Mountain belt and the Falklands Plateau to the south. Coastal glaciers left striated rock pavements in their wake and, with seasonal melting, discharged icebergs and huge quantities of glacial debris (forming layers up to a kilometre thick) into the sea.

Climatic change and an accompanying rise in sea levels at the end of the early Permian led to the collapse of the ice sheet, until by the late Permian only small highland ice caps persisted in southern Africa and Antarctica.

Common organisms: Glacial sediments are devoid of macrofossils, but may contain a few reworked plant spores

299 million years ago
Late Carboniferous to early Permian

- ① ophiacodontid synapsid
- ② embedded dropstone
- ③ glacial pavement

◑ OPHIACODONTID (1) This group of synapsids arose in the Middle Carboniferous, surviving the ice age into the early Permian, when it diversified and expanded to dominate land communities as the ice receded and the climate warmed. *Archaeothyris* from Nyrany belongs to this group, as does the 2m-long (6.6ft) *Ophiacodon* from the late Carboniferous and early Permian of New Mexico.

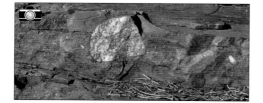

◐ DROPSTONE (2)
Evidence for extensive glaciation extending beyond the land is provided by dropstone boulders, carried by floating ice and found in fine-grained water-lain sediments.

◑ GLACIAL SURFACE (3)
The 290-million-year-old striated rock surfaces of the Permo-Carboniferous ice age seen in the Karoo region of South Africa are some of the best preserved in the world.

STRUGGLE TO RULE THE LAND
TEXAS REDBEDS, TEXAS, USA

Climate: equatorial with seasonal monsoon rains
Biota: terrestrial and freshwater aquatic

Latitude then: 0°

Latitude now: 35°N

Sea level: high (+200m/650ft)

Original environment: delta rivers and ponds

Deposits: red and grey muds and sands

Status: numerous localities across the southern USA

Preservation: includes some spectacular complete skeletons

↻ Earth c.295ma

↻ Fossil site today

As the late Carboniferous glaciation waned, global temperatures rose, and a new struggle to dominate the land was fought out in the equatorial river deltas, ponds, and banks of early Permian times. Tetrapods thrived in hot regions deep within a Pangean supercontinent that now stretched from pole to pole.

For the first time, reptiles were big enough to threaten the amphibian tetrapods on land, if not in the water. But there were still plenty of big temnospondyl amphibians around such as *Eryops*, a crocodile-like top predator 2m (6.6ft) long, which fed on fish and small tetrapods that it ambushed on land. Other amphibians had

reduced limbs, and were adapted to permanent life in the water, such as the bizarre nectridean *Diplocaulus* with its broad, arrow-shaped skull.

Among the reptiles, the 3m-long (10ft) synapsid pelycosaurs were most striking. *Edaphosaurus* and *Dimetrodon* both had sail-like structures on their backs that may have acted as heat exchangers, helping the animals to heat up in the morning and lose heat at midday. *Edaphosaurus* was one of the first reptiles adapted for plant eating, with a battery of grinding teeth on its lower jaw and a palate for macerating tough plant materials. *Dimetrodon*, in contrast, was a sharp-toothed predator.

Common organisms:
tetrapods, fish, and horsetail plants

295 million years ago
Asselian to Sakmarian stages of the Cisuralian Epoch

1. *Meganeuropsis*
2. *Araeoscelis*
3. *Dunbaria*
4. *Captorhinus*
5. dipnoan
6. *Dimetrodon*
7. *Diplocaulus*
8. *Eryops*
9. *Edaphosaurus*

➲ **ARAEOSCELIS (2)** One of the earliest known diapsid reptiles, this lizard-like genus was probably insectivorous.

↻ **DUNBARIA (3)** With dragonfly-like wings up to 3cm (1.2in) across, and extra winglets on the prothorax in front, this extinct group of insects had sucking mouthparts and probably fed on plant juices. They are known from the Elmo strata of Kansas.

➲ **DIMETRODON (6)** This large synapsid reptile is placed in a group known as the sphenacodontid pelycosaurs. Its sharp, blade-like teeth and powerful jaw suggest it was an active top predator.

99

A RARE VIEW OF UPLAND LIFE
BROMACKER, GERMANY

Climate: equatorial and strongly seasonal
Biota: freshwater and terrestrial tetrapods

Latitude then: 20°N

Latitude now: 50.5°N

Sea level: high (+200m/650ft)

Original environment: upland river plain

Deposits: sands and muds

Status: a single locality discovered in the 1980s

Preservation: well-preserved entire skeletons

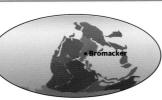

⊙ Earth c.290MA

⊙ Fossil site today

This rich early Permian site was discovered in the 1980s near Bromacker in what was then East Germany. Abundant and well-preserved tetrapods were uncovered from the sand and mud deposits of what had been, 290 million years ago, an upland alluvial plain crossed by small rivers. A strongly seasonal equatorial climate alternated between hot, wet seasons with full rivers and ponds, and hot, dry seasons when the ponds and streams dried out and many of the aquatic creatures perished.

Such upland environments are rarely preserved, and many of the species are unique to this locality – although they belong to clearly identifiable groups. There are also more cosmopolitan animals such as *Diadectes* and *Seymouria*, both of which are known from the early Permian of the USA's southern states. The presence of such animals is evidence for a continuity in terrestrial environments between North America and Europe at the time, as part of the Pangean supercontinent.

One of the most interesting of the Bromacker animals is *Eudibamus*, the oldest known bipedal reptile, which could run on its long hind legs. Since *Eudibamus* was a plant eater, this adaptation was almost certainly to aid escape from its enemies.

Common organisms: land-living tetrapods

290 million years ago
Sakmarian Stage of the Cisuralian Epoch

❶ *Diadectes*
❷ *Eudibamus*
❸ *Dimetrodon*
❹ *Seymouria*
❺ *Syscioblatta*

⦿ SYSCIOBLATTA (5) This cockroach is a spiloblattinid, characterized by extensively coloured, easily preserved wings that are common fossils in these equatorial deposits. The rapid evolution of this group is reflected in changes in wing vein patterns.

➔ ⦿ EUDIBAMUS (2)
This enigmatic 25cm-long (10in) reptile belongs to the small extinct group of bolosaurids, with rounded bulbous teeth, apparently used for feeding on tough plant material.

➔ SEYMOURIA (4) The seymouriamorphs were a small group of reptile-like tetrapods (reptiliomorphs) that included terrestrial and aquatic forms, both of which were carnivorous predators. *Seymouria* (60cm-long/24in) had powerful limbs to lift its body off the ground, and was evidently terrestrial.

101

THE GLOSSOPTERID FORESTS
QUEENSLAND AND NEW SOUTH WALES, AUSTRALIA

Climate: tropical, high global summer temperatures
Biota: swamp forest

Latitude then: 60°S

Latitude now: 22°S

Sea level: falling (+80m/260ft)

Original environment: swamps

Deposits: coals and associated sediments

Status: numerous coal mine localities in Queensland and New South Wales

Preservation: plant structures isolated but well preserved

⊕ Earth *c.*265 MA

⊕ Fossil site today

Australia mines more than 350 million tonnes of coal a year, mostly from late Permian deposits in Queensland and New South Wales. Before compression, these 30m (100ft) coal seams would have been 300m (1000ft) deep, taking around 300,000 years to form. High latitudes, crustal stretching, and widespread subsidence provided ideal conditions for continued growth of these cold-climate swamps and peatlands, whose closest present-day equivalent may be the boreal taiga of Canada and Siberia.

The deciduous swamp forest was dominated by tree-sized seed ferns – *Glossopteris* and *Gangamopteris* – along with some conifers, clubmosses, and horsetails. Despite abundant vegetation, there were few land tetrapods, and the insects were mostly small scorpion-flies and sap-sucking bugs. Bony fish such as *Ebenequa* swam in the swamp waters, but the expansion of terrestrial life seems still to have been limited to equatorial regions at this time.

Glossopterids are found in the Permian of Australia, South America, southern Africa, Madagascar, India, Antarctica, and New Zealand. In the 1920s German meteorologist Alfred Wegener argued that this distribution could only be explained if these regions had once been united in a single supercontinent.

Common organisms: tree-sized seed ferns
Volcanism: major explosive event (259ma) with widespread metre-thick tephra

267–260 million years ago
Capitanian Stage of the Guadalupian Epoch

- 1. *Glossopteris*
- 2. *Phyllotheca*
- 3. *Glossopteris linearis*
- 4. *Plumsteadia*
- 5. *Ebenequa*
- 6. *Sphenophyllum*
- 7. *Austraglossa*
- 8. *Dictyopteridium*
- 9. *Sphenopteris*

GLOSSOPTERIS (1) Leaves of this mid-sized seed fern (up to 4m/13ft tall) are elongated, with a prominent midrib.

PHYLLOTHECA (2)
This horsetail, up to 3m (10ft) tall, was widespread in glossopterid forests, but had a cosmopolitan distribution up to Cretaceous times. It is distinguished by a whorl of thin leaves extending from a disc-shaped base.

AUSTRAGLOSSA (7)
One group of glossopterids had an unusual fruiting body that produced a relatively small number of narrow-winged seeds attached as a group by a short stem.

COMPLEX TERRESTRIAL FOOD CHAINS
KAROO BASIN, SOUTH AFRICA

Climate: southern tropical

Biota: diverse terrestrial plant-and meat-eating tetrapod

Latitude then: 55°S

Latitude now: 31°S

Sea level: falling (+80m/260ft)

Original environment: lakes and rivers

Deposits: sands and muds

Status: numerous localities within South Africa's Karoo Basin

Preservation: often well-preserved skulls and some complete skeletons

• Karoo Basin

⊕ Earth *c.*260MA

• Karoo Basin

⊕ Fossil site today

By the late Permian, life on land was growing far more diverse – in terms of both variety of species and their ecological interactions. The great Karoo Basin of southern Africa provides a window onto this period, revealing over 70 new kinds of tetrapods, including the cynodont ancestors of today's mammals.

Some 1500 km (nearly 1000 miles) wide, the Basin was crossed by rivers bearing huge volumes of sediment from the Antarctic mountains to the south. Floodplains, lakes, and ponds were fringed with plant life, including horsetails, ferns, and glossopterids. Monsoon rains brought seasonal floods, drowning many creatures whose remains were preserved as fossils in the dry wadi deposits left as the waters evaporated in the hot sun. Life adapted to dramatic seasonal changes – small dicynodonts and even lungfish took shelter in specialized burrows.

Complex terrestrial food chains evolved, as plant-eating niches were filled by over 20 species of small tetrapods, especially dicynodonts and a few larger procolophonids. Carnivores ranged from small lizard-like insectivores (*Youngina*) to dog-sized therocephalians (*Moschorhinus*), and cat-sized cynodonts (*Procynosuchus*). The top predators were the first sabre-toothed meat-eaters – gorgonopsians such as *Cyanosaurus*.

260 million years ago
Wuchiapingian Stage of the Lopingian Epoch

Common organisms: small dicynodonts were the most common tetrapods

1. *Owenetta*
2. *Diictodon*
3. *Moschorhinus*
4. *Lystrosaurus*
5. *Youngina*
6. *Procynosuchus*
7. *Cyanosaurus*

➜ DIICTODON (2) Remains of the successful and widespread dicynodonts, *Diictodon* have been found within deep helical burrows, showing that its 45cm-long (18in) stocky body, short tail, and small, broad limbs were adaptations for burrowing. It had a complex horny beak to shred roots.

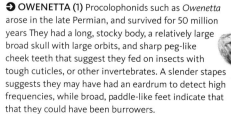

➜ OWENETTA (1) Procolophonids such as *Owenetta* arose in the late Permian, and survived for 50 million years They had a long, stocky body, a relatively large broad skull with large orbits, and sharp peg-like cheek teeth that suggest they fed on insects with tough cuticles, or other invertebrates. A slender stapes suggests they may have had an eardrum to detect high frequencies, while broad, paddle-like feet indicate that that they could have been burrowers.

➜ PROCYNOSUCHUS (6) This 60cm-long (24in) primitive cynodont displays a number of mammalian skull features such as the beginnings of a secondary palate, differentiation of the teeth into incisors and canines, and a backbone with distinct thoracic and lumbar regions. Its long tail suggests that it may have been semi-aquatic, with an otter-like lifestyle.

THE END OF THE PERMIAN WORLD
VOLGA BASIN, RUSSIA

Climate: tropical, high global summer temperatures
Biota: terrestrial

Latitude then: 30°N

Latitude now: 57°N

Sea level: low (-25m/-80ft)

Original environment: subtropical landscapes with seasonal rivers

Deposits: red sands and muds

Status: numerous localities scattered across the Volga and Moscow basins

Preservation: many skulls and some complete skeletons

Volga Basin •

⊕ Earth *c.*260 MA

Volga Basin •

⊕ Fossil site today

In 1841, British geologist Roderick Murchison recognized a distinctive series of red strata above Carboniferous rocks between Russia's Volga river and the Ural Mountains to the east. He named them as the Permian System after the regional capital Perm, and dedicated the ensuing map and memoir to Tsar Nicholas I. The Permian has since become an internationally recognized period of geological time.

Fossil vertebrates, believed to be mammals, were already known from these copper-rich Russian strata. These animals do indeed have mammalian features, but are now seen as belonging to an extinct synapsid group known as the dinocephalian therapsids. Explorations in the 20th century by Russian paleontologist Ivan Efremov uncovered hundreds of skeletons, including an array of amphibians and reptiles such as the seymouriamorph *Kotlassia* and the sabre-toothed gorgonopsian *Inostrancevia*.

The evolution of more and more reptiles allowed the first true herbivore-based ecosystem to develop (and also allow the strata to be correlated with others worldwide). Large herbivores such as the 3m (10ft) armoured *Scutosaurus* were closely followed by large predators such as *Inostrancevia*, which was equipped with 5cm (2in) long canines.

1. *Microphon*
2. *Arctotypus*
3. *Dicynodon*
4. *Kotlassia*
5. *Inostrancevia*
6. *Scutosaurus*
7. *Dvinia*

↪ ARMOURED GIANT (6)

Scutosaurus was the largest herbivore of its time, but still had an extensive shield of bony "osteoderms". In some areas, such as the shoulder and pelvis, these were sutured together.

↰↪ SABRE-TOOTHED JAWS (5)

The 2m-long (6.6ft) gorgonopsian *Inostrancevia* was the largest carnivore and top predator of its time. A sabre-toothed form, it had canines up to 5cm (2in) long, and jaws that could open to more than 90 degrees to produce a stabbing bite, which could cripple its prey. Carnivorous therapsids were more diverse than the herbivores – they included the gorgonopsians, therocephalians, and cynodonts.

THE GREATEST EXTINCTION
PUTORANA PLATEAU, SIBERIA, RUSSIA

Climate: seasonal – moist, temperate

Biota: apparently low diversity, but biased by small sample

Latitude then: 60°N

Latitude now: 68°N

Sea level: low (-25m/-82ft)

Original environment: widespread plateau basalts from fissure eruptions

Deposits: lavas

Status: both surface and buried outcrops across a huge area of Arctic Siberia

Preservation: well-preserved basaltic lavas

Putorana Plateau •

⊕ Earth *c.*251 MA

Putorana Plateau •

⊕ Fossil site today

The Paleozoic ended with an unprecedented extinction of life on Earth – perhaps as many as 90 per cent of all species died out. Although many geologists have looked to a major asteroid impact as a possible explanation for such a catastrophe, none has been found.

However, there is evidence for another cataclysmic event at this time – very large-scale outpourings of basalt lavas over some 7 million sq km (2.7 million sq miles) of Siberia around 252.2–251.1 million years ago. Such volcanism releases huge quantities of greenhouse gases into the atmosphere, resulting in global warming. And geochemical evidence suggests

that there was another source of warming – a massive release of methane from ocean-floor sediments, perhaps triggered by the volcanism. Ocean waters became anoxic, killing off much of the marine food chain, while warming of the atmosphere drastically affected life on land.

Between 50 and 60 per cent of tetrapod families (around 90 per cent of species) disappeared, with large herbivores and carnivores worst affected by the collapse of complex food webs. Some small diapsids, such as *Archosaurus* from Russia's late Permian, survived – their descendants would come to dominate the Mesozoic Era.

Volcanism: extensive eruptions of plateau basalts

252–251 million years ago
Changhsingian Stage of the Lopingian Epoch

❶ *Archosaurus*

➲ BASALT OUTCROPS
Today the Siberian trap outcrops only cover some 675,000 sq km (260,000 sq miles), but estimates suggest their original extent was much greater. Some individual flows were hundreds of metres thick and one tuff layer averaging 20m (66ft) thick has been traced over 30,000 sq km (11,600 sq miles). The region of volcanic material covered much of Siberia from the Urals east to Lake Baikal and south to Kazakhstan, an area almost equivalent to that of continental America.

⊙ ARCHOSAURUS (1) This metre-long predator with short limbs and sprawling posture is one of the earliest archosaurs with characteristic features such as a skull opening between the nostril and eye socket.

RUSSIAN SURVIVORS
MOSCOW BASIN, RUSSIA

250.5 million years ago
Induan Stage of the Early Triassic Epoch

Latitude then: 30°N

Latitude now: 60°N

Sea level: low (-25m/-82ft)

Original environment: semi-arid with seasonal rains

Deposits: sands and muds

Status: numerous localities throughout the Moscow Basin

Preservation: Three-dimensional skeletal remains with some good skulls and complete skeletons

⊕ Earth *c.*250.5ᴍᴀ

⊕ Fossil site today

Life was radically changed by the end-Permian extinction. Global average temperatures soared to 22°C (72°F) compared with 12–15°C (53–59°F) in the early Permian, and the polar ice caps disappeared completely. The changes on land are best seen in the earliest Triassic age strata of European Russia, where land environments became more arid but seasonal rains filled extensive river channels and lakes.

Here, the vast majority (over 90 per cent) of fossils are aquatic temnospondyl amphibians, such as *Benthosuchus*, mostly no more than a metre long (40in). The previously abundant reptiles were badly hit. A few, such as

Contritosaurus, did survive, but they were rare. Semi-aquatic reptiles such as *Lystrosaurus* were even rarer. The fish were also drastically reduced in number and diversity, with the lungfish *Gnathorhiza* being the most common survivor. Their fossil remains have been found in burrows for, like many lungfish, *Gnathorhiza* could survive drought conditions by aestivating in a moist, mucus-lined shelter.

Overall, Russia's Early Triassic shows the last remnants of some formerly worldwide tetrapod groups. The only remaining cosmopolitan connection was with India, where *Lystrosaurus* also survived.

Common organisms: freshwater amphibians with some lungfish and land-living tetrapods

Climate: northern hemisphere subtropical
Biota: terrestrial vertebrates

❶ *Benthosuchus*
❷ *Wetlugasaurus*
❸ *Lystrosaurus*
❹ *Contritosaurus*
❺ *Chasmatosuchus*

↩ **LYSTROSAURUS (3)** These semi-aquatic dicynodonts were the largest of Russia's Early Triassic survivors. They fed on drought-resistant horsetails and *Dicroidium* plants, but the largest species, *L. maccaigi*, (up to 1.5m/5ft long) may have specialized in glossopterids.

↪ **CONTRITOSAURUS (4)**
One of the most primitive of the procolophonid parareptiles, *Contritosaurus* was very small, with a skull only 2cm (0.8in) long, but had large eyes and small peg-like teeth. It probably specialized in feeding on insects.

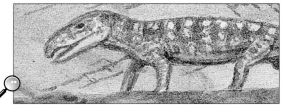

↪ **CHASMATOSUCHUS (5)**
This early diapsid archosaur was somewhat crocodile-like in appearance, and is placed in the proterosuchid family. These predatory carnivores were taking over the niches previously occupied by the gorgonopsids.

LIFE RECOVERS
GRAPHITE PEAK, ANTARCTICA

250 million years ago
Induan Stage of the Early Triassic Epoch

Latitude then: 78°S	
Latitude now: 85°S	
Sea level: (-25m/-82ft)	
Original environment: evergreen lowland forest	
Deposits: semi-arid stream deposits including sands and floodplain muds	
Status: inaccessible and inhospitable location in the Transantarctic Mountains	
Preservation: flattened but relatively well-preserved plant and animal remains	

⊕ Earth *c.*250 MA

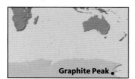

⊕ Fossil site today

One reason for the slow recovery of terrestrial life following the end-Permian extinction was a breakdown in terrestrial ecosystems following the loss of staple food plants at the base of the food chain. The extinction caused an abrupt change in vegetation from Permian broadleaf deciduous, and glossopterid swamp floras, to Early Triassic evergreen needle-leaf, humid, and cool-temperate forests.

These changes are particularly well seen in the rocks of Graphite Peak, where Antarctica's first Early Triassic vertebrates were discovered in 1968. Here, extensive river systems flowed through a 200,000 sq km (nearly 80,000 sq miles) basin filled with lowland *Voltziopsis* forests and flanked by a volcanic mountain range. The site's fossil soils contain abundant logs, stumps, and plant remains, but surprisingly no coals – part of a worldwide "coal gap" in strata of the time. The fossils may represent a post-apocalyptic state, with poor global floras of *Voltzia* and *Voltziopsis*, but abundant fungai – as if the destruction of the forests and the *Lystrosaurus* biota was followed by a period of massive decay. However, the sparse tetrapod fauna of the time shows some important evolutionary innovations, especially among the prolacertiforms and the cynodonts.

Common organisms: terrestrial plants

Climate: high-latitude cool and damp
Biota: terrestrial peat-swamp flora with some tetrapods

1. *Thrinaxadon*
2. *Lystrosaurus*
3. brachyopoid
4. *Voltziopsis*
5. *Prolacerta*
6. *Procolophon*

⮕ **THRINAXODON (1)** This cat-sized cynodont, found in South Africa and Antarctica, shows important developments in the evolution of the mammal condition, including modifications to the skull, jaw, and back, and a slender tail.

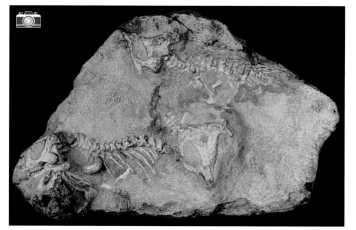

⮌ ⮍ **PROCOLOPHON (6)** This Early Triassic procolophonid, up to 40cm (16in) long, was part of a genus that survived the late Permian extinction. *Procolophon* skeletons are frequently found clustered together.

A MID-TRIASSIC CLOSE-UP
GRES A VOLTZIA, FRANCE

242 million years ago
Anisian Stage of the Middle Triassic Epoch

Latitude then: 20°N

Latitude now: 48°N

Sea level: as present

Original environment: coastal delta

Deposits: sands and muds

Status: numerous localities scattered through the northern Vosges hills

Preservation: flattened but entire fossils with soft tissue preservation in the mudrocks

Grès à Voltzia

↷ Earth *c.*242 MA

Grès à Voltzia

↷ Fossil site today

Despite the end-Permian extinction event, life was well on the road to recovery by the Middle Triassic. The Grès à Voltzia strata of northeastern France give a detailed insight into life in a coastal delta 242 million years ago. Ponds, cut off from the delta channels, dried out periodically, trapping and eventually killing their aquatic inhabitants. Bacterial mats grew over the putrefying remains and preserved them in astonishing detail.

The brackish waters of the delta teemed with small actinopterygian fish, shrimps, and crayfish, along with marine king crabs that came ashore in hordes to mate in the breeding season. Amphibious temnospondyls such as *Eocyclotosaurus* were top predators.

In addition to aquatic life, the pond muds preserve plants from the interchannel islands, such as the conifer *Voltzia*, along with ferns and horsetails that grew in abundance around the delta. A surprising diversity of modern-looking arthropods lived among the plants. They ranged from scorpions and millipedes to the earliest known funnel spider, *Rosamygale*, and are accompanied by host of insects – one collection contains over 5300 specimens, representing about 200 species, along with some of their eggs and larvae.

145.5

65.5

Common organisms: aquatic invertebrates, predominantly marine but with some terrestrial organisms including plants

Climate: tropical northern hemisphere

Biota: mixed terrestrial and aquatic

1. *Rosamygale*
2. blattodean
3. *Voltzia*
4. scorpionid
5. myriapod
6. *Antrimpos*
7. *Progonionenus*
8. *Anomopteris*
9. *Limulitella*
10. *Dipteronotus*
11. *Voltziaephemera*

⬆ ROSAMYGALE (1) The muds preserve soft-bodied chelicerates such as this early spider, 6mm long.

↻ VOLTZIA (3) The Grès à Voltzia is named after *Voltzia*, a bushy conifer that grew in dense thickets between delta distributaries along with other gymnosperms such as *Albertia*. *Voltzia* bore pollen and seeds in cones at the tips of branches with needle-like leaves, resembling modern conifers.

⬆ LIMULITELLA (9) The unique conditions of Grès à Voltzia preserve soft-bodied chelicerates such as this limulid, a 6cm-long (2.5in) king crab. *Limulitella* and its trackways are common fossils here. Although limulids are essentially marine, these remains show that they entered the delta's brackish waters to breed.

MARINE REPTILES DIVERSIFY
MONTE SAN GIORGIO, SWITZERLAND

233 million years ago
Ladinian Stage of the Middle Triassic Epoch

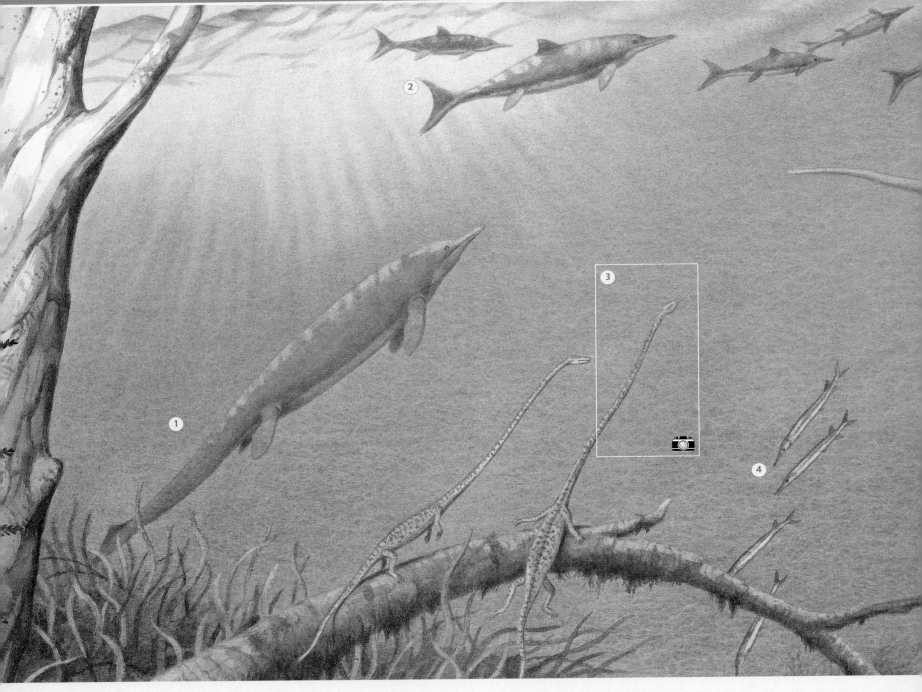

Latitude then: 15°N

Latitude now: 45.5°N

Sea level: +25m (82ft)

Original environment: stratified basin with anoxic bottom waters

Deposits: black bituminous shales and laminated dolomitic limestones

Status: quarrying has been necessary to excavate the fossiliferous strata

Preservation: flattened but often entire skeletons with some fossilized soft tissues

Monte San Giorgio •

⊕ Earth c.233 MA

Monte San Giorgio

⊕ Fossil site today

An extraordinary window on the marine life of the tropical Tethys Ocean first opened in the late 19th century, when remarkably well-preserved fossils were discovered in shales and dolomitic limestones near Besano in the southern Alps during quarrying for bitumen.

The outcrop of Middle Triassic strata straddles the Italian–Swiss border, and has surrendered hundreds of complete marine reptile skeletons over the last 150 years, along with thousands of fish (including 30 ray-finned species), clams, and ammonoids.

All these animals lived in the quiet waters of a marine embayment off the main Tethys Ocean.

Here abundant fish and ammonoid cephalopods were hunted by a variety of marine reptiles such as ichthyosaurs (eg *Mixosaurus*) and nothosaurs (eg *Ceresiosaurus*). Specialist predators such as the placodonts (eg *Paraplacodus*) had large flat teeth for crushing shellfish.

The strata also contain the skeletons of reptiles such as the bizarre *Tanystropheus*, which grew to 6m (20ft) long with a stiff, elongated neck twice the length of its body. Such an animal must have been essentially marine, but the related *Macronemus* was only 80cm (31.5in) long and may have spent more time on land.

Common organisms: ammonites and clams

Climate: tropical–equatorial
Biota: marine fish and reptiles

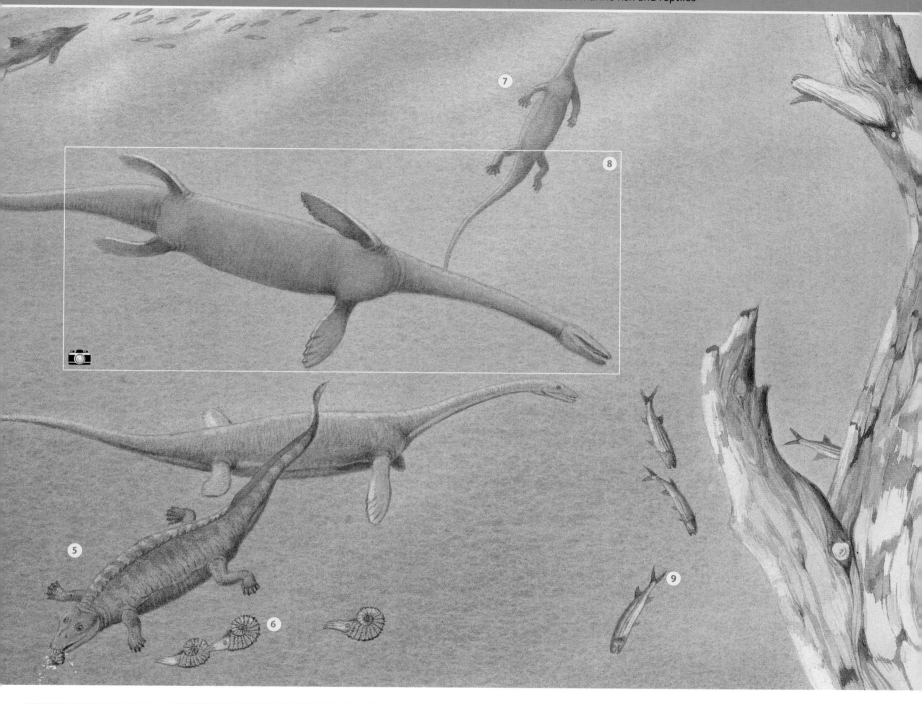

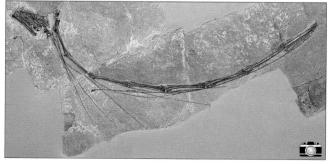

1 *Shastasaurus*
2 *Mixosaurus*
3 *Tanystropheus*
4 *Saurichthys*
5 *Paraplacodus*
6 *Eoprotrachyceras*
7 *Askeptosaurus*
8 *Ceresiosaurus*
9 *Birgeria*

↑ TANYSTROPHEUS (3) Three genera of archosauromorphs have been found at Monte San Giorgio, including the strange *Tanystropheus* with its extremely long neck, produced by elongation of between 12 and 24 individual vertebrae, depending on the species.

↑ CERESIOSAURUS (8) The most abundant reptiles at Monte San Giorgio were the amphibious nothosaurs. They were probably fish feeders, and may have laid eggs on land. They include *Ceresiosaurus* (up to 3m/10ft long), *Paranothosaurus*, *Lariosaurus*, and the tiny *Neusticosaurus* (up to 30cm/12in long), of which several hundred specimens have been found.

INSECTS DIVERSIFY WITH NEW PLANTS
FERGANA VALLEY, KYRGYZSTAN

228 million years ago
Ladinian to Carnian stages of the Middle–Late Triassic epochs

Latitude then: 38°N

Latitude now: 41°N

Sea level: +50m (165ft)

Original environment: forested floodplains with rivers, lakes, and swamps

Deposits: river and lake sands, silts, clays, and coal seams

Status: localities lie within a remote and largely inaccessible desert region of Kyrgyzstan

Preservation: flattened with some soft tissue preservation

⊕ Earth: *c.*228 MA

⊕ Fossil site today

Some of the richest insect-bearing rocks in the world are the Middle Triassic Madygen strata of Kyrgyzstan's Fergana Valley. They are particularly important because earlier Triassic insects are so rare. Some 15,000 fossils record increasing insect diversity, with new aquatic forms, and a disappearance of previous insect families. The transition from a Permian flora of lycopsids, ferns, cordaites, and pteridosperms, to a Triassic one of cycads, ginkgos, conifers, and bennittaleans clearly had a major impact.

Madygen's ancient river and lake deposits preserve discontinuous coal seams and muds with an astonishing abundance of plant and insect remains. Most spectacular are the giant predatory titanopterans, with wingspans up to 40cm (16in) and forelegs armed with stout spines. At rest, the wings were held flat over the abdomen and carried very large organs that probably produced a deep, bullfrog-like call. Like cicadas, titanopterans were very vocal and possibly territorial.

Madygen is also famous for two small but intriguing diapsid reptiles – *Longisquama* and *Sharovipteryx*. Both were discovered in 1965 and are claimed to have been aerial gliders. However, the exact nature of *Longisquama* remains highly problematic.

Common organisms: plants and insects

Climate: northern hemisphere humid temperate

Biota: freshwater and terrestrial

1 *Notocupoides*
2 *Hadeocoleus*
3 *Saurichthys*
4 *Longisquama*
5 *Podozamites*
6 *Sharovipteryx*
7 *Axioxyela*
8 *Madygenia*
9 *Gigatitan*

⮎ **LONGISQUAMA (4)** The 10cm-long (4in), feather-like modified scales on the back of this reptile may have assisted it in gliding.

⬆ **LONGISQUAMA (4)** Detailed examination of the so-called "scales" has raised questions about whether they could have functioned for gliding. One alternative explanation is that they may be plant leaves, fossilized with the skeleton purely by chance.

⬆ **GIGATITAN (9)** This giant insect not only had a wingspan of 40cm (16in), but also carried a stridulatory organ on the wings that was rubbed with the hind legs to make a loud noise, rather like the sounds made by modern crickets.

DINOSAUR BEGINNINGS
VALLEY OF THE MOON, LA RIOJA PROVINCE, ARGENTINA

227 million years ago
Carnian Stage of the Late Triassic Epoch

Latitude then: 40°S

Latitude now: 30°S

Sea level: +50m (165ft)

Original environment: volcanically active floodplain basin with rivers and lakes

Deposits: river-lain sands and muds

Status: World Heritage Site within the Ischigualasto/Talampaya Natural Parks since 2000

Preservation: disarticulated but three-dimensional bones

Valley of
the Moon

⊕ Earth: c.227MA

Valley of the Moon

⊕ Fossil site today

The discovery of some of the earliest dinosaur-like reptiles in the late 1950s has promoted the barren hilly landscapes of the Valley of the Moon, in Argentina's La Rioja and San Juan provinces, to international fame. These Late Triassic strata were laid down on a river floodplain abundant with 40m (130ft) giant conifers, ferns, and horsetails during a period of strongly seasonal climates.

Most of the fossil tetrapods are rhynchosaurs and cynodonts – dinosaur-related fossils are very rare. But importantly, two predatory genera, *Herrerasaurus* and *Eoraptor* are some of the earliest dinosaur-like

fossils (known as dinosauromorphs). As "basal saurischians" at the root of one major branch of dinosaurs, they are paralleled by the roughly contemporary ornithischian *Pisanosaurus*, found in the same region of Argentina.

Most importantly, these genera prove that, while dinosaurs are monophyletic (that is, they share a single common ancestor), the evolutionary split into ornithischian and saurischian groups (named for the distinctive structures of their of hipbones and pelvises) occurred even earlier than previously thought – indeed, *Pisanosaurus* is the very earliest known ornithischian.

145.5

65.5

Climate: highly seasonal temperate southern hemisphere
Biota: terrestrial

Common organisms: plants, cynodonts, and rhynchosaurs

❶ *Hyperodapedon*
❷ *Exaeretodon*
❸ *Herrerasaurus*
❹ *Saurosuchus*
❺ *Protojuniperoxylon*
❻ *Eoraptor*

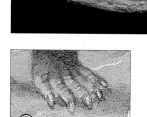

↪ **HYPERODAPEDON (1)**
Strong claws on the hind feet of this 2m-long (6.6ft) plant-eating rhynchosaur were probably used for digging out nutritious tubers and roots.

⬆ **HERRERASAURUS (3)** Large eyes and sharp teeth, including unusually long canines, show that this 4m-long (13ft) bipedal dinosaur was a highly active predator. Indeed, the remains of a pig-like plant-eating rhynchosaur have been found in the stomach area of one fossil specimen.

⬆ **EORAPTOR (6)**
Although only 1m (40in) long, this basal saurischian was also an active bipedal predator, with a combination of primitive and more advanced theropod characters. The eyes were large, the long muscular legs well adapted for fast running, and the arms, with long-fingered hands, ideal for catching and grasping prey. The teeth include curved, serrated, and more leaf-shaped forms, suggesting a mixed diet of small prey, insects, and possibly plants.

121

SCOTLAND'S DESERT TETRAPODS
EPOCH ELGIN, SCOTLAND

220 million years ago
Carnian Stage of the Late Triassic

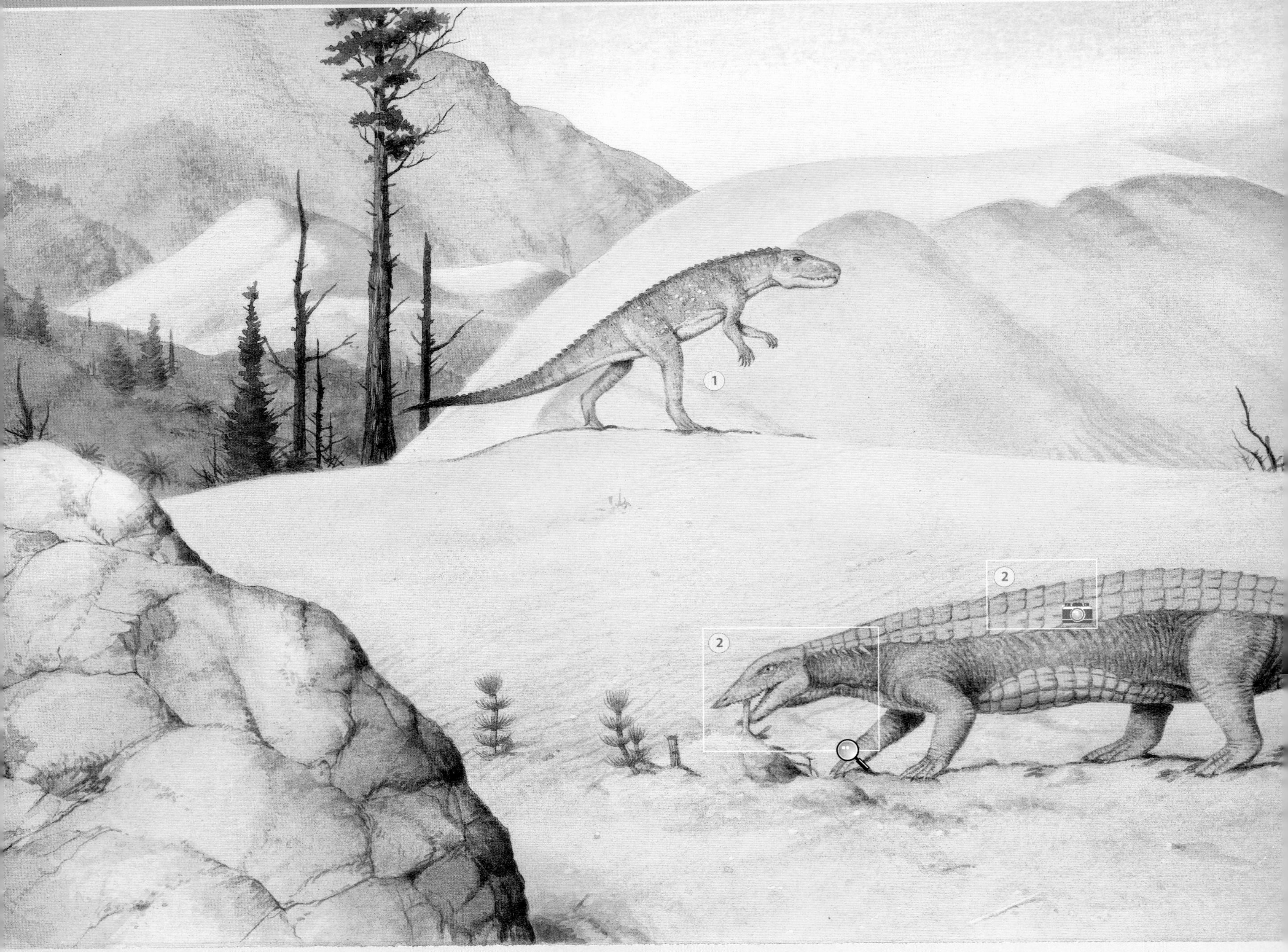

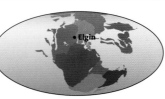

Latitude then: 35°N

Latitude now: 57°N

Sea level: +50m (165ft)

Original environment: desert but with some nearby vegetation

Deposits: wind-blown dune sands

Status: old stone quarries with rare fossils

Preservation: three-dimensional hollow moulds of bones and some articulated skeletons

Earth: *c.*220MA

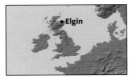

Fossil site today

Tetrapod remains found at Elgin in Scotland document the major transition from older Triassic faunas to the newly emerging dinosaurs. The fossils are preserved in sandstones that were originally wind-blown dune sands, and they are very difficult to extract since the bone is either soft or completely dissolved away to leave a hollow mould. Fortunately, modern casting techniques allow replication of the original bone shape, from which the detailed anatomy of several genera, such as *Ornithosuchus*, *Stagonolepis*, *Scleromochlus* and *Hyperodapedon*, has been obtained. These fossil remains are particularly important as many of the fossils preserve more-or-less complete, articulated skeletons.

The environment at Elgin is something of a puzzle. The animals died and were preserved in dune sands, but how did they actually survive in this apparently arid region, and were the dunes part of a coastal dune system or a larger desert? The variety of tetrapods, which includes some large plant-eaters, shows that there was a well-vegetated area relatively nearby, although no plant fossils are found in the sandstones. It is possible that the dunes were used for breeding and egg laying, and that the feeding grounds lay some way off.

Climate: dry continental interior with seasonal rains
Biota: terrestrial

Common organisms: various tetrapods but none common

- ❶ *Ornithosuchus*
- ❷ *Stagonolepis*
- ❸ *Hyperodapedon*
- ❹ *Scleromochlus*

↻ ↺ STAGONOLEPIS (2)
Growing to 3m (10ft) long, this aetosaur lacked teeth at the front of its mouth and had leaf-shaped cheek teeth for eating plant stems and tubers, which it rooted up with its blunt snout. With bony scales, its body was well armoured against predators.

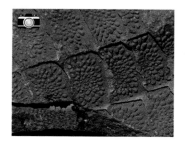

↑ ↻ HYPERODAPEDON (3) This stout-ribbed and barrel-chested rhynchosaur (represented here by a fossil of its spine and ribs, and its skull) grew up to 2m (6.6ft) long, and had a toothless "beak" and rows of palatal teeth for slicing and shredding vegetation.

↻ SCLEROMOCHLUS (4)
This 17cm-long (7in) archosaur had long hind legs for bipedal running or hopping like a jerboa, but relatively short arms and neck. Its taxonomy is controversial.

EVOLVING ARCHOSAURS
HAYDEN QUARRY, NEW MEXICO, USA

216 million years ago
Late Carnian to Norian stages of the Late Triassic Epoch

Latitude then: 10°N

Latitude now: 36°N

Sea level: +50m (165ft)

Original environment: forested floodplains with braided streams

Deposits: alluvial sands and silts

Status: one of several quarries in the Ghost Ranch locality – recovery of fossils required large-scale excavation

Preservation: disarticulated but otherwise well-preserved three-dimensional bones

Hayden Quarry

⊕ **Earth:** c.216 MA

• Hayden Quarry

⊕ **Fossil site today**

It was once thought that the first dinosaurs rapidly ousted the other Late Triassic archosaurs, but new fossils discovered close to the famous Ghost Ranch site (see over) show that the transition may been a longer process.

The Late Triassic Chinle Formation strata lie some 65m (210ft) below the later site. Since 2006, some 1300 vertebrate specimens have been recovered from Hayden Quarry, where the original environment was one of floodplains, braided streams, and a waterside open forest of the large conifer *Araucarioxylon*.

The new fossils include an intriguing mix of surviving Late Triassic archosaurs, such as the

plant-eating aetosaur *Typothorax*, dinosaur precursors, such as *Dromomeron*, and early true dinosaurs, such as the saurischian *Chindesaurus* and some coelophysoid theropods. Interestingly, dinosaurs known from high-latitude sites in Europe and Gondwana are absent here and across North America – perhaps a sign of real environmental variations , but perhaps explicable through bias in the type of sites that have been excavated. Whatever the truth, this new site shows that the evolutionary transition from archosaurs to dinosaurs was a gradual one, taking place over some 15–20 million years.

Common organisms: dinosaurs and other archosaurs

Climate: terrestrial and freshwater
Biota: tropical humid

- ❶ *Dromomeron*
- ❷ pseudopalatine archosaur
- ❸ *Chindesaurus*
- ❹ *Eucoelophysis*
- ❺ *Postosuchus*
- ❻ *Typothorax*

↪ DROMOMERON (1) This 1m-long (3.3ft) animal was a long-legged biped capable of fast running. Detailed analysis shows that it is closely related to the Middle Triassic dinosauromorph *Lagerpeton*, and as a result the two are placed together in a sister group to other dinosauromorphs.

↪ CHINDESAURUS (3) This 20cm-long (8in) femur is characteristic of *Chindesaurus*, a basal saurischian known from a partial skeleton and isolated bones.

↪ COELOPHYSOID (4) These Late Triassic theropods, growing up to 3m (10ft) long, are known only from a few disarticulated bones. This contrasts strikingly with the plentiful skeletons of a later member of the same group, *Coelophysis* itself, found in the slightly later strata of Ghost Ranch (see over).

↑ TYPOTHORAX (6) The formidable body armour of this 5m (16.5ft) long aetosaur consisted of heavy bony plates set into the skin. This was the plant-eater's only defence against predators such as *Postosuchus*. The small skull and short jaw only have small teeth, while the pig-like snout was entirely toothless.

DEATH AMONG THE DINOSAURS
GHOST RANCH, NEW MEXICO, USA

*c.*213 million years ago
Norian Stage of the Late Triassic Epoch

In 1947, one of paleontology's most remarkable fossil assemblages was found in the Whitaker Quarry at Ghost Ranch, New Mexico. It consisted of hundreds of articulated skeletons tangled together in a great jumble – and most belonged to a single genus, a small predatory theropod dinosaur called *Coelophysis*.

How all these meat-eaters were brought together is still something of a mystery – were they perhaps feeding or breeding? Three clear body forms are present in roughly equal numbers – robust (female?), slender (male?), and juvenile, but since there is no evidence of eggshells or nests found at the site, it seems unlikely that they were gathering

to breed. Remains of last meals, including the small crocodile relative *Hesperosuchus*, can be found among the preserved stomach contents of some animals. These were previously identified as juvenile *Coelophysis*, and taken as indications of cannibalism, but closer examination has revealed the truth.

One theory to explain this wealth of fossils is that animals may have died in large numbers during a drought. They were subsequently washed into the river channel, perhaps by a flash flood, and swept downstream to the place where they were finally buried. A notable puzzle is the lack of plant-eaters – even the non-coelophysoid remains at the site are mostly those of small predators.

⊕ COELOPHYSIS (1) A long, low skull tapering towards a narrow snout is typical of the coelophysoids. The animal's lower jaw also has unique features such as slightly heterodont (specialized) teeth – with some enlarged more than others. Postmortem crushing of the lightly built skulls unfortunately obscures some intriguing details, such as the structure of the internal braincase.

145.5

65.5

Climate: humid tropical

Common organisms: theropod dinosaurs

Biota: terrestrial and freshwater

1. *Coelophysis*
2. *Araucarioxylon*
3. *Semionotus*
4. *Chinlea*
5. *Hesperosuchus*
6. *Rutiodon*

⊙ COELOPHYSIS (1) This is one of the many complete skeletons from the Whitaker Quarry. They often display this typical posture, with the neck curved back sharply on itself. However, what looks like a sign of the animal's death throes is in fact a result of muscles contracting after death.

⊙ HESPEROSUCHUS (5) This metre-long (40in), lightly built sphenosuchian predator may have adapted a "high walk" posture (similar to that seen in modern crocodiles) for running, with the limbs extended and brought closer to the body than in normal resting mode. Parallel rows of protective scutes may have helped strengthen the backbone for such energetic activity, while the animal's belly was protected with rib-like structures.

251   199.6

JURASSIC SEAWORLD
LYME REGIS, DORSET, ENGLAND

195 million years ago
Hettangian to Sinemurian stages of the Early Jurassic Epoch

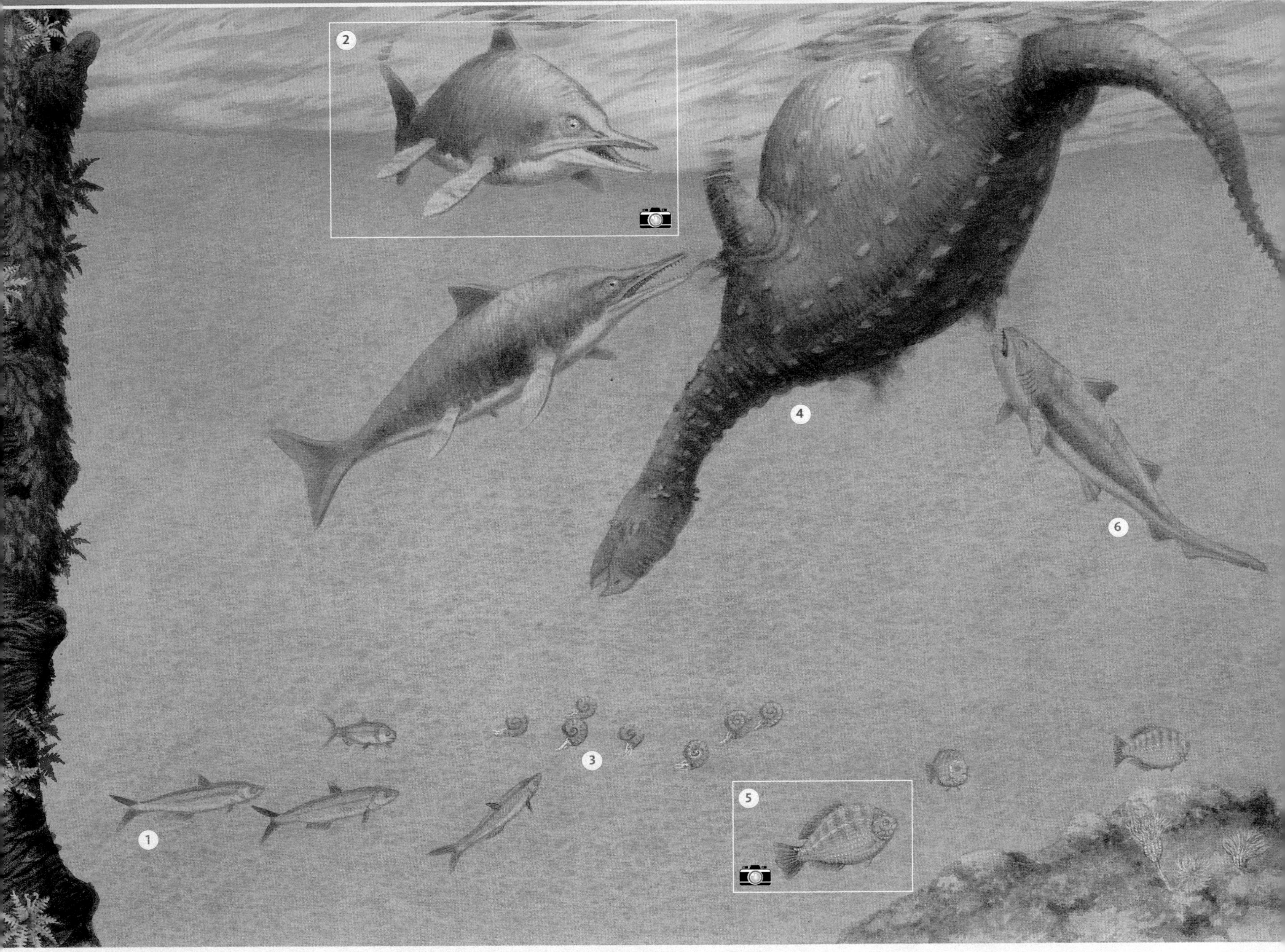

Latitude then: 36°N

Latitude now: 50°N

Sea level: +25m (82ft)

Original environment: nearshore waters

Deposits: low-oxygen seabed muds with carbonate nodules and limestone layers

Status: World Heritage Site and Site of Special Scientific Interest

Preservation: mostly flattened, but some three-dimensional fossils and with some soft tissue preservation

⊕ Earth: c.195 MA

⊕ Fossil site today

The Early Jurassic coastal cliffs around Lyme Regis in Dorset are one of the most famous fossil localities in Britain, and are now a World Heritage Site. For more than two centuries, fossils found by local collectors in the limestone and shale strata have found their way into museum collections around the world. They range from rare dinosaurs to some 50 species of fish and countless ammonites.

In the early 19th century, the Anning family recovered and prepared some of the earliest known and best preserved specimens of marine reptiles – including some of the finest ichthyosaurs and plesiosaurs, which were sold to major British museums. Lyme's Early Jurassic reptiles are still the most abundant and diverse of this age known in the world: some 14 species have been recovered, of which nine are unique to the site. As well as ichthyosaurs and plesiosaurs, they include a flying pterosaur (*Dimorphodon*) and the armoured thyreophoran dinosaur *Scelidosaurus*, of which a remarkably complete specimen has been found in recent years.

Among the fish are many ray-finned genera and some complete cartilaginous sharks and chimaeras, making Lyme Regis one of the most important fossil fish sites in the British Isles.

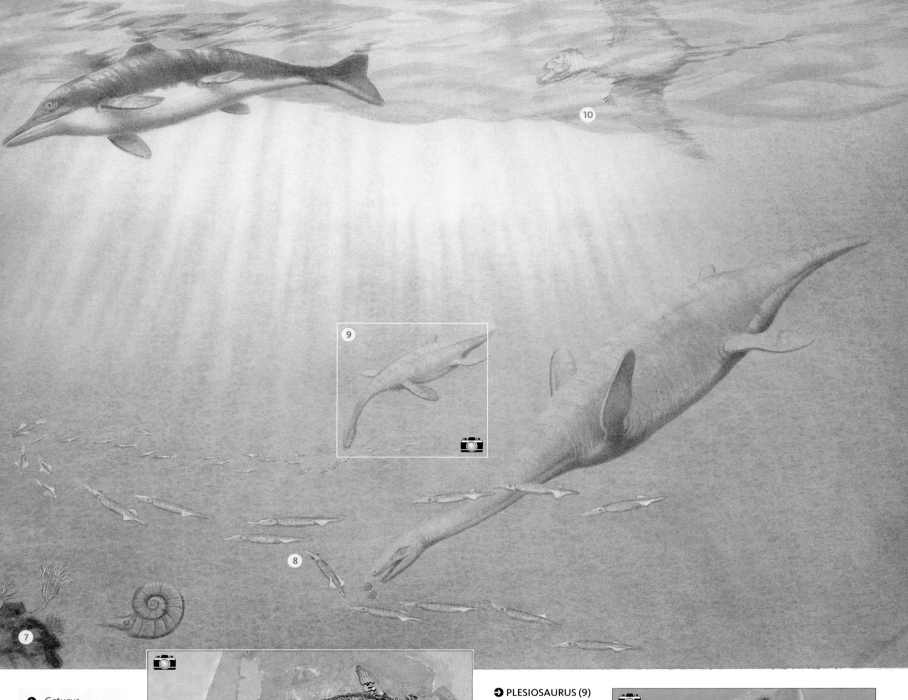

Common organisms: molluscs (ammonites, belemnites, and bivalves) and fish

Climate: mid-latitude warm humid
Biota: marine, mostly nektonic

❶ *Caturus*
❷ *Ichthyosaurus*
❸ *Microderoceras*
❹ *Scelidosaurus*
❺ *Dapedium*
❻ *Hybodus*
❼ *Pentacrinites*
❽ *Passaloteuthis*
❾ *Plesiosaurus*
❿ *Dimorphodon*

➲ **ICHTHYOSAURUS (2)** Four species of this large, dolphin-like, marine predatory reptile, up to 9m (30ft) long, have been found at Lyme Regis. To judge from fossilized stomach contents and faeces, their varied diet included fish and invertebrates.

➲ **PLESIOSAURUS (9)**
Growing up to 5m (16.5ft) long, this small-headed, long-necked plesiosaur was found alongside large-headed, short-necked pliosaurids such as *Eurycleidus*.

➲ **DAPEDIUM (5)**
Small fins on this deep-bodied actinopterygian fish suggest that it could not swim very quickly – it may have fed on invertebrates such as clams.

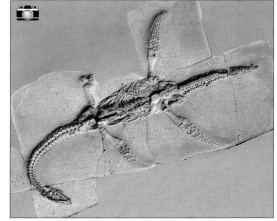

129

251   199.6

REPTILES RULE THE SEAS
HOLZMADEN, GERMANY

182 million years ago
Toarcian Stage of the Early Jurassic Epoch

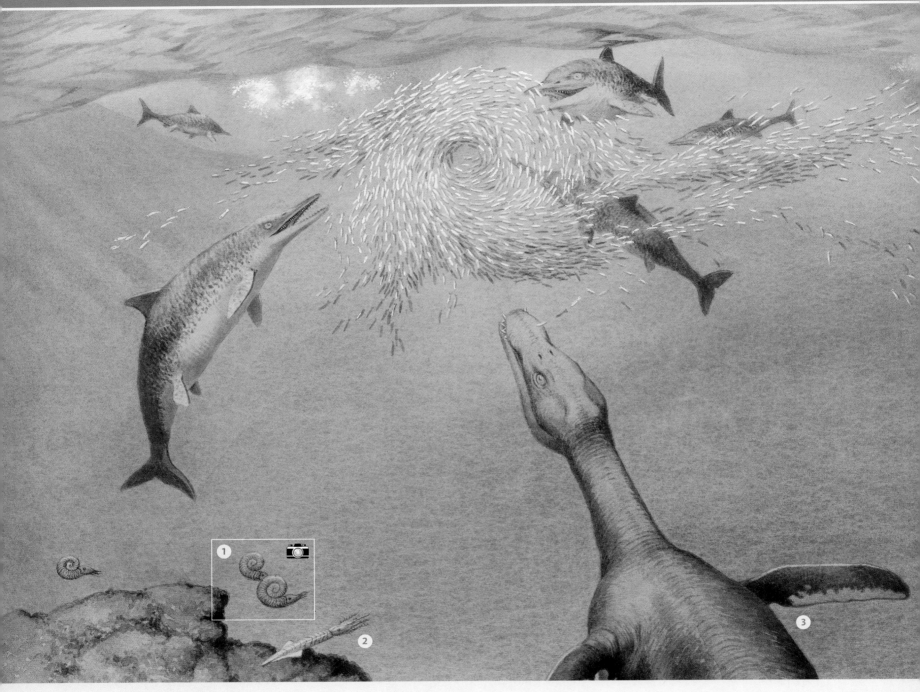

Latitude then: 30°N

Latitude now: 48°N

Sea level: +50m (165ft)

Original environment: deep shelf sea

Deposits: bituminous shales and limestones

Status: mostly exposed in commercial quarries, with a few open to collectors

Preservation: numerous articulated skeletons with body outlines preserved as black organic films that replicate soft tissues

⊕ Earth: *c.*182 MA

⊕ Fossil site today

More than five hundred almost complete skeletons of dolphin-like ichthyosaurs have been recovered from Early Jurassic bituminous limestones and shales at Holzmaden in southern Germany, where strata have been quarried for roofing, building, and oil production since the 16th century,

Some of the Holzmaden ichthyosaurs preserve an outline of skin and body profile replicated by a bacterial mat, providing proof that they had a dorsal fin and an upper lobe to the tail, neither of which normally fossilize.

Almost as spectacular are the 70 specimens of marine teleosaur crocodiles found at the

same site, especially the long-snouted, fish-eating *Steneosaurus*, some 3m (10ft) long, which had powerful limbs and was amphibious.

The Holzmaden strata were laid down in one of several deep basins within the shallow sea that covered much of northern Europe at the time. At certain periods, the deeper basin waters were starved of oxygen and became stagnant. This restricted life on the seabed to a few bivalves and crustaceans, but it also provided ideal conditions for the preservation of many animals that had lived in the more oxygen-rich waters closer to the surface, including some long-stemmed sea lilies.

Climate: mid-latitude warm humid
Biota: marine nektonic

Common organisms: cephalopods, fish, and reptiles

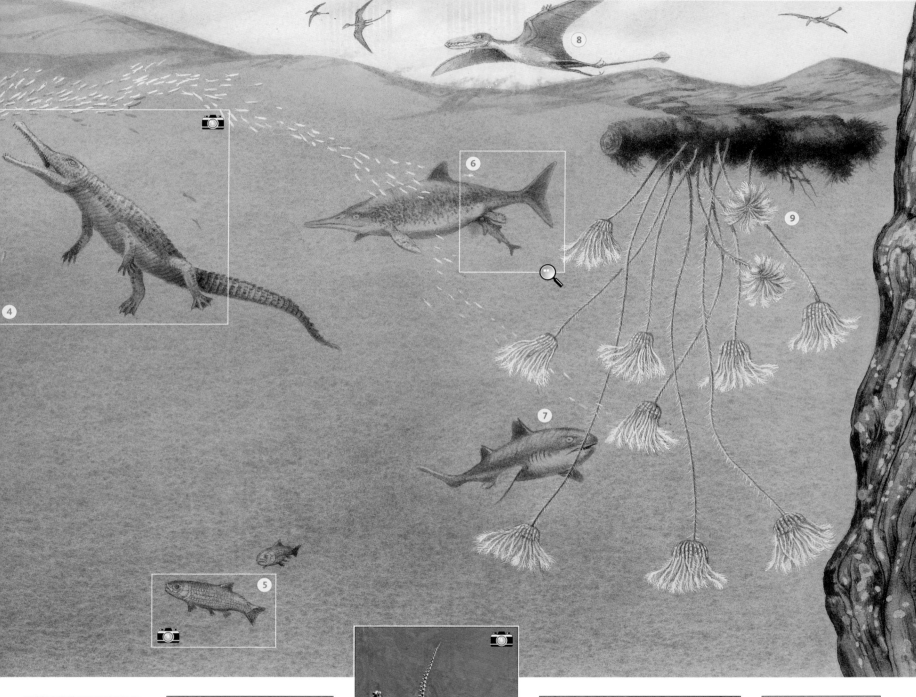

1. *Harpoceras*
2. *Passaloteuthis*
3. *Rhomaleosaurus*
4. *Steneosaurus*
5. *Lepidotes*
6. *Stenopterygius*
7. *Palaeospinax*
8. *Dorygnathus*
9. *Pentacrinus*

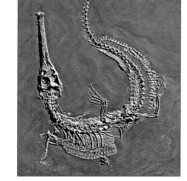

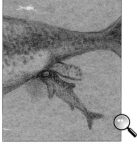

HARPOCERAS (1) This laterally flattened and thin-shelled ammonite, 20cm (8in) in diameter, was widespread in Early Jurassic seas. Its streamlined form indicates that it was free-swimming. At Holzmaden some ammonites are found with their aptychi (plates that could seal the shell) still in place.

STENEOSAURUS (4) With a long, narrow snout and upward-facing eyes, this 4.8m (16ft) crocodile probably attacked shoals of fish from below.

LEPIDOTES (5) This heavily built holostean fish had thick and heavy, diamond-shaped "ganoid" scales, grew to around a metre (40in) in length, and was widespread throughout Early Jurassic seas. The marginal teeth were peg-like, but the inner teeth were rounded and specialized for crushing mollusc shells.

STENOPTERYGIUS (6) Fossils of this 3m (10ft) ichthyosaur include some pregnant females, revealing that they bore live young.

131

MAMMALS TAKE TO THE AIR AND WATER
DAOHUGOU, INNER MONGOLIA, CHINA

171–164 million years ago
Bajocian to Bathonian stages of the Middle Jurassic Epoch

Latitude then: 60°N

Latitude now: 41°N

Sea level: +75m (246ft)

Original environment: lakes, streams, and their floodplains

Deposits: fine-grained mudrocks

Status: remote locality in Inner Mongolia

Preservation: flattened but with soft tissues preserved

• Daohugou

⊕ Earth: c.168MA

• Daohugou

⊕ Fossil site today

An extraordinary series of fossils has recently been uncovered from the Middle Jurassic freshwater lake and stream deposits of Daohugou in Inner Mongolia. Previously, Jurassic mammals were seen as tiny rodent-like creatures that avoided reptilian predators by nocturnal or burrowing lifestyles. Although dating of the site remains controversial, and some argue that it is in fact Early Cretaceous, these fine-grained mudrocks preserving soft tissues have the potential to revolutionize our view of early mammal evolution and diversity.

Among the highlights is the 50cm-long (20in) swimming *Castorocauda* (meaning

"beaver tail"). It is the largest early mammal known and may push back the record of mammalian conquest of water by more than a hundred million years. A whole new family of mammals has been created to accommodate the 14cm (6in) *Volaticotherium*, a primitive gliding form. A new burrowing monotreme, *Pseudotribos*, might seem less unusual, but has a unique arrangement of cheek teeth. Other Daohugou fossils include *Chunerpeton*, the earliest known salamander, and the oldest known feathered dinosaurs – the large *Pedopenna* and the pigeon-sized and extremely bird-like (though flightless) *Epidexipteryx*.

Climate: mid-latitude warm temperate
Biota: terrestrial and freshwater aquatic

Common organisms: arthropods aquatic and terrestrial, amphibian tadpoles

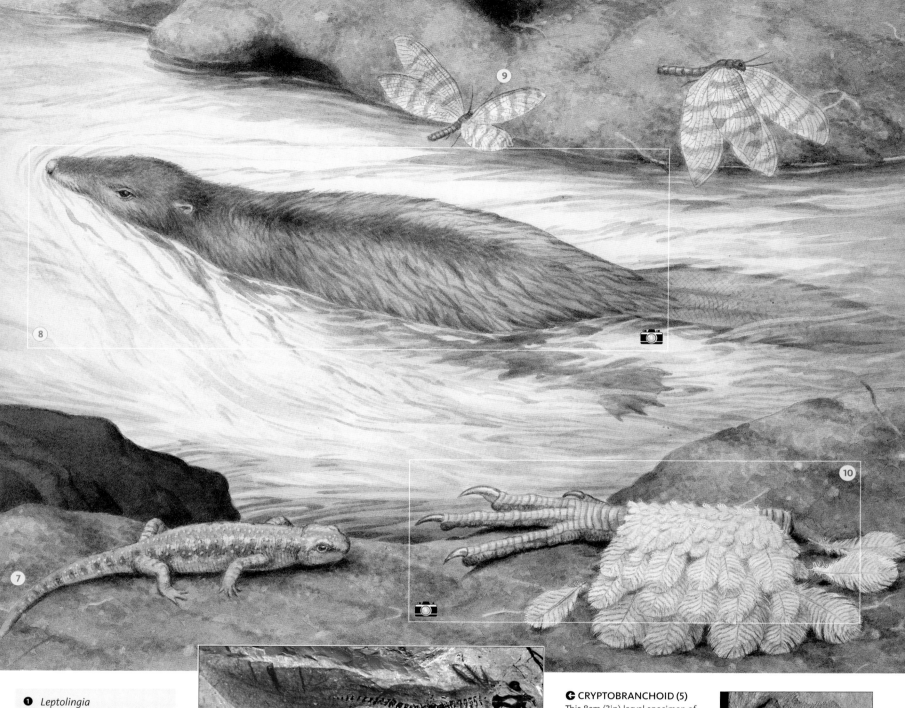

① *Leptolingia*
② *Pseudotribos*
③ *Volaticotherium*
④ *Mongolbittacus*
⑤ cryptobranchoid
⑥ *Quadraticossus*
⑦ *Chunerpeton*
⑧ *Castorocauda*
⑨ *Grammolingia*
⑩ *Pedopenna*

↻ CASTOROCAUDA (8) Combining features of platypus, otter, and beaver, *Castorocauda* was a semi-aquatic docodont mammal with a flat, scaly tail.

↺ CRYPTOBRANCHOID (5)
This 8cm (3in) larval specimen of the earliest known salamander, a primitive member of the cryptobranchids, preserves eyes, gill filaments, tail keel, notochord, limb outlines, and a gut filled with conchostrachan brachiopods.

↪ PEDOPENNA (10) The 12cm-long (5in) foot bones of this maniraptoran dinosaur are similar to those of troodontids and dromaeosaurids. They are surrounded by impressions of long symmetrical feathers overlain by shorter ones.

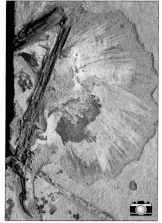

THE OLDEST OCTOPUS
LA VOULTE-SUR-RHONE, FRANCE

163 million years ago
Early Callovian Stage of the Middle Jurassic Epoch

Latitude then: 30°N

Latitude now: 44°N

Sea level: rising – marine transgression

Original environment: low-oxygen seabed below wave-base

Deposits: fine muds with carbonate nodules

Status: last extensively collected in the 1980s

Preservation: flattened in shales and three-dimensional in nodules, with some soft tissue preservation

La Voulte-sur-Rhône •

⊕ Earth: c.163 MA

◦ La Voulte-sur-Rhône

⊕ Fossil site today

The stagnant marine muds of La Voulte-sur-Rhône are one of the few fossil deposits in the world that preserve soft-bodied cephalopods such as the octopus. Active animals without skeletons, like the octopus, normally stand little chance of being preserved because their remains are scavenged after death, but unusual seabed conditions at La Voulte have fossilized a number of specimens. This is the oldest known fossil record of a true octopus.

The original environment was just below the wave-base, with an inhospitable soft mud seabed on which few animals could survive because of low oxygen levels. A few specially adapted creatures were present in very large numbers, such as the small clam *Bositra* and the brittlestar *Ophiopinna*, which sometimes occur in swarms of up to 3000 per sq metre (280 per sq foot). Both animals may have had some swimming ability, but evidently they were periodically smothered by inflows of mud. Most of the fossils are flattened, but where dead animals were quickly surrounded by carbonate nodules, they are preserved in three dimensions, with some remarkable soft tissue preservation. Fish and decapod crustaceans that swam in the more oxygenated surface waters are also well preserved here.

Common organisms: bivalves and brittlestars

Climate: mid-latitude warm humid
Biota: low-diversity mostly free-swimming marine

1. *Dollocaris*
2. *Rhomboteuthis*
3. *Aeger*
4. *Ophiopinna*
5. *Pholidophorus*
6. *Proteroctopus*
7. *Bositra*
8. *Eryma*

➲ **OPHIOPINNA (4)** The presence of some paddle-shaped spines on the arms of this brittlestar suggests that it may have been able to swim, like living comatulid sea lilies.

➲ **PROTEROCTOPUS (6)** If this 15cm-wide (6in) octopod-like cephalopod is a true octopus, as it appears to be, then it is the oldest known member of the Palaeoctopodidae.

☾ **BOSITRA (7)** This small clam probably survived stagnant, muddy bottom waters by using its thin, flat shells to swim. It is often found with the shells open in a "butterfly" position.

➲ **ERYMA (8)** Fossils of this 11cm-long (4.5in), shrimp-like decapod crustacean, with slender but powerful pincers, are occasionally preserved whole within carbonate nodules.

THE FIRST GIANT DINOSAURS
TENDAGURU, TANZANIA

Climate: subtropical with seasonal rains
Biota: terrestrial, reptile-dominated

Latitude then: 35°S

Latitude now: 10°S

Sea level: 200m (650ft) above present

Original environment: coastal riverine plains

Deposits: muds and sands

Status: remote outcrops in rural Tanzania

Preservation: well-preserved three-dimensional bones, including some articulated skeletons

Tendaguru

⊕ Earth: *c.*152 MA

Tendaguru

⊕ Fossil site today

Between 1907 and 1931, more than 200 tonnes of fossils excavated from Tendaguru in southeastern Tanzania were shipped back to Germany. Most were the bones of very large sauropod dinosaurs, and in August 1937 Berliners queued to see the mounted skeleton of one of them unveiled at the Natural History Museum. The *Brachiosaurus*, 11.87m (39ft) high and 22.65m (74ft) long, was the largest complete fossil of a land animal on display anywhere in the world.

With individual leg bones measuring more than 2m (6.6ft) long, the excavation of such large dinosaurs from such a remote region was a significant feat, and further work at Tendaguru has revealed a remarkable view of subtropical life in the Late Jurassic. The giant herbivores must have coexisted by exploiting an abundance of different plants and foliage types growing at different heights and perhaps in different situations within a complex and varied environment of coastal plains, estuarine river channels, and water bodies of varying salinity. In their quest for lush vegetation some of these huge sauropods became mired in swamps – their fossil legbones have been found still stuck vertically into the ancient swamp mudrocks.

145.5

65.5

152 million years ago
Kimmeridgian Stage of the Late Jurassic Epoch

Common organisms: plant pollen and herbivorous dinosaurs

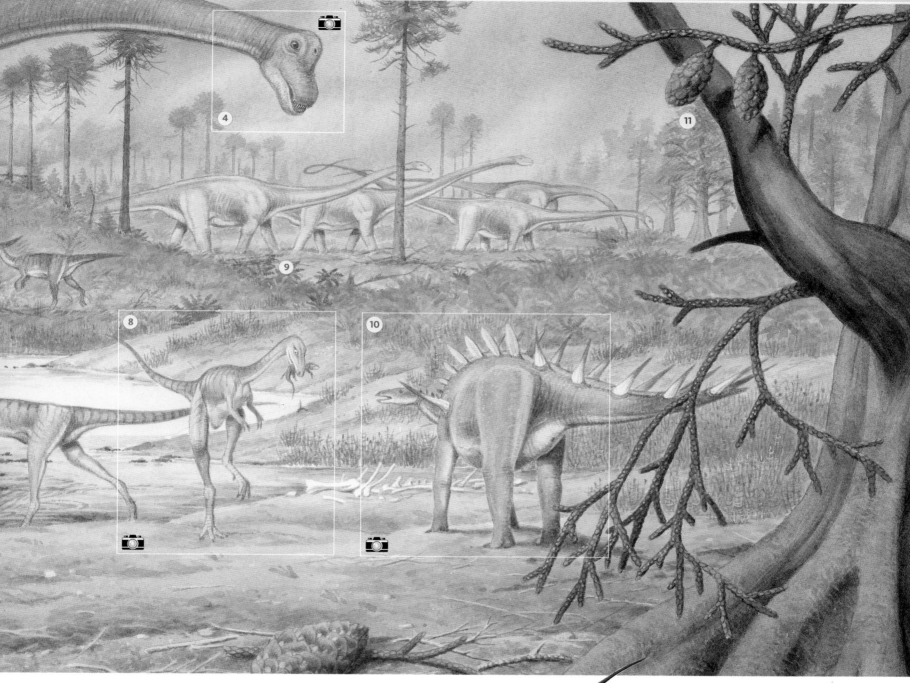

1. paramacellodid lizard
2. *Dicraeosaurus*
3. gleicheniacean fern
4. *Brachiosaurus*
5. *Rhamphorhynchus*
6. *Tendagurutherium*
7. *Dryosaurus*
8. *Elaphrosaurus*
9. *Barosaurus*
10. *Kentrosaurus*
11. cheirolepidacean conifer

◐ BRACHIOSAURUS (4)

The long-snouted skull of this giant brachiosaurid has chisel-shaped teeth for raking and chopping its tough plant food. The nostrils were set on top of a domed skull 70cm (28in) long, which may also have acted as a sound-resonating chamber.

◑ ELAPHROSAURUS (8)

This 6m-long (20ft) ceratosaurian theropod, known from a single partial skeleton, was a medium-sized bipedal carnivore adapted for fast running.

◑ KENTROSAURUS (10)

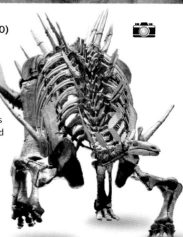

This relatively small, 5m-long (16.5ft), plant-eating stegosaur lived in large herds at Tendaguru. With smaller forelimbs than hind ones, it was perhaps able to rear up on its hind limbs – though not for long as the body was bulky. It was well armoured against predators such as *Elaphrosaurus*, with bony spines and plates on its back, tail, and shoulders.

WHO RULED THE JURASSIC FORESTS?
GUIMAROTA, PORTUGAL

Climate: humid subtropical
Biota: terrestrial and aquatic

Latitude then: 20°N

Latitude now: 39°N

Sea level: high (+125m/410ft)

Original environment: coal measure forest swamps

Deposits: shales, coals, and limestones

Status: disused coal mine

Preservation: partially flattened, but with some entire skeletons

⊕ Earth: c.152 MA

⊕ Fossil site today

A rare glimpse into a Late Jurassic swamp forest in Portugal reveals a very different view of life among the dinosaurs. Tens of thousands of fossils have been recovered from among the coals of the dense forest vegetation, and their associated brackish lagoon limestones. They show a great diversity of terrestrial and aquatic organisms, including shellfish, insects, fish, salamander-like amphibians, turtles, flying pterosaurs, dinosaurs, primitive birds, small mammals, and even a marine crocodile.

The fauna suggests that these forests were not ruled by the few small dinosaurs, such as 60cm-long (24in) *Compsognathus* and the

2m (6.6ft) *Aviatyrannis*, but rather by the numerous small mammals (over 26 species have now been described), abundant amphibians (over 9000 fossils recovered), and lizards. Among the mammals, *Henkelotherium*, a mouse-sized, tree-climbing insectivore with a long tail, was the first articulated Jurassic mammal to be found and is the oldest known representative of the therians, a group that includes the marsupial and the placental mammals.

Among numerous fossil teeth found here are some from *Archaeopteryx* – the only fossils of this primitive bird found outside Germany.

145.5

65.5

152 million years ago
Kimmeridgian Stage of the Late Jurassic Epoch

Common organisms: molluscs, insects, amphibians, and lizards

1. *Saurillodon*
2. *Haldanodon*
3. *Compsognathus*
4. *Celtedens*
5. *Phlebopteris*
6. *Archaeopteryx*
7. *Henkelotherium*
8. *Pagiophyllum*
9. *Baiera*
10. *Klukia*
11. *Lycopodium*
12. paulchoffatiid
13. *Machimosaurus*
14. *Rhamphorhynchus*

↻ HALDANODON (2) The fossil remains of this small docodont mammal are mostly jaws no more than 3cm (1.2in) long, and teeth. However, a partial skeleton from Guimarota shows that it was a mole-sized semi-aquatic burrower, similar to the living desmans, with massive limb bones, short dense fur, small eyes, and no ear opening.

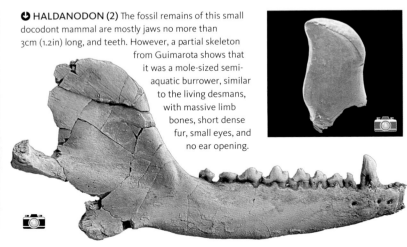

↻ ARCHAEOPTERYX (6) Small teeth (1.6mm long) found at Guimarota have the distinctive sigmoidal shape and finely serrated cutting edge characteristic of *Archaeopteryx*, though they are somewhat larger.

↻ HENKELOTHERIUM (7) This mouse-sized arboreal insectivore with a long tail was the first articulated Jurassic mammal to be found, and is the earliest known member of the Theria.

FIRST FOSSIL EVIDENCE FOR EVOLUTION
SOLNHOFEN, GERMANY

Climate: semi-arid tropical monsoonal
Biota: mixed marine and terrestrial

Latitude then: 25°N

Latitude now: 49°N

Sea level: stable (+200m/650ft)

Original environment: coastal lagoon with islands

Deposits: fine-grained calcareous muds

Status: a few commercial quarries still work the limestones

Preservation: flattened but well-preserved complete fossils with soft tissues occasionally preserved as impressions

⊕ Earth: c.151 MA

⊕ Fossil site today

A single fossil feather discovered at Solnhofen in 1860 offered the first hint that birds existed as early as the Late Jurassic. The following year, a complete fossil was indeed found – named *Archaeopteryx* (meaning "ancient wing"), the specimen displayed a curious mixture of bird and reptile features, including arms and clawed finger bones surrounded by the impressions of asymmetrical flight feathers and a long bony tail. The evolutionist Thomas Henry Huxley realized that *Archaeopteryx* provided the first good fossil evidence to support Darwin's evolutionary argument for transitions (often known as "missing links")

between the major classes of animals such as reptiles and birds.

Today, some ten *Archaeopteryx* specimens are known. At least two species existed, and recent discoveries show that the foot had a hyperextendible second toe like those seen in deinonychosaurs. *Archaeopteryx* probably spent most of its life on the ground, only occasionally perching in trees, since it lacked the reversible first toe that helps modern birds to grip branches. The form of one of its skull bones is closer to non-avian theropods than birds. *Archaeopteryx* therefore presents a true mix of dinosaur-like and bird-like features.

151 million years ago
Kimmeridgian Stage of the Late Jurassic Epoch

Common organisms: shrimps, fish, sea lilies

➲ **COMPSOGNATHUS (2)**
One specimen of this fast-running, metre-long (40in), theropod has been found with its prey – a gekkotan lizard (*Bavarisaurus*) – in its stomach.

① *Aegirosaurus*
② *Compsognathus*
③ *Bavarisaurus*
④ *Archaeopteryx*
⑤ *Rhamphorhynchus*
⑥ *Ginkgoites*
⑦ *Mesolimulus*
⑧ *Leptolepides*

➲ **ARCHAEOPTERYX (4)**
The earliest bird fossil is now classified as a theropod dinosaur along with other small-feathered dinosaurs in a group known as the "avialian maniraptorans".

↻ **MESOLIMULUS (7)**
This horseshoe crab or limulid has been found where it died at the end of its track, probably expiring from lack of oxygen or some toxin in the lagoon waters.

141

LIFE AND DEATH IN A JURASSIC LAGOON
SOLNHOFEN, GERMANY (CONT.)

Climate: semi-arid tropical monsoonal
Biota: mixed marine and terrestrial

The fine-grained calcareous muds of Solnhofen, quarried for centuries as lithographic stone, were originally deposited in island-studded coastal lagoons swept by frequent monsoon cyclones. The storms drowned a variety of land-living animals, and their remains were well preserved in the anoxic muds alongside marine organisms. Thousands of fossils from over 600 different species have now been recovered from several quarries excavated in these 150-million-year-old deposits.

Apart from the famous *Archaeopteryx*, the Solnhofen organisms range from marine algae and shrimps (such as *Aeger*), fish (*Caturus,*

Aspidorhynchus), and squid (*Acanthoteuthis*), to terrestrial insects (*Tarsophlebia*) and pterosaurs (*Pterodactylus*).

In 1784, Italian naturalist Cosimo Collini described and accurately illustrated a fossil *Pterodactylus* specimen, found in limestone at Eichstatt near Solnhofen, and then in the collection of Karl Theodor, Elector Palatione. Collini drew no conclusion on the animal's identity, but a few decades later, French anatomist Georges Cuvier studied his illustrations and recognized the creature's reptilian nature, coining the name "pterodactyle" (from the Greek words for "wing" and "finger").

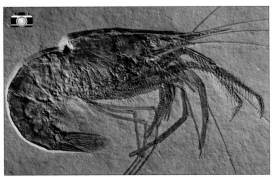

🎧 **AEGER (2)** This 6cm (2.5in) shrimp-like decapod crustacean is one of the commonest fossils at Solnhofen and had a worldwide distribution in Jurassic times.

151 million years ago
Late Kimmeridgian Stage of the Late Jurassic Epoch

Common organisms: shrimps, fish, sea lilies

1. *Tarsophlebia*
2. *Aeger*
3. *Pterodactylus*
4. *Caturus*
5. *Acanthoteuthis*
6. *Cycleryon*
7. *Aspidorhynchus*

PTERODACTYLUS

(3) This short-tailed flying pterosaur had a wingspan of 30cm (12in), and numerous sharp pointed teeth – it may have been a fish-eater.

CYCLERYON (6) Well-preserved speciments of this seabed-living decapod crustacean, growing up to 25cm (10in) long, have been found at Solnhofen. The closest living relatives of these extinct animals are modern spiny, furry, and slipper lobsters.

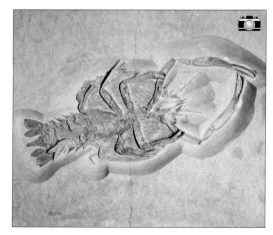

143

THE REAL JURASSIC PARK
MORRISON FORMATION, WESTERN USA

Climate: humid temperate with seasonal droughts

Biota: terrestrial and freshwater aquatic

Latitude then: 20–40°N

Latitude now: 35–42°N

Sea level: high (+125m/410ft)

Original environment: deserts to coal measure swamps, rivers, and lakes

Deposits: diverse, from desert sands to organic muds

Status: extensive outcrops with some rich fossil-bearing horizons

Preservation: three-dimensional bones, some articulated skeletons

• Morrison Formation

⊕ Earth: c.150 MA

• Morrison Formation

⊕ Fossil site today

The Late Jurassic Morrison Formation is a complex system of strata that outcrops across 1.5 million sq km (0.6 million sq miles) along the Rockies from Montana south to New Mexico. It preserves a wide variety of environments from deserts in the south to coal measure swamps in the north. Flood-prone rivers and lakes throughout this region preserved abundant, spectacular fossils, especially of the giant sauropods for which the region is renowned. The discovery of these large dinosaurs, such as *Diplodocus*, sparked an intense 19th-century rivalry between American collectors Edward Cope and Othniel Marsh.

As the shallow sea that covered much of western North America in the Middle Jurassic retreated north, it left vast plains covered with meandering rivers and numerous swamps. Cyclic droughts every few years or decades concentrated thirsty animals around remaining waterholes, where many eventually died. With the coming of the rains, flash floods swept the carcasses away and partially disarticulated them before dumping them in large quantities. One site in Utah has yielded remains of *Camarasaurus*, *Diplodocus*, *Apatosaurus*, *Camptosaurus*, *Stegosaurus*, *Allosaurus* (some 44 individuals), and two smaller theropods.

155–148 million years ago
Kimmeridgian to Tithonian stages of the Late Jurassic Epoch

Common organisms: plants, reptiles, amphibians, and fish

❶ *Nilssonia*
❷ cycadalean
❸ *Czekanowskia*
❹ *Goniopholis*
❺ *Diplodocus*
❻ osmundacean
❼ *Fruitafossor*
❽ *Camarasaurus*

➲ **OSMUNDACEAN (6)** These unique ferns had a conical stem, thickened towards the top by a persistent mantle of leaf bases and roots. When fossilized, these conical structures are known as *Osmunda*.

↻ **FRUITAFOSSOR (7)**
With an echidna-like skeleton and peg-shaped teeth, this 10cm-long (4in) mammal was well adapted to digging up insects for food and perhaps even burrowing underground.

➲ **CAMARASAURUS (8)**
This common 20m-long (66ft) sauropod had a relatively large skull but a very small brain. Its 52 chisel-shaped teeth were well suited to tackling tough coniferous foliage.

145

THE MORRISON ECOSYSTEM
MORRISON FORMATION, WESTERN USA (CONT.)

Climate: humid temperate with seasonal droughts
Biota: terrestrial and freshwater aquatic

Early explorations of the Morrison Formation strata inevitably focussed on the big dinosaurs and tended to overlook other fossils. Since then, however, a more complete assessment has been made of the fossil biota and the environments. For instance, the Dry Mesa Dinosaur Quarry in Colorado preserves not only the remains of 23 kinds of dinosaurs (including the enormous *Supersaurus* and *Ultrasaurus*) but also the fossil remains of flying pterosaurs, aquatic crocodiles, turtles, amphibians, and lungfish, as well as a diversity of small mammals.

Plant-bearing horizons contain the remains of terrestrial bryophytes, horsetails, ferns, cycads, ginkgos, and conifers. Of these, the conifers were the dominant plants in both dry and wet locations, forming the most abundant vegetation of the Middle to Late Jurassic, from North America across into Europe.

Numerous mammal fossils were found, including some of the earliest recorded from North America, but until 1998 they largely consisted of isolated teeth and jawbones from animals that seemed mostly to be small, primitive insect-eaters. However, the recent discovery of *Fruitafossor* shows that some early mammals had developed a burrowing lifestyle. In fact, *Fruitafossor's* skeleton is very similar to the living Australian echidna – another animal that spends some of its life below ground.

❶ *Camptosaurus*
❷ *Stegosaurus*
❸ *Fruitafossor*
❹ *Glyptops*
❺ *Diplodocus*
❻ *Coniopteris*
❼ *Apatosaurus*
❽ *Ceratodus*
❾ *Allosaurus*

➲ STEGOSAURUS (2) This small-brained, 9m-long (30ft), plant-eating ornithischian was common across the western USA in Late Jurassic times. It is easily distinguished by the two rows of large bony plates along its back and its spiked tail. Individual plates could be up to a metre (40in) tall, and may have had a role in thermoregulation, display, or defence. The sharp tail spikes some 60cm (24in) long were certainly for protection.

145.5

65.5

155–148 million years ago
Kimmeridgian to Tithonian stages of the Late Jurassic Epoch

Common organisms: plants, reptiles, amphibians, and fish

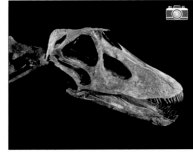

⊙ **DIPLODOCUS (4)** The teeth of this 29m-long (95ft) sauropod have a simple pencil-like shape, ideal for raking and stripping foliage from ferns and other low-growing plants.

⊙ **ALLOSAURUS (9)** This 12m-long (40ft) theropod is one of the best known dinosaurs, and a top predator in the Morrison ecosytem. Its strong skull and teeth could deliver a powerful bite.

HISTORIC BELGIAN DINOSAURS
BERNISSART, BELGIUM

Climate: northern mid-latitude warm humid
Biota: terrestrial

Latitude then: 35°N

Latitude now: 51°North

Sea level: rising (+200m/650ft)

Original environment: continental near coast

Deposits: freshwater lake muds

Status: deposits excavated more than 300m (1000ft) underground in 19th-century coal mine, now flooded

Preservation: generally excellent, with entire skeletons embedded in clay deposits

Bernissart •

⊕ Earth: *c.*127MA

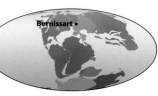

• Bernissart

⊕ Fossil site today

Around 120 million years ago, the Early Cretaceous landscapes of Bernissart, in what is now southern Belgium, hid a dark and treacherous secret. Herds of large, plant-eating *Iguanodon* prospered on the abundant plant life, unaware that the lush vegetation of their surroundings was a potential death trap. Local lake waters would have seemed innocuous enough – full of fish, turtles, and small crocodiles that posed no threat to these muscular ornithopod dinosaurs.

But every so often, the swollen torrents of a flash flood caught the *Iguanodon* unawares. Sedimentary evidence suggests that at least three flood events drowned numerous animals. The carcasses were swept into the lake, and as they sank into the deep, cold, bottom sediments they were preserved almost intact.

The discovery of the Bernissart skeletons in 1878 revolutionized ideas about what the dinosaurs might have looked like, for these were some of the first complete skeletons ever found. Initially, the 11m (36ft) skeletons were reconstructed with a kangaroo-like stance, but they are now known to have used all four limbs with their tails extended behind them, switching between quadrupedal and bipedal gaits with ease when necessary.

CRETACEOUS

145.5   65.5

128–125 million years ago
Barremian to Aptian stages of the Early Cretaceous Epoch
Common organisms: dinosaurs

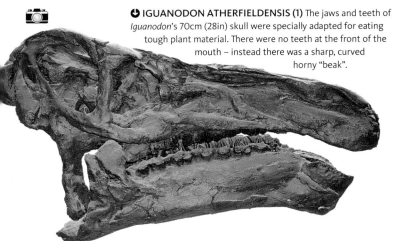

IGUANODON ATHERFIELDENSIS (1) The jaws and teeth of *Iguanodon*'s 70cm (28in) skull were specially adapted for eating tough plant material. There were no teeth at the front of the mouth – instead there was a sharp, curved horny "beak".

1. *Iguanodon atherfieldensis*
2. *Ornithocheirus*
3. *Iguanodon bernissartensis*

ORNITHOCHEIRUS (2) This abundant and widespread pterosaur, with a wingspan of 2.5m (8.5ft), had a global range from Europe to South America.

IGUANODON BERNISSARTENSIS (3) *Iguanodon*'s hand displays the bony core of a prominent thumb spike. When it was originally found in isolation by Gideon Mantell, he compared its position with the protuberance on the nose of living iguana lizards, and so gave *Iguanodon* the nose spike which featured on many early reconstructions.

149

THE MODERN BIOTA EMERGES
LIAONING, CHINA

Volcanism: frequent eruptions with ashfalls

Common organisms: some insects, fishes, salamanders, choristodere reptiles, and birds

Latitude then: 39°N

Latitude now: 41°N

Sea level: high (+150m/490ft)

Original environment: continental with volcanoes

Deposits: freshwater lake muds and volcanic ashfalls

Status: actively being excavated in many localities across a wide region of Liaoning and northern China

Preservation: generally excellent – flattened but with some soft tissue preserved

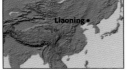

⊕ Earth: c.128ᴍᴀ

⊕ Fossil site today

The Early Cretaceous forests of Liaoning in northern China were a strange and exciting evolutionary backwater. Their remarkably well-preserved fossil riches have only been explored in detail over the last few decades, revealing wonders of diversity, anatomical detail, and behaviour. The landscapes, their flora, and fauna were under constant threat from highly active volcanoes, which produced showers of ash and gas that periodically killed off life and allowed its remains to accumulate, especially in the lake sediments.

Finds from the region include feathered dinosaurs that nested like birds, many new kinds of birds themselves, and even mammals that ate baby dinosaurs. The thousands of fossils now recovered have revolutionized our view of the link between dinosaurs and birds, and thrown new light on the evolution of mammals and plants as well.

The exceptional preservation and plentiful fossils of Liaoning provide a window to the surprising and exotic diversity of Early Cretaceous life. By contrast, many of the plants are familiar cosmopolitan forms, such as conifers, ginkgos, and ferns, although there are surprises to be found here too, including some of the earliest flowering plants.

Climate: seasonal – semi-arid to moist temperate
Biota: diverse terrestrial and freshwater aquatic

128 million years ago
Barremian Stage of the Early Cretaceous Epoch

◉ SINOSAUROPTERYX (4)
Found in the mid-1990s, this basal coelurosaur was one of the first important fossils to be discovered in the Liaoning region. Its skeleton is very similar to that of the small theropod dinosaur *Compsognathus*, but preserved soft tissues showed that it is covered with short, "fuzzy", hair-like feathers. The first feathered dinosaur to be found, it offered strong evidence that birds and some dinosaurs share a common ancestor.

❶ iguanodontid
❷ *Callobatrachus*
❸ *Ischnidium*
❹ *Sinosauropteryx*
❺ *Confuciusornis*
❻ *Protarchaeopteryx*

◉ CONFUCIUSORNIS (5)
Thousands of specimens of this early beaked bird, one of the oldest and most primitive known, have now been found in Liaoning. Its distinctive long tail feathers were probably only present in the males, and would have been used for mating display.

151

A DINOSAUR-EATING MAMMAL
LIAONING, CHINA (CONT.)

Volcanism: frequent eruptions with ashfalls
Common organisms: some insects, fishes, salamanders, choristodere reptiles, and birds

The extraordinary diversity of the Liaoning fossils and their importance to the fossil record of life in the Early Cretaceous is largely due to their rapid burial within fine-grained lakeshore and bottom muds. Here, low oxygen levels in the sediments deterred scavenging of the corpses, and even delayed natural decomposition, helping to preserve some soft tissues – particularly tough, keratinous materials such as fur, feathers, and related structures. As a result, the Liaoning fossils revealed for the first time the extent to which many small dinosaurs were in fact partially or entirely covered with such structures.

The Liaoning fossils also prove conclusively that some dinosaurs not only nested like modern flightless birds, but that they also took care of their offspring after hatching. One specimen of *Psittacosaurus* has been found still sitting on a nest of some 40 hatchlings – the entire family must have been killed in some sudden catastrophe, perhaps smothered by a volcanic ashfall.

What is more, the recent discovery of the primitive mammal *Repenomamus* has shown one reason why the dinosaurs had to look after their hatchlings – evidently this badger-sized animal was an active carnivore that predated upon defenceless dinosaur young. We know this because a remarkable fossil of *Repenomamus* has been found with the remains of a *Psittacosaurus* hatchling in its stomach.

⊙ **MICRORAPTOR (4)** This tiny 7.7cm (3in) feathered dromaeosaurid is unique in having feathers on both its legs and arms – it may have been capable of gliding flight.

145.5

65.5

Climate: seasonal – semi-arid to moist temperate
Biota: diverse terrestrial and freshwater aquatic

128 million years ago
Barremian Stage of the Early Cretaceous Epoch

1. cicadomorph
2. *Manchurochelys*
3. *Jinzhousaurus*
4. *Microraptor*
5. *Ephemeropsis*
6. *Repenomamus*
7. *Psittacosaurus*
8. *Jeholopterus*
9. *Psittacosaurus* hatchlings
10. *Beipiaosaurus*
11. *Dilong*

REPENOMAMUS (6) The discovery of this unusually large, badger-sized, and predatory triconodont is further evidence that early mammals were more diverse than previously thought.

BEIPIAOSAURUS (10) This bizarre-looking therizinosaur, with 10cm-long (4in) claws, was part of a theropod group that had adapted to a herbivorous lifestyle.

PSITTACOSAURUS (7) The skull of this dinosaur had a parrot-like horny beak for slicing plant matter, with strong cheek teeth and a sliding lower jaw for chewing.

153

DINOSAURS TAKE TO THE AIR
LIAONING, CHINA (CONT.)

Volcanism: frequent eruptions with ashfalls
Common organisms: some insects, fishes, salamanders, choristodere reptiles, and birds

Thanks largely to fossil discoveries made at Liaoning, feathers can longer be seen as a unique attribute of birds. Instead, their presence in some theropod dinosaurs reinforces the view that birds did indeed evolve from small theropods. Some ornithischians, such as *Psittacosaurus*, had bristle-like structures on their tails that may or may not be related to feathers. Among theropods, relatively simple filamentous structures are known in primitive coelurosaurs such as *Sinosauropteryx*, and the evolution of feathers can be followed through increasingly complex forms, as seen in *Caudipteryx* and *Protarchaeopteryx*.

Feathers were clearly "preadapted" for flight – in other words, they evolved first in flightless dinosaurs.

Filamentous "protofeathers" such as those of *Sinosauropteryx* probably provided thermal insulation, like fur on modern mammals. More complex forms, such as the arm feathers of *Protarchaeopteryx*, may have been used for mating displays, or as camouflage. Only *Microraptor* and the true birds had asymmetrical feathers that could generate lift for flight.

However, it is important to realize that these Early Cretaceous fossils do not represent the true "road to flight". Feathered flight had already evolved by the Late Jurassic, as seen from *Archaeopteryx*. Instead, the various feathered dinosaurs from China and elsewhere are simply those that had developed their own unique uses for feathers, or retained them from earlier times.

⊙ **EOMAIA (1)** The teeth of this 10cm (4in) fossil show it is the earliest known placental mammal, and its pelvic structure suggests it may have been a marsupial. The limb proportions are typical of modern tree-climbers.

Climate: seasonal – semi-arid to moist temperate
Biota: diverse terrestrial and freshwater aquatic

128 million years ago
Barremian Stage of the Early Cretaceous Epoch

1. Eomaia
2. Sapeornis
3. Protarchaeopteryx
4. Caudipteryx
5. Archaefructus
6. Sinodelphys
7. Tetraphalerus
8. Czekanowskia
9. Sinornithosaurus
10. Sinosauropteryx
11. Jeholodens

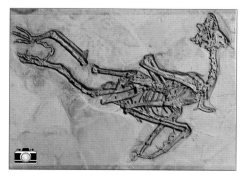

SAPEORNIS (2) This large (30cm or 12in long) and primitive Early Cretaceous bird probably ate seeds and fruit. Its elongated forelimbs suggest it may have been a soaring or gliding form rather than an able flier.

ARCHAEFRUCTUS (5)
This is perhaps the earliest known angiosperm (flowering plant) with male and female reproductive organs on the same shoot.

JEHOLODENS (11)
This 8cm-long (3in) triconodont mammal is known from an entire skeleton, showing that it retained a relatively primitive hindlimb structure but had a more advanced forelimb.

THE RISE OF THE LITTLE MAMMALS
KHOOVOR, OVORHANGAI, MONGOLIA

Climate: northern hot arid belt
Biota: terrestrial

Latitude then: 25 degrees North

Latitude now: 46 degrees North

Sea level: high (+175m/575ft)

Original environment: terrestrial rivers, deltas, and lakes

Deposits: waterlain sands, silts, and muds

Status: a remote locality in the Mongolian desert

Preservation: three-dimensional and disarticulated small mammal bones plus some entire dinosaur remains

Khoovor •

⊕ Earth: *c.*112 MA

Khoovor •

⊕ Fossil site today

The Early Cretaceous Khoovor fauna was discovered by Soviet–Mongolian expeditions that recovered thousands of bones from waterlain sediments in this locality. These sediments were deposited in rivers and their deltas as they flowed into lakes. Altogether, some 500 isolated small mammal fossils – mostly symmetrodont and triconodont jaw fragments and teeth – were recovered by sieving and washing the sediments. The mammals must have drowned during intermittent flood events, their remains washed downstream until they were deposited when the current slowed as it entered a lake.

Some 17 different kinds of mammal have been identified from the fossils, ranging from primitive forms to more advanced ones such as *Prokennalestes*, one of the earliest accepted eutherians, and some strangely specialized types. A eutriconodont very similar to one found in Montana indicates a faunal connection between the continents, and further evidence of intercontinental connections is provided by other cosmopolitan animals, such as the plant-eating dinosaur *Iguanodon bernissartensis*, found in Belgium (see p.148), and the smaller *Psittacosaurus,* also found at Liaoning (see p.152).

Common organisms: small mammals and dinosaurs

112 million years ago
Aptian to Albian stages of the Early Cretaceous Epoch

1. *Nilssoniopteris*
2. eutriconodontid
3. *Balera*
4. *Kielantherium*
5. *Araucaria*
6. *Shamosaurus*
7. *Iguanodon*
8. *Psittacosaurus*
9. *Sparganium*
10. *Prokennalestes*
11. *Pterophyllum*
12. *Lycoptera*

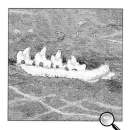

⬆ **KIELANTHERIUM (4)**
Known only from fossil
fragments at this locality,
this tiny Early Cretaceous
"tribosphenidan" is close to
the origin of the therian
mammals.

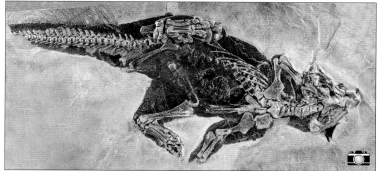

⬆ **PSITTACOSAURUS (8)** Growing up to 2m (6.6ft) long, this early ceratopsian is only
known from Asian sites. It is estimated that it grew to a maximum weight of 20kg (44lb) in
9 years – four times faster than living reptiles, but four times slower than living marsupials.

⟳ **LYCOPTERA (12)** This primitive
10cm-long (4in) "osteoglossomorph" teleost
may be the earliest skeletal fossil that can
be classified within a living group of
teleost fish.

AN EARLY CRETACEOUS LAKE ENVIRONMENT
CRATO FORMATION, ARARIPE BASIN, BRAZIL

Climate: southern hot arid belt

Biota: terrestrial and aquatic

Latitude then: 12°S

Latitude now: 7°S

Sea level: high (+175m/575ft)

Original environment: predominantly freshwater lagoon with river deltas

Deposits: muds, silts, sands, and carbonates

Status: proposed as a potential World Heritage Site in 1999

Preservation: flattened but with soft tissue preservation

↻ Earth: *c.*112 MA

↻ Fossil site today

Over the last two decades, some 200 new species have been discovered in the Early Cretaceous Crato strata of Brazil. They include more than 100 insect species, 11 arachnids, and 9 species of fish, with rarer turtles, lizards, pterosaurs, plants, a bird, and feathers that may belong either to dinosaurs or birds.

The extraordinary preservation by phosphate reveals anatomical details of soft tissues such as muscles and gills in 110-million-year-old fossils. Most of the known fossils from the site have been found through commercial quarrying or trade among collectors rather than through organized excavation, but such

are the potential riches that the region was proposed for World Heritage Site status in 1999.

The discovery of the fossil beds, however, dates back to the early 19th century, when German naturalists Johann Spix and Carl Friedrich Martius first found fish-bearing nodules in the northeastern region of Ceara.

The fossils are found in strata exposed around the flanks of the 800m-high (2600ft) Chapada do Aripe plateau, part of the much larger sedimentary Araripe Basin. From here Louis Agassiz, the Swiss expert on fossil fish, described seven new species in the 1840s and recognized their Cretaceous age.

Common organisms: insects, plants, fish

112 million years ago
Late Aptian Stage of the Early Cretaceous Epoch

1. *Welwitschiostrobus*
2. *Irritator*
3. *Santanmantis*
4. *Ruffordia*
5. *Tapejara*
6. *Baeocossus*
7. nymphaealean
8. *Cretofedtschenkia*
9. belostomatid
10. *Lindleycladus*
11. *Cretaraneus*
12. *Tettagalma*
13. *Ludodactylus*
14. *Dastilbe*

◐ NYMPHAEALEAN (7)
This common water plant
with large leaves and
creeping rhizomes belongs to
one of the oldest angiosperm
(flowering plant) clades.

↥ CRETARANEUS (11) This
Crato spider was the first to
be described, and belongs to
the araneoids, now known
from hundreds of specimens
found here.

↥ LUDODACTYLUS (13) The leaf fragment jammed
into the lower jaw of this 4m-wingspan (13ft) pterodactyl
may have been responsible for its death.

FLOWERING PLANETS AND INSECTS CO-EVOLVE
CRATO FORMATION, ARARIPE BASIN, BRAZIL (CONT.)

Climate: southern hot arid belt
Biota: terrestrial and aquatic

The Crato strata were laid down on the bed of a shallow, stagnant, freshwater lake, surrounded by thick vegetation that included early flowering plants and a great diversity of terrestrial and semi-aquatic creatures. Some remains of this fauna and flora are found in the deposits alongside the fish.

Today the fossil-bearing deposits are worked from two horizons – the lower and older Crato Formation, 50–60m (165–200ft), thick and the higher and younger Santana Formation. Their valuable fossils, especially from the lower laminated mudstones of the Crato Formation, are recovered from countless small quarries. The strata are used for paving, and it is the commercial business of splitting the rocks that has revealed many of the world's most remarkable Cretaceous insect fossils.

The preservation of so many insects alongside the newly diversifying angiosperms (flowering plants) allows an insight into their co-evolution. Increased insect diversity seems to have been linked to the exploitation of new food supplies – especially nectar, which was in turn linked to new methods of pollination and a number of strategies developed by the plants to attract insects and encourage them to transfer pollen from male to female reproductive organs. For example, the nectar drops and distinctive odour carried by the male flowers of living welwitschioids may be primtive traits developed in Early Cretaceous times.

As a whole, Crato provides one of our clearest "windows" onto a Cretaceous ecosystem in the southern hemisphere Gondwanan supercontinent.

🔊 **BAISOPARDUS (3)** This antlion with a 10-cm (4-in) wingspan was a predatory myrmeliontiform, one of the largest neuropterans.

Common organisms: insects, plants, fish

112 million years ago
Late Aptian Stage of the Early Cretaceous Epoch

1. myrmeliontid
2. ephedroid (female)
3. *Baisopardus*
4. *Cladocyclus*
5. *Protoischnurus*
6. *Tapejara*
7. *Cratoraricrus*
8. *Britopygus*
9. *Araripeliupanshania*
10. *Susisuchus*
11. ephedroid (male)
12. *Arariphyrnus*
13. *Ruffordia*

CLADOCYCLUS (4) Growing to a metre (40in) long, this fish was the commonest of the extinct teleost ichthyodectiformes. It had a worldwide distribution between the Late Jurassic and the Late Cretaceous, and could tolerate reduced salinities.

SUSISUCHUS (10)
This metre-long (40in) neosuchian probably lived around rather than in the Crato lagoon. One of the earliest crocodilians, it has dermal armour similar to modern crocodiles. The presence of similar basal eusuchians in Middle Cretaceous Australia, which was connected to South America via Antarctica, suggests that modern crocodilians arose in Gondwana.

DINOSAURS OF THE SUBPOLAR WINTER
DINOSAUR COVE, VICTORIA, AUSTRALIA

Climate: southern high-latitude temperate humid belt
Biota: predominantly terrestrial vertebrates

Latitude then: 75°S

Latitude now: 38°S

Sea level: rising (+175m/575ft)

Original environment: large river channels and floodplains

Deposits: sandstones and bone conglomerates

Status: the bone-bearing conglomerate has now been entirely excavated from the site

Preservation: three-dimensional but disarticulated bone material

Dinosaur Cove

⊕ Earth: c.110MA

Dinosaur Cove •

⊕ Fossil site today

In 1904, a small theropod dinosaur claw was found in Early Cretaceous coastal sandstones 200km (125 miles) southwest of Melbourne, but no more remains were found at the site now known as Dinosaur Cove until the late 1970s. Following its rediscovery, this ancient river channel was excavated by tunnelling into the cliff. Several thousand bones were recovered, mostly from small plant-eating ornithopod dinosaurs. They included two new genera, *Leaellynasaura* and *Atlascoposaurus*, along with more remains of *Fulgurotherium*, found previously at Lightning Ridge in New South Wales.

At the time, Dinosaur Cove lay well inside the Antarctic Circle, and as a result there was little daylight and low temperatures for much of the winter – oxygen isotopes suggest mean annual temperatures between -6 and +5°C (21–41°F). Some have suggested that dinosaurs were able to survive these conditions because they may have been warm-blooded.

The last known amphibian temnospondyl, *Koolasuchus*, may also have survived longer here than elsewhere because of geographical isolation and a tolerance of cold water. Living salamanders, as a modern parallel, can be active in water as cold as -2°C (28.5°F).

CRETACEOUS

145.5

65.5

Common organisms: plants

110 million years ago
Albian Stage of the Early Cretaceous Epoch

❶ sphenopterid
❷ *Bishops*
❸ *Gingkoites australis*
❹ *Leaellynasaura*
❺ sphenopterid
❻ *Koolasuchus*
❼ *Taeniopteris*

➲ **BISHOPS (2)** The primitive jaw structure of this recently discovered Gondwanan australosphenid mammal suggests that it might represent an early divergent clade that gave rise to the egg-laying monotremes.

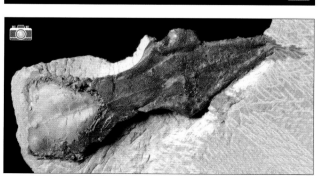

🔄 **LEAELLYNASAURA (4)** This metre-long (40in) basal euornithopod is known from isolated bones and a partial juvenile skull. It has an enlarged optic lobe – possibly an adaptation for coping with the months of darkness that it experienced annually.

🔄 **KOOLASUCHUS (6)** This 4m-long (13ft) temnospondyl has a 50cm (20in), flat and broad head. Its jaws were armed with a hundred or so teeth, each up to 10cm (3.5in) long. With eyes on top of the skull, it was probably an ambush predator lying hidden in riverbed mud awaiting its prey.

LIFE IN A GREENHOUSE POLAR LANDSCAPE
HUGHENDEN, QUEENSLAND, AUSTRALIA

Climate: southern, high-latitude, temperate humid
Biota: coastal marine and terrestrial

Latitude then: 66°S	
Latitude now: 20°S	
Sea level: high (+200m/650ft)	
Original environment: coastal marine and terrestrial	
Deposits: shallow marine and riverine	
Status: isolated sites scattered across western central Queensland	
Preservation: many bones have been found in carbonate nodules	

⊕ Earth: *c.*105MA

⊕ Fossil site today

Despite its subpolar location, dinosaurs roamed the lushly vegetated Australian landscape in the Early Cretaceous. Rising sea levels saw shallow waters flooding the interior of the continent, restricting dry land to southern Western Australia, the Kimberley region, and Queensland to the east.

At their maximum extent, 117 million years ago, these seaways were patrolled by spectacular predatory marine reptiles, such as the 13m-long (43ft) pliosaur *Kronosaurus*, which fed upon newly evolving teleost bony fish. However, by 99 million years ago the seas had withdrawn.

Australia at the time was still part of the fragmenting Gondwanan supercontinent, and attached to Antarctica. Even close to the Antarctic Circle, climates were temperate, and ranged from freezing to 12°C (54°F) in southern Australia and across Antarctica, with a high annual rainfall of some 750–1150mm (30–46in)

This humidity produced large rivers and lakes occupied by turtles and lungfish, such as *Neoceratodus*, surrounded by vegetation. Here araucarian pines, gingkos, and podocarps, with an understorey of mosses, ferns, and cycads fed plant-eating dinosaurs such as *Minmi* and *Muttaburrasaurus*.

Volcanism: offshore island volcanoes
Common organisms: plants and fish

105 million years ago
Albian Stage of the Early Cretaceous Epoch

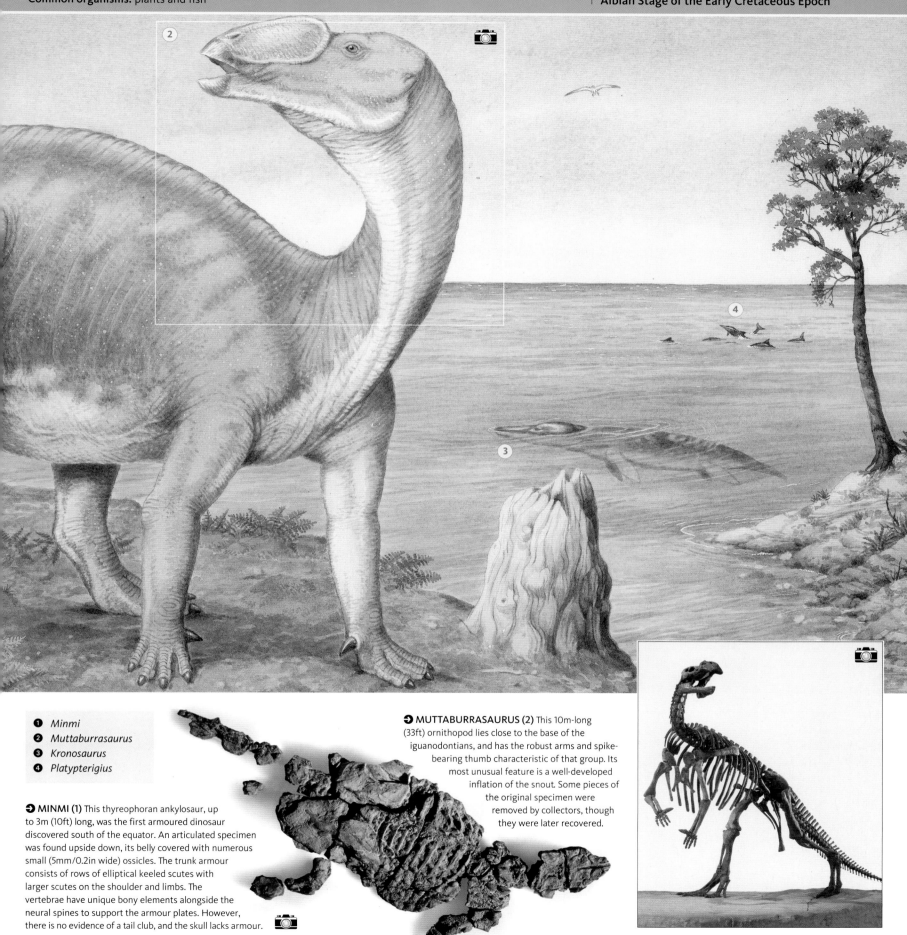

❶ *Minmi*
❷ *Muttaburrasaurus*
❸ *Kronosaurus*
❹ *Platypterigius*

➲ **MINMI (1)** This thyreophoran ankylosaur, up to 3m (10ft) long, was the first armoured dinosaur discovered south of the equator. An articulated specimen was found upside down, its belly covered with numerous small (5mm/0.2in wide) ossicles. The trunk armour consists of rows of elliptical keeled scutes with larger scutes on the shoulder and limbs. The vertebrae have unique bony elements alongside the neural spines to support the armour plates. However, there is no evidence of a tail club, and the skull lacks armour.

➲ **MUTTABURRASAURUS (2)** This 10m-long (33ft) ornithopod lies close to the base of the iguanodontians, and has the robust arms and spike-bearing thumb characteristic of that group. Its most unusual feature is a well-developed inflation of the snout. Some pieces of the original specimen were removed by collectors, though they were later recovered.

AFTER THE GONDWANAN BREAKUP
NEUQUEN, PATAGONIA, ARGENTINA

Volcanism: subduction-related volcanism to the west
Common organisms: small to medium-sized reptiles

Latitude then: 55°S

Latitude now: 39°N

Sea level: high (+200m/650ft)

Original environment: terrestrial with rivers, deltas, and lakes

Deposits: fluvial sandstones, silts, and muds

Status: although relatively isolated the region now attracts both dinosaur hunters and tourists

Preservation: abundant tetrapod remains in places include complete skeletons

Neuquén

⊕ Earth: c.98 MA

Neuquén

⊕ Fossil site today

In recent decades, Patagonian South America has emerged as one of the most exciting regions for new discoveries of terrestrial Mesozoic life. A sequence of deposits in the Neuquén Basin preserves the remains of abundant new tetrapods, ranging from exquisitely articulated skeletons of small and medium-sized reptiles such as sphenodontids, crocodiliforms, and primitive limbed snakes, through to mammals and rarer dinosaurs.

The latter include enormous tetrapod predators called carcharodontosaurids, rebbachisaurid and titanosaur sauropods, and a new dromaeosaurid. The unique nature of these fossils reinforces the separation between the Gondwanan and Laurasian biotas at this time, and they continued to evolve in parallel through the rest of the Cretaceous.

The end of Cenomanian times saw a marked turnover of many Gondwanan dinosaur species, especially among saurischians, where the effect seems to have been global. All the carcharodontosaurids of the Neuquén Basin belong to the "giganotosaurine" group that seems to have been endemic to South America. They did not spread to Africa even when there was a land connection, and became extinct by about 90 million years ago.

98 million years ago
Cenomanian Stage of the Late Cretaceous Epoch

Climate: southern warm semi-arid to humid
Biota: terrestrial

❶ *Buitreraptor*
❷ *Andesaurus*
❸ *Giganotosaurus*
❹ *Anabisetia*

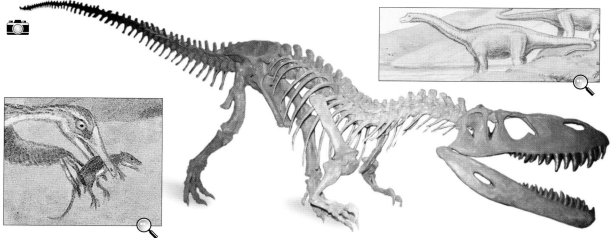

❯ BUITRERAPTOR (1)

This metre-long (40in) dromaeosaurid had a *Velociraptor*-like muzzle but a lack of serrated teeth. It was part of a distinctly Gondwanan lineage of dromaeosaurs – another example of the isolation from Laurasia at this time.

↻ ANDESAURUS (2)

This giant 30m-long (100ft) titanosaur, known from a single partial skeleton, has unusually high dorsal spines on each vertebra, one and a half times as high as each vertebra is long.

↻ GIGANOTOSAURUS (3)

This large predatory carcharodontosaurid had a massive 1.6m (5ft) ornamented skull, with long neural spines along its backbone and robust hands and hind legs.

167

BRINGING UP BABY – A DINOSAUR NURSERY
NEUQUEN, PATAGONIA, ARGENTINA (CONT.)

Volcanism: subduction-related volcanism to the west
Common organisms: dinosaurs

Thousands of sauropod egg clutches have been found scattered over several square kilometers of the Auca Mahuevo region of the Neuquén Basin. Although most are fragmented, some eggs have been found with exquisitely preserved dinosaur embryos. Identified as titanosaurs, these astonishing fossils give us an insight into the nesting behaviour of large sauropods.

The 85m-thick (280ft) sands, silts, and mudrocks of the Anacleto Formation contain four distinct egg-bearing "horizons". They were laid down intermittently across a wide river floodplain during a period when the climate veered sharply between semi-arid and subhumid conditions. Geological activity to the west produced volcanoes and rising

mountains that isolated the region from the Pacific and rivers running off these rising highlands shed large amounts of sediment into the basin.

Most egg clutches contain between 20 and 40 eggs, apparently laid in piles with no particular structure, although some were evidently laid in excavated hollows 1m (3ft) wide. The density of the clutches suggests that, on each occasion, a large number of gregarious females laid eggs at the same time and in the same nesting ground. The sheer size of the adults and proximity of the clutches suggest that little or no parental care took place once the eggs were laid – there simply would not have been room for all the parents to tend their young at the same time.

🔵 **NEST EGG (1)** At one locality more than 500 clustered eggs, each 12–15cm (5–6in) across, have been found in an area of just 65 square metres (700 sq ft).

Climate: southern warm semi-arid to humid
Biota: terrestrial

84–80 million years ago
Campanian Stage of the Late Cretaceous Epoch

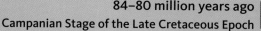

❶ *Argentinosaurus*
❷ *Aucasaurus*

❷ EMBRYO SKIN (1)
Impressions from the hide of the unborn infants show non-overlapping tubercular patterns of scales, similar to those found on other dinosaur skin fossils.

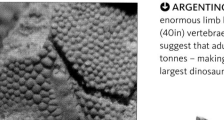

⟳ ARGENTINOSAURUS (1) The enormous limb bones and metre-sized (40in) vertebrae of this giant sauropod suggest that adults weighed up to 90 tonnes – making them perhaps the largest dinosaurs of all.

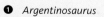

⬆ HATCHLINGS (1) From measurements of embryos still within their eggs, the baby dinosaurs are though to have been about 25cm (10in) long when they hatched – in adulthood they grew to around 40m (130ft).

⬆ AUCASAURUS (2) Found 25m (82ft) above the egg-bearing layer, this 4m-long (13ft) theropod was an abelisaurid, related to *Carnotaurus*.

LATE CRETACEOUS LIFE IN MONGOLIA
UKHAA TOLGOD, MONGOLIA

Common organisms: lizards, mammals, and ceratopsian dinosaurs

Latitude then: 10°N

Latitude now: 44°N

Sea level: very high
(+225m/730ft)

Original environment:
semi-arid desert with ephemeral
rivers

Deposits: dune sands, silts, and
muds

Status: scattered locations in
remote desert region

Preservation: three-dimensional
with many entire skeletons

Ukhaa Tolgod

⊕ Earth: *c.*80 MA

• Ukhaa Tolgod

⊕ Fossil site today

The fossil treasures of Mongolia's Gobi Basin
were first revealed to the wider world in the
1920s, when Roy Chapman Andrews led a
series of expeditions from the American
Museum of Natural History. Since then,
international digs have quarried numerous
sites in these Late Cretaceous deposits for
dinosaurs and early mammals. Ukhaa Tolgod,
in the Nemegt Basin, has revealed some 200
dinosaur skeletons along with an extraordinary
diversity of small mammals, and about 1400
lizards. The thousand or so mammal fossils
include entire skulls and skeletons, with the
now-extinct multituberculates outnumbering

the therians (ancestors of modern marsupials
and placental mammals) by nine to one.

The sediments show a site largely
dominated by slowly migrating dunes, with an
occasional oasis along an ephemeral river
floodplain. Small trees and shrubs grew along a
river bank leading to a marshy lake. In places,
mammals and small reptiles burrowed into
stabilized dunes, overgrown with plants. Many
skeletons were buried in loose wind-blown
sand that avalanched from high dunes into the
intervening gullies where the animals lived.
Fossil footprints have also been found on the
surfaces of what were once active dunes.

Climate: northern hot arid belt
Biota: terrestrial tetrapods

80 million years ago
Campanian Stage of the Late Cretaceous Epoch

❶ *Pinacosaurus*
❷ *Zalambdalestes*
❸ *Kryptobataar*
❹ *Protoceratops*
❺ *Estesia*
❻ *Mononykus*
❼ *Saurornithoides*

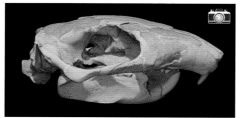

⮡ **ZALAMBDALESTES (2)**
This 24cm (9.5in) mammal has epipubic bones like marsupials and monotremes, but a placental-like skull and jaw (shown here). Its long legs were adapted for running.

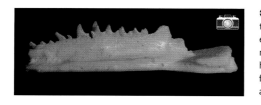

⟲ **KRYPTOBATAAR (3)** This 25cm (10in) multituberculate mammal is known from more than 150 specimens. Its skeleton retains primitive features, most notably a sprawling stance that gave it an asymmetric gait with a steep jumping action.

⮡ **PROTOCERATOPS (4)** Known from more than a hundred skeletons, this 2m (6.6ft) plant-eating ceratopsian walked on four legs. It was more advanced than *Psittacosaurus*, but did not have the enlarged nostrils and nasal horns found in the larger ceratopsian that flourished at the very end of the dinosaurs' reign.

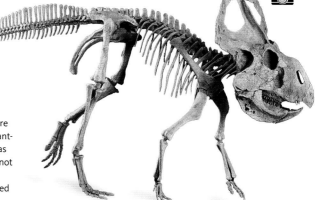

171

DINOSAUR PARENTING SKILLS
UKHAA TOLGOD, MONGOLIA (CONT.)

Common organisms: lizards, mammals, and ceratopsian dinosaurs

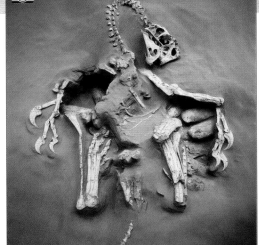

Recent discoveries at Ukhaa Tolgod have overturned the interpretation of one of Roy Chapman Andrews' most famous finds from another Mongolian locality (Bayn Dzak, otherwise known as the Flaming Cliffs). Here, Andrews found a nest of dinosaur eggs that was thought to have been laid by the plant-eating *Protoceratops*, the most common dinosaur at the site. Nearby, Andrews found the remains of a small, bird-like theropod, soon named *Oviraptor* ("egg thief"), in the belief that it had been stealing the eggs.

In 1993 another expedition from the American Museum of Natural History found a new nest of eggs – and one of these still contained the tiny bones of an embryo. Nearby lay the skulls of two sharp-toothed predatory theropods. Detailed examination showed that the embryo was not the expected *Protoceratops*, but was in fact an *Oviraptor*. Far from *Oviraptor* being an egg-stealer, it now seems possible that the animal Andrews found may in fact have been defending its nest against a thieving *Protoceratops*.

The little theropods, meanwhile, turned out to be dromaeosaurids, and may have been provided as food for *Oviraptor* hatchlings. Further evidence that the "egg-thief" has had a bad press came with the discovery of the fossil remains of an adult *Oviraptor* that died on the nest while incubating its eggs, apparently just like a modern chicken.

Climate: northern hot arid belt
Biota: terrestrial tetrapods

80 million years ago
Campanian Stage of the Late Cretaceous Epoch

❶ *Oviraptor*
❷ *Velociraptor*
❸ *Nemegtbataar*

↻ **OVIRAPTOR (1)** At up to 2m (6.6ft) long, this small, long-legged, and fast-moving coelurosaurian theropod was very bird-like, with a highly modified, short and deep toothless skull equippped with a strong, parrot-like beak. This famous skeleton preserves an adult that apparently died on the nest.

↻ ↥ **VELOCIRAPTOR (2)** This 2m (6.6ft) dromaeosaurid was a highly active predator with long muscular legs and sharply clawed toes and fingers. The long, light skull and jaw was strong and lined with inwardly curving sharp teeth. One of the most remarkable finds from the Ukhaa Tolgod excavations was a *Velociraptor* skeleton that was killed while locked in mortal combat with a *Protoceratops*.

THE DINOSAURS OF "CRETACEOUS PARK"
JUDITH RIVER, MONTANA, USA AND ALBERTA, CANADA

Climate: northern mid-latitude warm humid
Biota: terrestrial

Latitude then: 58°N

Latitude now: 50°N

Sea level: very high (+225m/735ft)

Original environment: coastal lowlands

Deposits: river and lake sands, silts, and muds

Status: World Heritage Site since 1979

Preservation: many complete and partial skeletons

• Judith River

⊕ Earth: c.77 MA

• Judith River

⊕ Fossil site today

Hundreds of dinosaur remain found amid the badlands of southern Alberta's Red Deer Valley, including some 300 articulated skeletons, have been protected since 1955 within the Dinosaur Provincial Park (now recognized as a World Heritage Site). The fossils here are embedded in Late Cretaceous strata, laid down in an eastward-pointing wedge of sediments, between the mountains in the west and a lost seaway to the east that once linked the Arctic Ocean to the Gulf of Mexico.

Here, large rivers meandered across coastal plains with shallow lakes and swampy peatlands, through coastal dunes, lagoons, and deltas, into the seaway itself. The freshwaters were full of fish, shellfish, salamander-like amphibians, frogs, turtles, and crocodilians.

The lush, humid surroundings were home to a huge range of plants and animals. Almost 200 plant and fungi species have been identified, including many species of the newly evolving flowering plants, such as oaks, lilies, and sunflowers. The dinosaurs included members of well-known groups such as tyrannosaurids, ceratopsians, and hadrosaurs. Giant azhdarchid pterosaurs flew overhead, and at least 20 species of small mammals and several lizards have also been found.

79–74 million years ago
Campanian stage of the Late Cretaceous Epoch

Common organisms: plants, fish, and dinosaurs

1. *Struthiomimus*
2. *Quetzalcoatlus*
3. *Chasmosaurus*
4. *Deltatheridium*
5. *Stegoceras*
6. *Gorgosaurus*
7. *Lambeosaurus*
8. *Troodon*

➲ GORGOSAURUS (6) This relatively small (9m or 30ft) tyrannosaurid, known from many partial skeletons and a dozen skulls, was the area's most common large predator.

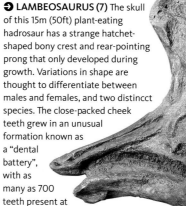

➲ LAMBEOSAURUS (7) The skull of this 15m (50ft) plant-eating hadrosaur has a strange hatchet-shaped bony crest and rear-pointing prong that only developed during growth. Variations in shape are thought to differentiate between males and females, and two distincct species. The close-packed cheek teeth grew in an unusual formation known as a "dental battery", with as many as 700 teeth present at any one time.

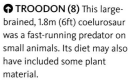

➦ TROODON (8) This large-brained, 1.8m (6ft) coelurosaur was a fast-running predator on small animals. Its diet may also have included some plant material.

175

THE BEAST OF MAASTRICHT
ST PIETER'S MOUNT, MAASTRICHT, THE NETHERLANDS

Climate: northern mid-latitude warm humid
Biota: marine

Latitude then: 40°N

Latitude now: 51°N

Sea level: high (+175m/575ft)

Original environment: deep continental shelf seas

Deposits: carbonate mud (chalk)

Status: underground tunnels within the chalk hills around Maastricht

Preservation: well-preserved three-dimensional shells and bones

⊕ Earth: *c.*68 MA

⊕ Fossil site today

Discovered in the late eighteenth century, the historic Maastricht mosasaur played a key role in the growing understanding of the prehistoric past. The beast's monstrous metre-long (40in) jaws were speculated to belong to an extinct crocodile or a toothed whale. Such was their fame that the fossils were looted by Napoleon's forces and taken to Paris, where they survive in the National Museum of Natural History. In 1808, famed anatomist Georges Cuvier concluded that the skull came from a creature related to the varanid lizards.

During the last 25 million years of the Cretaceous some 20 genera of marine mosasaurs evolved, diversified, spread worldwide, and became extinct. They were the dominant marine predators, but had to share the seas with other carnivores, notably sharks.

Mosasaur stomach contents have revealed a varied diet of fish, plesiosaurs, birds, and other mosasaurs. They also seem to have attacked the large marine turtle *Allopleuron*. However, evidence from healed bite marks shows that mosasaurs were themselves sometimes attacked by sharks. Other bite marks, found on mosasaur ribs, are thought to be the result of scavenging, probably by *Squalicorax*, a relatively small Cretaceous shark.

145.5

65.5

68 million years ago
Maastrichtian Stage of the Late Cretaceous Epoch

❶ *Mosasaurus*
❷ *Belemnitella*
❸ *Scaphites*
❹ *Squalicorax*
❺ *Marsupites*
❻ *Temnocidaris*
❼ *Hyotissa*
❽ *Plagiostoma*

⊙ **MOSASAURUS (1)** This ferocious marine predator's huge jawbones have conical, slightly recurved teeth. The entire animal grew up to 10m (33ft) long.

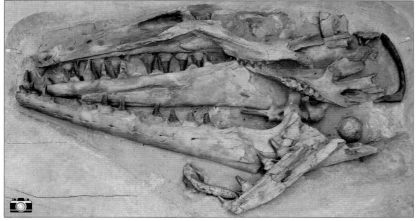

⊙ **SQUALICORAX (4)** Like the living tiger shark, this neoselachian shark had broad, triangular teeth topped with finely serrated crowns that were ideal for scavenging off large carcasses. It was widespread in shallow seas of the Late Cretaceous, and grew to a maximum size of 3m (10ft).

LAST DAYS OF THE CRETACEOUS
HELL CREEK, MONTANA, DAKOTAS, AND WYOMING, USA

Climate: northern mid-latitude warm humid belt
Biota: terrestrial

Latitude then: 45°N

Latitude now: 47°N

Sea level: high (+175m/575ft)

Original environment: river floodplains

Deposits: river sands, silts, and muds with swampy peats

Status: numerous sections exposed in dry creeks within the badland topography

Preservation: many partial skeletons and a few entire ones

• Hell Creek

⊕ Earth: c.66 MA

Hell Creek •

⊕ Fossil site today

The Hell Creek strata in the badlands of eastern Montana, Dakotas, and Wyoming are among the most intensely studied in the world, because they are well exposed and straddle the boundary between the Cretaceous and the Paleogene. Laid down by meandering rivers on a floodplain in a humid climate, they preserve a rich record of plants and animals during the two million years that saw the extinction of the dinosaurs and many other organisms.

Famously, the end Cretaceous extinction event is represented here by a thin iridium-enriched layer that is also found around the world, and this "iridium spike" has been linked to a major impact from space at the time. The succession of animals and plants below the boundary has been closely examined to discover which organisms were most affected.

Fossil pollen reveals hundreds of plant species, with angiosperm woodlands largely replacing the earlier cycad–fern–palm meadows. The plants also preserve evidence of newly evolving relationships with insects. A hundred or more small mammal species actually outnumber the dinosaurs by about three to one, but it is the dinosaurs, including *Tyrannosaurus rex*, that have grabbed the popular imagination.

Common organisms: plants, small mammals, and dinosaurs

66 million years ago
Maastrichtian Stage of the Late Cretaceous Epoch

❶ *Magnolia*
❷ *Edmontonia*
❸ *Cimolestes*
❹ *Edmontosaurus*
❺ *Triceratops*
❻ *Tyrannosaurus*

➜ **CIMOLESTES (3)** This 15cm-long (6in) eutherian mammal may have been ancestral to the modern Carnivora order. Fossils reveal long canines and shearing crests on the molar teeth that indicate a carnivorous diet

↻ **TRICERATOPS (5)** This 9m-long (30ft) ceratopsian plant-eater defended itself against the even bigger predators of the time with large neck frills and long horns on a massive skull..

↻ **TYRANNOSAURUS (6)** The most famous and largest coelurosaurian predator grew to 12m (40ft) long. Its 1.5m (5ft) skull and massive jaws could deliver a bone-crushing bite to its victims.

THE MAMMALIAN REVOLUTION BEGINS
CRAZY MOUNTAIN, MONTANA, USA

62 million years ago
Torrejonian Stage of the Paleogene Period

Latitude then: 46°N

Latitude now: 46°N

Sea level: rising (+200m/650ft)

Original environment: river channel and surrounding woodland

Deposits: river and floodplain sands, silts, and muds

Status: numerous sites over a wide region are actively being excavated for their fossils

Preservation: disarticulated bones and, mostly, teeth

⊕ Earth: c.62MA

⊕ Fossil site today

The demise of the dinosaurs left vast regions virtually devoid of large land animals, and vacant niches were soon occupied by newly evolving and diversifying mammals and birds. But questions about the evolutionary origins of these mammals have proved difficult to resolve, and it now seems there may have been several waves of origination and extinction.

The best known continental deposits of the period are in western North America. In the very earliest Paleogene strata, representatives of three mammal groups – multituberculates, marsupials and placentals – are found in changing proportions. The marsupials suffered the greatest decline, while multituberculates flourished and placentals experienced an explosive radiation. By Torrejonian times, there were more than 88 genera of placentals belonging to 29 families.

The Torrejonian mammal age had a high extinction rate, and half of all genera were extinct by the Tiffanian (beginning 60.9 MA). Their equally diverse replacements were mostly placentals, as multituberculates were now in decline. But again more than half of the 94 known Tiffanian genera were extinct by 56 MA – a sign of complex struggles for supremacy among the rising mammals.

Common organisms: plants and small mammals

Climate: northern subtropical humid
Biota: terrestrial

1. *Conoryctes*
2. *Platanus*
3. *Ptilodus*
4. *Stilpnodon*
5. *Rhamnus*
6. *Plesiadapis*
7. *Didymictis*
8. *Vitis*
9. *Chriacus*
10. *Taxodium*
11. *Pantolambda*
12. *Prodiacodon*

CONORYCTES (1) This tusk-like tooth belonged to a taeniodont. These pig-like animals also had clawed feet, probably for digging up plants.

PTILODUS (3) Relatively large for a multituberculate, *Ptilodus* was a squirrel-sized animal no more than 50cm (20in) long. Its laterally placed eyes, coupled with its long tail and the structure of its legs and feet, suggests that it was a good climber well adapted for an arboreal lifestyle.

PTILODUS TOOTH (3) The distinctive serrated form of this multituberculate cheek tooth, coupled with the shape of its broad, short-snouted skull, suggests that these animals had rodent-like habits.

PLESIADAPIS (6) This lemur-like plesiadapiform weighed up to 5kg (11lb). Its teeth are superficially rodent-like, with rectangular, blunted cusps to the molars indicating an omnivorous or fruit-rich diet. Plesiadapiforms were at one time thought to be early primates.

181

1. LIFE IN THE LATE CRETACEOUS 100-65.5ma

Recent fossil discoveries, especially in China and Mongolia, are transforming ideas about life in Late Cretaceous times. Terrestrial environments are no longer seen as simply dominated by reptilian groups. Mammals and birds were already diversifying along with the flowering plants and the insects they depended upon. There were even carnivorous mammals that preyed upon dinosaur hatchlings. Nevertheless, many dinosaur groups, such as the hadrosaurs, were thriving and show no sign of decline in diversity or numbers prior to their extinction. By contrast, the pterosaurs were in terminal decline, perhaps due to competition with the birds. By the End-Cretaceous they were reduced to a single family, albeit the most spectacular of all – the enormous azhdarchids.

2. THE EYE OF THE STORM

A major impact such as the one that occurred at Chicxulub would have had an immediate and devastating local effect, with tsunamis and wildfires spreading much further in the hours and days that followed. But to cause a significant global loss of plant and animal life in the seas and on land takes time and a global mechanism such as a drastic and lasting climate change. Scientists are still arguing about whether the impact itself could have been the prime cause of such a change,

ANGIOSPERMS
Prior to the extinction, the angiosperms were already well established alongside other terrestrial plant groups.

◀ MAGNOLIA

MAMMALS
By the Late Cretaceous, mammals were diverse and widespread, though mostly small in size.

◀ CIMOLESTES

NON-AVIAN DINOSAURS
Many non-bird dinosaur groups, including diverse ornithischians, continued to thrive right up until the extinction event.

◀ PROTOCERATOPS

BIRDS
During the Late Cretaceous, several other major bird groups flourished alongside the modern Neornithes.

◀ SAPEORNIS

AMMONITES
The ammonites were already diminishing in diversity – the extinction event wiped out the last surviving families.

◀ SCAPHITES

PTEROSAURS
Most of the smaller pterosaur families were extinct by the Late Cretaceous, leaving only the giant azhdarchids.

◀ QUETZALCOATLUS

FISH
Teleost fish originated in the early Mesozoic, and had already gone through a number of evolutionary changes by the Late Cretaceous.

◀ CATURUS

HEXAPODS
The hexapods had a long history stretching back into the Paleozoic, but benefited from the development of flowering plants in the Cretaceous.

◀ BAISOPARDUS

DISCOVERING MASS EXTINCTIONS

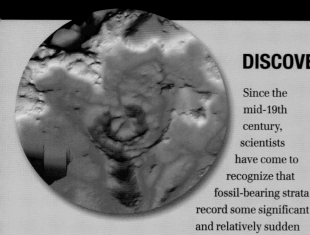

▲ **GROUND ZERO** Canada's Mancouagan crater, from the Late Triassic, is one of Earth's most impressive impact scars.

Since the mid-19th century, scientists have come to recognize that fossil-bearing strata record some significant and relatively sudden changes in the overall pattern of past life. John Phillips, Professor of Geology at Oxford around 1856, identified three major phases in the history of life – the Paleozoic, Mesozoic, and Cenozoic – which still give their name to major geological eras. The Paleozoic, meaning "ancient life", begins with the Cambrian explosion and is best known for extinct marine arthropods such as the trilobites and many groups of extinct jawless fish. It finishes at the end of the Permian Period, but which point life had made the move to land and diversified into a variety of forms including early reptiles and the ancestors of mammals. The Mesozoic, meaning "middle life", begins with the Triassic Period and finishes at the end of the Cretaceous. It is famous for its extinct reptiles such as the dinosaurs and pterosaurs. The Cenozoic, meaning "recent life", begins with the Paleocene and records the rise of the flowering plants and placental mammals.

These three major eras in the history of life are now known to correspond to three phases of evolutionary radiation and diversification, punctuated by major extinction events. The largest of these, at the end of the Permian, brought the Paleozoic Era to an end, while the best known is that which terminates the Cretaceous Period and the Mesozoic Era. A further three major extinctions have also been identified – at the end of the Ordovician, in the late Devonian, and at the end of Triassic times – and together these are known as the "big five".

To compare the scale of these extinctions, we need a common system of measurement, and scientists have settled on counting the number of major families of certain commonly fossilized marine organisms found on either side of each extinction. Accordingly, the end of the Ordovician saw 26 families become extinct, including many graptolites, trilobites and brachiopods; in the late Devonian some 22 families became extinct, especially corals, ammonoids, trilobites, and jawless fish; the cataclysmic End-Permian event saw 51 families die out, including nearly all the remaining corals, and many echinoderms and brachiopods; the End-Triassic event saw many ammonite families become extinct, along with brachiopods and bivalves. Finally, the End-Cretaceous event saw the ammonites, belemnites disappear completely, along with many bivalves and foraminiferans.

OTHER CAUSES?

Over the three decades since the so-called "Crater of Doom" hypothesis was first put forward, a great deal of research has gone into all aspects of the extinction question. Although the possibility of an extraterrestrial impact certainly made the biggest splash in the headlines, none of the other extinction events has been conclusively linked to a similar impact, and serious questions have been raised over how a single event, on this scale could have had such a global impact and whether its repercussions could have killed organisms so selectively.

Other mechanisms have been sought for the other extinctions, with various experts arguing for their own favourite cause. Large-scale volcanic events have been linked to three of the extinctions, climate change to another three, and sea level changes (coupled with deoxygenation of the oceans) have been associated with all five. To begin with sea level change was considered as a primary cause, but this has been partly displaced by the volcanic events that ended the Permian, Triassic and Cretaceous periods. The End-Permian flood basalts of the so-called Siberian Traps may have erupting some 1-2 million cubic kilometres (250,000-500,000 cubic miles) of lava and associated gases over a period of about a million years. Volcanic gases such as carbon dioxide are well known greenhouse gases, and could have created significant climate change, potentially giving rise to the measured anoxia of the ocean waters, and even the sea level changes.

Many non-geologists find it hard to believe that apparently simple questions about the nature of extinction events do not have simple answers, but there are significant problems with interpretations of the rock and fossil record, especially when it comes to the resolution of time. We tend to think of catastrophes as virtually instantaneous, but earthquakes, volcanoes, tsunamis and even impact events are localized, while extinction events occur on a truly global scale.

EVIDENCE

▶ **THE IRIDIUM ANOMALY** The boundary between Mesozoic and Cenozoic strata is marked around the world by sediments that are unusually rich in the rare element iridium. Although still measured in parts per trillion, iridium levels in these rocks are 30-130 times higher than normal, and this excess can only have come from an extra-terrestrial source. Other evidence includes shocked quartz that can only be produced in an impact event, and tsunami deposits in some areas.

◀ **THE "CRATER OF DOOM"** The Chicxulub crater in Mexico, formed by the End-Cretaceous impact, took years to find because it now lies buried beneath layers of younger strata in the Gulf of Mexico. Eventually unearthed identified by geophysicists in 1990, the structure is some 180km (110 miles) across, implying that the object that formed it had a diameter of at least 10km (6 miles)

EARTH'S

The evolution of life has been punctuated by several mass extinctions that have, at times, wiped out more than half of all living organisms. The best known of these – although not the largest – occurred at the end of the Cretaceous Period, famously marking the disappearance of the non-bird dinosaurs, flying pterosaurs and most marine reptiles. In the aftermath, surviving animal groups diversified rapidly and a new biota rose to prominence, setting the stage for the Cenozoic Era. Debate about the mechanisms behind such evolutionary crises is intense, but the evidence points to a range of different causes, each cataclysmic in its own right.

EXTINCTION
EVENTS

Debates over the cause of these catastrophic events became very public
in the 1980s when two American scientists, Walter and Luis Alvarez,
suggested that the extinction of the dinosaurs was caused by the impact
of a large object from space – a comet or asteroid – that scattered debris
around the world. The theory polarized geological opinion, and the
crucial discovery of a crater with the right age only came in the 1990s.

By this time it was clear that the End-Cretaceous extinction was by no
means the biggest suffered by life in the past – the end of the Permian,
for example, saw 70 percent or more of all life become extinct. Inevitably,
another impact event was suggested, but the tell-tale evidence was not
forthcoming. Instead, the End-Permian extinction seems to be linked to
extensive volcanism and sea-level changes – events that also occurred at
the end of the Cretaceous. Meanwhile, some paleontologists have
questioned just how an impact event could selectively kill off certain
groups of organisms while leaving others largely unscathed, and there
are also suggestions that the dinosaurs were in decline for some time
before their ultimate extinction. Whatever the truth, it seems that
extinction events have a complex range of causes.

4. MODERN LEGACY

While the end Cretaceous extinction certainly changed some important aspects of the global biota, the evolution of surviving organisms soon produced replacements for the organisms that died out, and it is these surviving groups that dominate the world's present-day flora and fauna. The flowering plants, insects, birds and mammals were all well established in Late Cretaceous times and were already co-evolving alongside one another as a result of the reproductive requirements of the plants and the food requirements of the animals. And of course even the dinosaurs are still with us today – albeit in the feathered form of birds. They have filled the niches vacated by the flying pterosaurs so successfully that their mammalian competitors, the bats, which are themselves very successful, have nevertheless had to occupy a rather specialised niche as nocturnal predators.

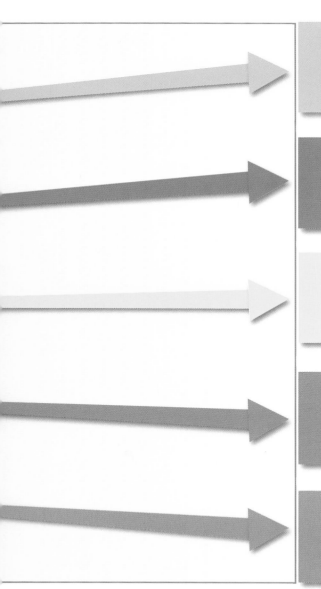

FLOWERING PLANTS

The angiosperms continue to be our planet's largest plant group today, aided by the extraordinary success of the grasses. They continue their intimate relationship with insect and other pollinators.

MAMMALS

From a relatively small range of superficially similar animals, mammals have evolved throughout the Cenozoic to occupy a wide range of ecological niches, on land, in water, and even in the air.

BIRDS

The neornithean birds continue to flourish in a wide variety of environments around the world. Since the extinction event they have branched into many distinct subgroups, including flightless forms.

TELEOST FISH

The advanced "neoteleost" fish, with mouth adaptations that improve their predatory ability, today number more than 12,000 species, making them the most successful vertebrate group.

HEXAPODS

The hexapods are the largest group of arthropods, and indeed of all life, on the planet today, with more than a million described species, and perhaps five times that number still unrecognized.

perhaps through the creation of a "nuclear winter" in which debris suspended in the atmosphere blocked heat and light from the Sun.

On a geological timescale, however, the effect was almost instantaneous – the disappearance of the non-avian dinosaurs and other reptile families, as well as the long-lived ocean ammonites and other groups including some plants. There is also some evidence that the extinction was followed by a short-lived burst of fungal growth in its aftermath.

3. LIFE RECOVERS   65-34ma

The End-Cretaceous extinction devastated food chains both on land and at sea, thanks to the loss of land plants and marine phytoplankton. As the effects spread along the food chain, herbivores of all kinds, intermediate carnivores, and eventually specialist top predators were affected – with the top predators suffering most. While some survivors at lower levels in the food chain rapidly adapted to fill ecological niches left by the extinction, the top-level vacancies were probably reoccupied rather

more slowly, since new top carnivores had to emerge from the multitude. In the seas the teleost (bony) fish underwent an explosive radiation, as did the flowering plants, insects, birds, and mammals on land. However, the early Cenozoic saw significant "turnover" within some of these groups, as new families of animals diversified, flourished, and then faded away to be replaced by others. As a result, few Early Cenozoic forms are immediately recognisable today.

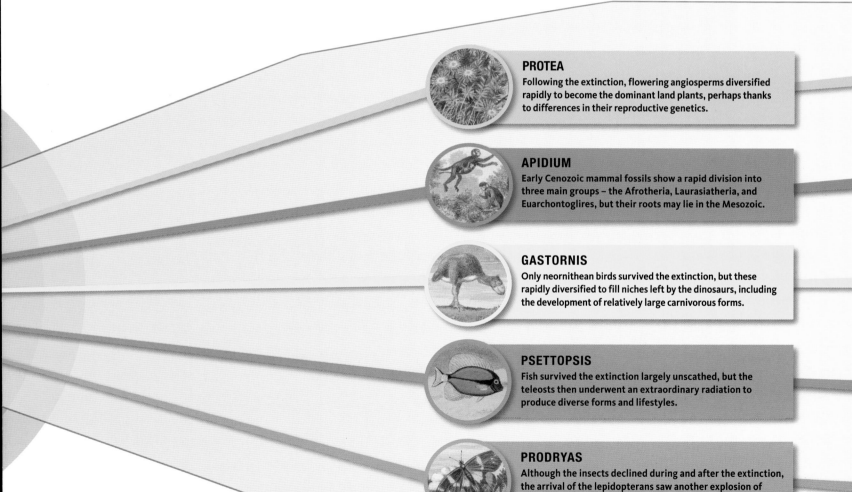

PROTEA
Following the extinction, flowering angiosperms diversified rapidly to become the dominant land plants, perhaps thanks to differences in their reproductive genetics.

APIDIUM
Early Cenozoic mammal fossils show a rapid division into three main groups – the Afrotheria, Laurasiatheria, and Euarchontoglires, but their roots may lie in the Mesozoic.

GASTORNIS
Only neornithean birds survived the extinction, but these rapidly diversified to fill niches left by the dinosaurs, including the development of relatively large carnivorous forms.

PSETTOPSIS
Fish survived the extinction largely unscathed, but the teleosts then underwent an extraordinary radiation to produce diverse forms and lifestyles.

PRODRYAS
Although the insects declined during and after the extinction, the arrival of the lepidopterans saw another explosion of diversity.

GLOBAL WARMING IN THE EARLY EOCENE
BIGHORN BASIN, WYOMING, USA

54 million years ago
Wasatchian Age of the Paleogene Period

Latitude then: 45°N

Latitude now: 45°N

Sea level: high

Original environment: terrestrial with small rivers

Deposits: river and floodplain gravels, sands, muds, and soils

Status: numerous localites scattered across Wyoming

Preservation: often well-preserved leaves and pollen, but the mammal remains are mostly teeth

• Bighorn Basin

⊕ Earth: c.54 MA

• Bighorn Basin

⊕ Fossil site today

Following the end-Cretaceous extinction, life took several million years to recover. Large plants that occupied dry land seem to have been particularly badly hit by the extinction – especially in the Americas, which were the nearest landmasses to the extraterrestrial impact in what is now the Gulf of Mexico.

However, the recovery was a complex one – North American strata show a moderate increase in plant diversity through the Paleocene, followed by a decrease to just 25 species at the Paleocene–Eocene boundary, around 55.8MA. The early Eocene saw a further increase to around 50 species.

Mammals, meanwhile, show a high turnover and rapidly increasing numbers across the Cretaceous–Paleogene boundary (from 10 to 50 genera), followed by a slight decline to 40 genera and then an increase to 75 genera by around 50MA. Contrary to what one might expect, the changes in plant and mammal diversity are distinctly out of step with one another. However, this may simply reflect different modes of response to the same problem – a phase of rapid global warming that saw mean annual temperatures rise up to 4°C (7°F) from the late Paleocene, reaching to 26°C (79°F) in the early Eocene.

Common organisms: plants and mammals

Climate: warm and dry, rising to warm humid
Biota: terrestrial

1. *Miacis*
2. *Arfia*
3. *Hyopssodus*
4. *Platanus*
5. *Didymictis*
6. *Cantius*
7. *Hyracotherium*
8. *Diacodexis*
9. *Celtis*
10. *Phenacodus*

PLATANUS (4) This distinctive leaf shape is typical of the deciduous platanaceans (plane trees), a group of large northern hemisphere angiosperms that grow to more than 30m (100ft) high. The group originated in early Cretaceous times, and typically grew beside streams and in wetlands.

DIACODEXIS (8) At around 50cm 20in) long, this small animal was one of the oldest known even-toed ungulates (artiodactyls), and fed on a variety of plant foods. A slender and long-limbed animal, it has the characteristic lower leg and ankle typical of the artiodactyls, which have movement restricted to the vertical plane. It probably moved by leaping and bounding.

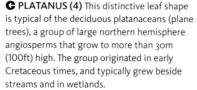

MIACIS (1) The teeth of this early placental mammal show that it was one of the first of its kind to include meat in its diet..

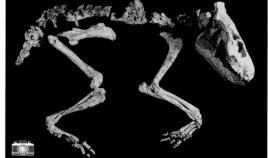

189

A TROPICAL EOCENE LAKE
GREEN RIVER, WYOMING, USA

54 million years ago
Wasatchian Age of the Paleogene Period

Latitude then: 40°N

Latitude now: 40°N

Sea level: rising (+200m/650ft)

Original environment: freshwater lake and surrounding forests

Deposits: fine-grained silts and muds

Status: rich strata are commercially quarried for their fossils, but collecting licences can be bought

Preservation: flattened but often entire, with some soft tissue preservation

⊕ Earth: *c.*54MA

⊕ Fossil site today

Over some 17 million years of late Paleocene to early Eocene times one of the largest accumulations (around 65,000sq km/25,000sq miles) of lake sediments built up to more than 2km (1.2 miles) thick within the Midwest of the USA as the Rocky Mountains were formed. The Green River strata were deposited in three lake basins under paratropical climates, with annual average temperatures of 15–20°C (59–68°F) and annual rainfall of 750–1000mm (30–40in). The fine-grained lake sediments preserve the remains of plants, insects, molluscs, fish, lizards, snakes, turtles, crocodiles, birds, and mammals that lived in and around the lakes.

Fish are particularly abundant, with a million or so fossils collected over the last 25 years. Despite their numbers, however, they are not very diverse, belonging to just 21 genera of mostly modern teleost families. They seem to have suffered from repeated mass mortalities due to lack of oxygen when the lake waters stagnated.

Rare mammals include early horses, tapirs, rhinoceroses, and the elephant-sized *Uintatherium*, the largest land animal of its time. The potential for important new finds is illustrated by the recent discovery of *Onychonycteris*, the most primitive bat known .

Common organisms: plants, fish, birds, and insects

Climate: warm paratropical
Biota: diverse freshwater lake and terrestrial

- ❶ *Sumac*
- ❷ *Limnofregata*
- ❸ *Sabalites*
- ❹ *Heliobatis*
- ❺ *Knightia*
- ❻ *Trionyx*
- ❼ *Borealosuchus*
- ❽ *Amia*
- ❾ *Gallinuloides*
- ❿ *Presbyornis*
- ⓫ *Typha*
- ⓬ *Icaronycteris*
- ⓭ *Onychonycteris*
- ⓮ *Boavus*
- ⓯ *Uintatherium*
- ⓰ *Ailanthus*

↪ **HELIOBATIS (4)** With a flat disc-shaped body and long barbed tail, *Heliobatis* was a typical batoid stingray that lived on the lake bed, burying itself slightly in sediment for camouflage. With teeth adapted for crushing, it fed on small crustaceans, clams, and small fish. Some grew to 90cm (35in) long.

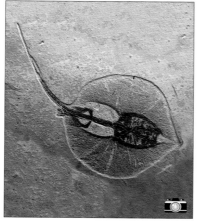

↪ **ONYCHONYCTERIS (13)** This primitive bat had wings developed for powered flight, but its ear lacked the adaptations for echolocation – instead it must have hunted insects in other ways.

↑ **BOAVUS (14)** This neotropical wood snake, *Boavus idelmani*, is the only one of its kind known from Green River, and was previously mistaken for a boa.

EOCENE LIFE IN TROPICAL LONDON
ABBEY WOOD, LONDON, ENGLAND

54 million years ago
Wasatchian Age of the Paleogene Period

Latitude then: 51°N

Latitude now: 51°N

Sea level: rising (+200m/650ft)

Original environment: coastal rainforest

Deposits: channel sands and pebbles

Status: legally protected site that can only be exposed by excavation

Preservation: small, fragmentary, and isolated vertebrate remains, especially teeth

Abbey Wood

⊕ Earth: c.54MA

• Abbey Wood

⊕ Fossil site today

Bulk sampling and sieving of channel sands from the early Eocene of suburban London has produced abundant tiny fossils. Mostly mollusc shells and various animal teeth, they represent life in a remarkably diverse terrestrial and low-salinity coastal environment, surrounded by mangrove swamps and rainforest. The waters and lush vegetation supported many different vertebrates, including more than 20 species of fish, turtles such as *Trionyx*, birds including *Marinavis*, and some 30 mammal species.

The mammals range from small marsupials and late-surviving multituberculates to rodents (*Paramys*), insect-eating bats, primates (the fruit-eating *Cantius*), and the primitive horse *Pliolophus*. They also include extinct groups such as plant-eating pantodonts (*Coryphodon*), insectivorous pantolestids (*Palaeosinopa*) and apatotherians (*Apatemys*), and carnivorous creodonts (*Palaeonictis* and *Oxyaena*), as well as various primitive plant-eating ungulates (condylarthans) and artiodactyls.

The abundance of plant-eaters shows that the vegetation provided varied and rich food for all these mammals. Nearby plant fossils show that there were more than 300 angiosperm species in a rainforest assemblage resembling those of modern lowland Asia.

Climate: warm paratropical
Biota: terrestrial and coastal marine

Common organisms: fish and mammals

❶ *Coryphodon*
❷ *Ficus*
❸ *Paramys*
❹ *Palaeosinopa*
❺ *Apatemys*
❻ lauracean
❼ *Marinavis?*
❽ *Ceriops*
❾ *Palaeonictis*
❿ *Cantius*
⓫ *Oxyaena*
⓬ *Pliolophus*

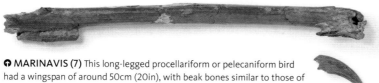

↻ **MARINAVIS (7)** This long-legged procellariform or pelecaniform bird had a wingspan of around 50cm (20in), with beak bones similar to those of the modern shearwater.

↻ **CANTIUS (10)** Fragments of teeth and skulls of *Cantius* represent the first adapiform primate fossils. Found in the early Eocene strata of North America and Europe, these small animals weighed around 3kg (6.6lb), and had unspecialized small incisors with non-shearing cheek teeth. They are likely to have been fruit-eaters.

↻ **PLIOLOPHUS (12)** This dog-sized primitive horse is known at Abbey Wood from jaw fragments, teeth, and other bones – the more complete skull shown here comes from sediments at nearby Harwich, where the same species has been found.

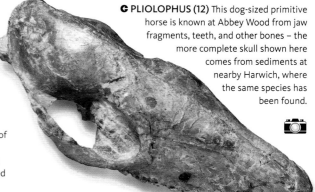

193

EOCENE FISH OF THE TETHYS OCEAN
MONTE BOLCA, ITALY

49 million years ago
Mid-Eocene Stage of the Paleogene Period

Latitude then: 42°N

Latitude now: 46°N

Sea level: +100m (325ft)

Original environment: backreef lagoon or silled basin

Deposits: fine-grained carbonate muds

Status: several of the quarries around Monte Bolca have been incorporated into a fossil park that is open to the public

Preservation: flattened, but including numerous entire animals with some soft tissue preservation

⊙ Earth: *c.*49MA

⊙ Fossil site today

Excavated since the 16th century, the mid-Eocene strata of Monte Bolca contain the remains of more than 500 species of terrestrial and marine organisms, including plants, jellyfish, polychaete worms, corals, molluscs, crustaceans, fish, and reptiles.

The first written description of Bolca fossil material dates to 1555, when Siennese doctor Andrea Mattioli reported seeing "some slabs of stone which, on being split in half, revealed the shapes of various species of fish, every detail of which had been transformed into stone". The first monograph on fossil fish, Giovanni Serafino Volta's *l'Ittiolitologia Veronese* (1796–

1808) and some of Louis Agassiz's pioneering fish studies were based largely on fossils from Monte Bolca. The fish fauna is now known to be extraordinarily diverse, with 250 species assigned to 140 genera and 90 families of which 80 per cent are still living. Altogether they are typical of the tropical waters of the Indo-Pacific and Atlantic Oceans.

The deposits here were laid down in a lagoon, landward of a barrier reef with surface waters open to the Tethys Ocean to the south. The bottom waters were subject to seasonal stagnation, perhaps promoted by algal blooms, resulting in accumulations of asphyxiated fish.

Climate: tropical

Biota: marine with terrestrial elements

Common organisms: fish

1. *Exellia*
2. *Lophius*
3. *Ceratoichys*
4. *Eomyrophis*
5. *Eoholocentrum*
6. *Psettopsis*
7. *Trygon*
8. *Eobothus*
9. *Mene*

EXELLIA (1) This extinct spadefish was distinguished by very long pelvic fins and a high dorsal fin rising above the eyes.

EOHIOLOCENTRUM (5) Resembling a modern beryciform squirrelfish, with pelvic and pectoral fins with spiny fin rays just behind the head, this fish's large eyes suggest that it might have been nocturnal.

TRYGON (7) This elasmobranch batoid, up to 80cm (32in) long, is one of many rays at Monte Bolca. Up to two thirds of its length was typically made up by the long, thin tail. Like the sharks, rays have a cartilaginous skeleton, and so their remains are only preserved in special conditions.

195

EOCENE LIFE IN A EUROPEAN RAINFOREST
MESSEL, NEAR DARMSTADT, GERMANY

48 million years ago
Lutetian Stage of the Eocene Epoch

Latitude then: 49 degrees North

Latitude now: 49 degrees North

Sea level: +100m (330ft)

Original environment: a maar lake and its surrounding forests

Deposits: lake-bed muds and silts

Status: Designated World Heritage Site

Preservation: largely flattened but often complete organisms, including soft tissue preservation

• Messel

⊕ Earth: c.48 MA

• Messel

⊕ Fossil site today

Today the early Eocene age oil shales of Messel are a World Heritage Site, preserved for the extraordinary richness and quality of their fossil remains. From dog-sized primitive horses to early primates, anteaters, birds, fishes, beetles and plants, many of Messel's fossils are preserved with their original hair, feathers and other soft tissues. They give an extraordinary insight into life in and around a rainforest lake in paratropical northwestern Europe during early Eocene times, some 48 million years ago.

The Messel fossils reveal a world without grass or grass-eating mammals, but one with warm climates and a wonderful diversity of life.

The deposits preserve a maar lake (a flooded volcanic crater) with fossils suggestive of habitats ranging from open water, swamp, bankside and damp forest, to drier elevated banks, and higher ground further away from the lake with pines, beech, chestnut and oaks.

Many of the mammals found at Messel originated outside Europe but migrated across the region in Eocene times. Amongst these migrants were early modern forms such as the rodents, horses, bats and primates. However, some mammals were insect-eating survivors from Mesozoic times – including early hedgehog-like animals.

Common organisms: fish and insects

Climate: paratropical humid
Biota: terrestrial and aquatic

1. *Palaeopython*
2. *Propalaeotherium*
3. *Paroodectes*
4. *Messelobunodon*
5. *Miacis*
6. *Formicium*
7. *Palaeoglaux*
8. cicada
9. *Archaeonycteris*
10. *Primozygodactylus*
11. *Darwinius masillae*
12. *Eomanis*
13. *Hyrachyus*

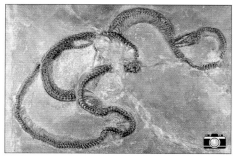

⊕ **PALAEOPYTHON (1)** At 2m (6.6ft) long, this is the largest snake found at Messel and belongs to the family Boidae. Snakes were the last major group of reptiles to evolve in Late Cretaceous times, probably developing from burrowing lizard ancestors.

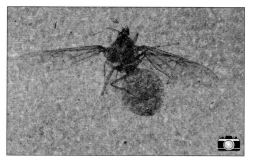

⊕ **FORMICIUM (6)** With a wingspan of around 6cm (2.4in) this winged female flying ant perished during her nuptial flight, some of the fossil queen ants found at Messel had wingspans of 16cm (6in) and would have weighed up to 10g (1oz) or so, the size and weight of a hummingbird.

⊕ **DARWINIUS MASILLAE (11)** – a new Messel treasure, this 58cm (22¾ in) long primate (nicknamed "Ida") has stirred debate over whether it is an early haplorhine primate and member of our human lineage or an adapiform on the lemur lineage.

DIVERSITY OF LIFE AT MESSEL
MESSEL, NEAR DARMSTADT, GERMANY (CONT.)

48 million years ago
Lutetian Stage of the Eocene Epoch

The dominant members of Messel's aquatic biota ranged from water lilies and insect larvae to bony fish (90 per cent of the vertebrates) but also included metre-long (40in) predatory gars, salamanders, frogs, turtles, and crocodiles up to 4m (13ft) long.

The terrestrial ecosystem, meanwhile, included ferns, palms, vines, citrus, and tea plants (some 176 species altogether), plus associated insects. Some 63 per cent of insect species were beetles, but there were also numerous ants, wasps, bugs, bees, spiders, cicadas, cockroaches, and moths, but dragonflies and butterflies are rarities.

Messel preserves an important record of Eocene lizards, iguanas, and snakes. Its birds are equally

significant and include representatives of both extinct and modern groups, ranging in size from the tiny *Messelirrisor*, with a wingspan of 17cm (7in), to the flightless *Diatryma*, some 1.7m (5.5ft) tall, along with ibises, rails, owls, swifts, and roller-like birds.

The mammals include tiny marsupial opossums, primitive insectivores (*Leptictidium*), a fish-eater (*Buxolestes*), hedgehogs (*Pholidocercus*), primates (*Propithecus*), pangolins (*Eomanis*), a possible anteater (*Eurotamandua*), carnivores (*Paroodectes, Miacis*), bats, rodents, even-toed ungulates (*Messelobunodon*), and four species of horse. These last show that horses entered North America via Europe, since the most primitive equid, *Hallensia*, is only known from Europe.

❶	*Palaeochiropteryx*	❿	*Gastornis*
❷	buprestid	⓫	*Eocoracias*
❸	*Rhynchaeites*	⓬	*Cephalotaxus*
❹	*Eurotamandua*	⓭	*Pholidocercus*
❺	*Zantedeschia*	⓮	*Buxolestes*
❻	*Eopelobates*	⓯	*Asiatosuchus*
❼	*Kopidodon*	⓰	*Eurohippus*
❽	*Leptictidium*	⓱	*Aegialornis*
❾	*Trionyx*	⓲	*Typha*

➲ **LEPTICTIDIUM (8)** This 75cm-long (30in) insect-eating bipedal mammal is one of the most remarkable found at Messel. As with its Cretaceous ancestors, the very primitive structure of its pelvis shows that it ran on its hind legs but did not hop or jump.

Common organisms: fish and insects

Climate: paratropical humid
Biota: terrestrial and aquatic

➲ **TRIONYX (9)** This long-necked and soft-shelled trionychid turtle had powerful jaws and an almost worldwide distribution. With lengths of up to 60cm (24in), it was the largest of the Messel turtles and like its surviving relatives was a predator in subtropical freshwaters and fed partly on fish.

⬆ **GASTORNIS (10)** This robust flightless bird was a predator with a large powerful beak and strong legs, standing 1.75m (5.75ft) tall. Its relatives were widespread across Europe and North America in Paleogene times.

OUR EARLIEST ANTHROPOID RELATIVES
SHANGHUANG, SOUTH JIANGSU PROVINCE, CHINA

Latitude then: 31°N

Latitude now: 31°N

Sea level: +100m (330ft)

Original environment: subtropical limestone karst

Deposits: karst fossil fissure fills

Status: found in commercial limestone quarries

Preservation: disarticulated vertebrate remains – mostly teeth and jawbones

⊕ Earth: c.45 MA

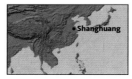

⊕ Fossil site today

Tiny teeth, jaw fragments, and foot bones from Eocene-age fissure fillings near Shanghuang, China, belong to a very early member of the anthropoid radiation, and fill a gap between more primitive primates and the more advanced animals from which monkeys, apes, and humans eventually evolved.

Named *Eosimias*, it has been placed in a new family (Eosimiidae) and provides evidence that primates originated even earlier in Paleogene times. Until this discovery the oldest known anthropoid came from Fayum in Egypt.

Eosimias was very different from the early Egyptian anthropoids. The teeth of this tiny primate, which weighed no more than 100g (3oz), show that it was very primitive and tarsier-like, with small spatulate incisors, enlarged canines, and an anthropoid-like jawbone. Its ankle anatomy has a combination of lemur-like prosimian and anthropoid traits, suggesting that *Eosimias* had a monkey-like foot posture with its soles face down rather than angled inwards. It seems these tiny primates probably moved through its forest habitat by climbing and walking on all fours along branches, in contrast to the typical prosimian preference for leaping and clinging over quadrupedal walking and running.

Common organisms: rodents and lagomorphs

Climate: subtropical humid
Biota: terrestrial

❶ eosimiid
❷ *Eosimias*
❸ *Adapoides*

↘ **EOSIMIAS (2)** This tiny (100g/3oz) primate's teeth, with small incisors and large canines set in an anthropoid-like jawbone, display a mixture of primitive and advanced features, suggesting that the animal was a very primitive, tarsier-like anthropoid.

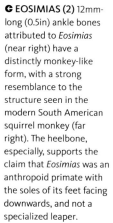

◖ **EOSIMIAS (2)** 12mm-long (0.5in) ankle bones attributed to *Eosimias* (near right) have a distinctly monkey-like form, with a strong resemblance to the structure seen in the modern South American squirrel monkey (far right). The heelbone, especially, supports the claim that *Eosimias* was an anthropoid primate with the soles of its feet facing downwards, and not a specialized leaper.

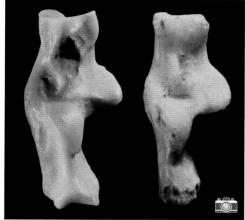

65.5

PRIMATES PROLIFERATE IN EOCENE EGYPT
FAYUM, EGYPT

34 million years ago
Rupelian Stage of the Oligocene Epoch

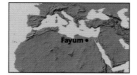

Latitude then: 30°N

Latitude now: 29°N

Sea level: falling (+75m/245ft)

Original environment:
forested swampy coastal floodplain

Deposits: muds, silts, and sands

Status: scattered outcrops
southwest of Cairo

Preservation: disarticulated
three-dimensional bony material
with some entire skeletons

Earth: *c.*34MA

Fayum •

Fossil site today

Fayum, on the edge of Egypt's modern Western Desert, was once a coastal floodplain. Some 33 million years ago, all kinds of vertebrates lived around the freshwater and brackish estuarine swamps. There were numerous and diverse aquatic birds, with big-footed jacanas walking over lily pads, while flamingos, storks, and herons flourished in the shallows alongside crocodiles, sirenians, and turtles. Further out, cormorants dived for fish while ospreys and fish eagles skimmed the surface to snatch prey.

The fossil mammals from the Fayum deposits include representatives of both extinct and modern groups. The former include large herbivorous embrithopods (the rhino-like *Arsinoitherium*, some 4m or 13ft long), and predatory hyaenodonts. Among the latter are the primitive proboscidean *Moeritherium*, 3m (10ft) long, sirenians, browsing hyracoids, rodents, and tree-dwelling early primates.

Late Eocene climate cooling had brought changes in mammal faunas, with primates far less widespread. The early adapids and tarsiiforms were nearly exterminated, but there followed a proliferation of anthropoids in North Africa, giving rise to the parapithecid *Apidium* and eight species of catarrhine propliopithecids including *Aegyptopithecus*.

Common organisms: shellfish, fish, and small mammals

Climate: subtropical humid

Biota: freshwater aquatic and terrestrial

1. *Ardea*
2. *Sarothrura*
3. *Apidium*
4. *Phiomia*
5. *Aegyptopithecus*
6. *Epipremnum*
7. *Arsinotherium*
8. *Pandion*
9. *Arctophilornis*
10. *Moeritherium*
11. *Nyctiocorax*
12. *Balaeniceps*
13. *Haliaetus*

⏻ ARSINOTHERIUM (7) One of the oddest of the extinct mammals that inhabited Fayum's swamps was this 4m-long (13ft) herbivore with a pair of side-by-side nasal horns and teeth specialized for browsing large quantities of tough, fibrous vegetation. Recent work suggests that these embrithopods were related to proboscideans and sirenians.

⏻ ➲ MOERITHERIUM (10) This 3m-long (10ft) plant-eater stood some 70cm (28in) high, and occupied the aquatic niche filled by hippos today. Its name means "the beast from Lake Moeris", and it belonged to an extinct group of moeritheres, related to the probiscidean gomphotheres. Its teeth suggest a diet of soft vegetation.

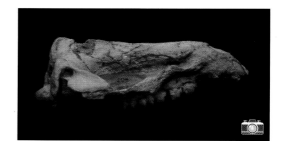

FLOWERS AND INSECTS EVOLVE TOGETHER
FLORISSANT, COLORADO, USA

34 million years ago
Rupelian Stage of the Oligocene Epoch

Latitude then: 39 degrees North

Latitude now: 39 degrees North

Sea level: falling (+50m/165ft)

Original environment: lacustrine

Deposits: muds and volcanic ash

Status: a legally protected National Monument

Preservation: flattened but exquisitely preserved biota often with fossilized soft tissues

Florissant •

⊕ Earth: c.34MA

Florissant •

⊕ Fossil site today

Florissant's wealth of petrified life reveals some of the most exquisite and delicate fossils ever found, including entire flowers and butterflies, still preserved with traces of their markings. The abundance of plants and some 1500 species of insects records critical details about how the two groups evolved together – without insect pollination, the modern flowering plants would simply not survive.

The setting was a lava-dammed lake beneath a volcano. Lethal pyroclastic flows periodically swept across the lake, bringing small land animals and plant debris that sank into the lakebed sediments. The most abundant vertebrates are fish – birds and mammals are very rare but include one of the oldest mole relatives, while amphibians and reptiles are unknown.

The remains of more than 100 species of upland plants have been found, including deciduous trees, palms, and stumps of *Sequoia*. Most abundant are the extinct *Fagopsis* and *Cedrelospermum*, along with *Ailanthus* (tree of heaven) and *Koeltreuteria* (goldenrain tree). This indicates a warm temperate to subtropical climate, and sufficient rainfall to support dense forests along riverbanks, with pines and oaks dominating open higher ground.

Climate: warm temperate to subtropical
Common organisms: plants and insects
Biota: terrestrial and lacustrine

❶ charadriid	⓮ Aphodius
❷ Mesohippus	⓯ Microstylum
❸ Herpetotherium	⓰ Koelreuteria
❹ Mahonia	⓱ Nephila
❺ Labiduromma	⓲ Merycoidodon
❻ "Bledius"	⓳ Myopodagrion
❼ Ephemera	⓴ Amelanchier
❽ Oligodonta	㉑ Vanessa
❾ Rosa	㉒ Palaeovespa
❿ Marquettia	㉓ Holcorpa
⓫ Prodryas	㉔ Megacerops
⓬ Heriades	㉕ Syrphus
⓭ Florissantia	

⚲ PRODRYAS (11) Discovered in 1878, this remarkably well-preserved brush-footed nymphalid butterfly, with a wingspan of 5cm (2in), was one of the first fossils to be found at Florissant. Similar to forms that live in Central and South America, it still preserves the original colour pattern.

⚲ FLORISSANTIA (13) Fossil flowers are exceedingly rare as they are essentially ephemeral and their tissues are normally very delicate. This 2cm-wide (0.8in) flower was toughened to aid wind dispersal of the maturing fruit, which helped ensure its chances of fossilization.

⚲ PALAEOVESPA (22) This 2cm-long (0.8in) fossil hornet, a member of the Vespidae family, is one of Florissant's most famous fossils. Like living hornets, it was probably a social insect that constructed a papery, multichambered nest for breeding. Several different species have been found.

65.5

THE FIRST SAVANNAH GRASS COMMUNITY?
SHAND GOL, MONGOLIA

Climate: northern hemisphere warm temperate
Biota: terrestrial vertebrates

Latitude then: 44°N
Latitude now: 44°N
Sea level: +50m (165ft)
Original environment: savannah grasslands?
Deposits: sands and silts
Status: desert location southwest of Ulanbataar
Preservation: mostly small isolated teeth and fragmentary bone material

Shand Gol •

⊕ Earth: c. 30MA

• Shand Gol

⊕ Fossil site today

One of the most spectacular finds of the American Museum of Natural History's expeditions to Central Asia in the 1920s was *Indricotherium*, the largest known land mammal, which grew to nearly 5m (16.5ft) tall at the shoulder. This giant rhino relative was a browser in open country, possibly the oldest grass-dominated community in the Old World.

Shand Gol's Oligocene mammal fauna is very different from the region's late Eocene fauna. Gone are the rhino-sized brontotheres, tapiroids, and the last of the bear-like hoofed mesonychids, although there were still some archaic survivors such as the pig-like

enteledonts, carnivorous hyaenodonts, and deer-like artiodactyls. Alongside them, several new groups appeared, including ruminant artiodactyls, rodents, several advanced carnivorans, and the first true cats, as well as cat-like nimravids and civets.

However, there were almost no medium-sized mammals – instead, smaller rodents and lagomorphs dominated, their high crowned teeth well adapted for eating gritty plants such as grass. Despite this, there is no continuity with later savannah-mosaic communities – instead evolution of the modern grassland mammal fauna probably happened in Africa.

1. *Plesictis*
2. *Hyaenodon*
3. *Palaeoprionodon*
4. *Amphicynodon*
5. *Indricotherium*
6. *Cricetops*
7. *Ochonta*
8. *Nimravus*
9. *Tupaiodon*

⊙ PALAEOPRIONODON (3) Early Oligocene times saw the evolution of a number of small animals similar to true cats. These archaic viverrid "feloids" resembled the living civets and genets. With a skull and teeth like the present-day genet, *Palaeoprionodon* probably lived off a similar diet of small vertebrates.

⊙ INDRICOTHERIUM (5) The largest land mammal ever known is also known as *Paraceratherium*, and was formerly also called *Baluchitherium*. It had a rhino-like skull that was some 1.4m (55in) long, with hyracodont-like teeth, including prominent upper incisors, adapted for browsing tough plant material. With long, relatively slender limbs, it stood nearly 5m (16.5ft) high at the shoulder, and its long neck allowed it to reach even higher into tree canopies to feed.

AN AUSTRALIAN "ARK"
RIVERSLEIGH, QUEENSLAND, AUSTRALIA

Climate: southern hemisphere humid subtropical
Biota: terrestrial and freshwater vertebrates

Latitude then: 19°S
Latitude now: 19°S
Sea level: as present
Original environment: lacustrine and lowland rainforest
Deposits: lakebed carbonate muds
Status: scattered limestone outcrops in arid northwest Queensland
Preservation: disarticulated but three-dimensional bones

Riversleigh •

⊕ Earth: c .23MA

Riversleigh •

⊕ Fossil site today

When Australia became isolated from the rest of the world, it preserved a phase of mammal evolution dominated by marsupials, which were later almost entirely replaced elsewhere by placental mammals. As a result, the 23-million-year-old lakebed muds of Riversleigh provide a window onto an array of unique late Oligocene and early Miocene mammals.

The largest of these are abundant hippo-like plant eaters (*Neohelos*) quite unlike anything alive today, whose presence indicates rich vegetation. The few predators include a cat-sized marsupial "lion" (*Priscileo*), a carnivorous kangaroo (*Ekaltadeta*), crocodiles, and snakes,

but their paucity allowed the survival of flightless birds such as *Bullockornis*. Above ground, the tree canopy was populated by six possum species and many songbirds, while the forest floor litter was constantly stirred by tiny wallabies and marsupial moles.

The lowland rainforest developed on limestone karst, where caves provided roosts for large numbers of leaf-nosed bats. High groundwater levels formed semi-permanent lakes and rivers teeming with turtles, frogs, eels, catfish, lungfish growing to 4m (13ft) long, and crocodiles – one of which was found with the skull of a small *Priscileo* in its mouth.

23 million years ago
Aquitanian Stage of the Miocene Epoch

Common organisms: fish, bats, birds and small marsupials

1. Litokoala
2. Namilamadeta
3. Priscileo
4. Bullockornis
5. Montypythonoides
6. Distioechurus
7. Hypsiprymnodon
8. Yalkaparidon
9. Burramys
10. Strigocuscus
11. Litoria
12. Nimbacinus
13. meliphagid
14. Paljara
15. paradisaeid
16. Pseudochirops
17. Ekaltadeta
18. Brachipposideros
19. Neohelos
20. Physignathus

☞ **PRISCILEO (3)** This marsupial "lion" was a thylacoleonid, related to living wombats. Its descendants evolved into leopard-sized predators in Pleistocene times before becoming extinct some 40,000 years ago.

☞ **YALKAPARIDON (8)** Unofficially known as the "thingodon", this 30cm (12in) marsupial was thought to be a rare placental mammal until the discovery of its typically marsupial lower jaw and teeth. It cannot yet be assigned to any known marsupial order, and may have been a specialized worm-eater.

☞ **EKALTADETA (17)** Modern kangaroos are all plant-eaters, but this small extinct form, known only from a couple of 18cm (7in) skulls, has teeth that indicate that it was either a carnivore or omnivorous, with lower incisors adapted for stabbing, premolars for holding and slicing, and molars for crushing. Marginal flanges prevented bone from piercing the gums – another carnivore characteristic.

EARLY APE EVOLUTION IN AFRICA
RUSINGA ISLAND, LAKE VICTORIA, KENYA

Climate: tropical with seasonal rains
Biota: terrestrial plants and vertebrates

Latitude then: 0°S

Latitude now: 0°S

Sea level: +10m (33ft)

Original environment: alluvial floodplains surrounded by volcanic highlands

Deposits: volcanic ashes

Status: island localities, now mostly cultivated land

Preservation: three-dimensional partial skeletons

⊕ Earth: *c.* 18MA

⊕ Fossil site today

Overlooked by a volcano, the woodlands and floodplains of Rusinga in Africa's Great Rift Valley were frequently showered with ash that preserved many of the plant and animal inhabitants. Unusually for terrestrial deposits, many fossils were preserved in woodland soils close to their original habitats, with nearly complete skeletons of early bovids, rodents, and apes such as *Proconsul*.

The primitive-looking proconsulids are classified as hominoids on account of their relatively large brains, tooth structure, and lack of a tail. Ranging in size between a large monkey and a female gorilla, skeletal analysis suggests they were tree-dwellers, sharing the generalized features of a quadrupedal, monkey-like ancestor. Sexual dimorphism in their canine teeth suggests that they did not live in monogamous social groups but in a larger hierarchical grouping consisting of a dominant male alongside several females and their infants.

Two *Proconsul* species have been found at Rusinga, together with smaller arboreal apes, such as *Dendropithecus*, lorises, flying squirrels, and animals adapted to life among the leaf litter of the forest floor, such as elephant shrews and water chevrotains.

23.03

18 million years ago
Burdigalian Stage of the Miocene Epoch

2.59   0

Common organisms: woodland plants and rodents

1. *Chalicotherium*
2. *Gymnurechinus*
3. *Proconsul*
4. *Paranomalurus*
5. *Masritherium*
6. *Rhynchocyon*
7. *Python*
8. *Gomphotherium*
9. *Dendropithecus*
10. *Hyainailourus*
11. *Dorcatherium*

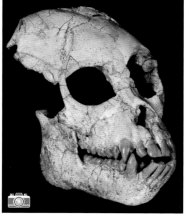

↺ ↻ ↻ PROCONSUL (3)
The best known of the proconsulid primates has four species ranging in weight from 17 to 50kg (37–110lb), and grew up to 1m (40in) long. The brain is similar in size to that of a large monkey, but the ear region is more like that of apes and cercopithecoid monkeys. Limb proportions are monkey-like and quite short relative to the body size – giving an overall a mixture of primitive ape-like and monkey features.

↻ DENDROPITHECUS (9) With long, slender limbs similar to those of a spider monkey, *Dendropithecus* was probably quadrupedal but also swung around by its arms. The 9kg (20lb) animal is known from numerous jaws and teeth that suggest a diet of fruit and leaves. There is striking sexual dimorphism in the teeth, but both sexes have long, sharp canines.

65.5

LIFE ON THE HIMALAYAN FLANKS
SIWALIK HILLS, INDIA, AND PAKISTAN

Climate: warm temperate with seasonal rains

Biota: terrestrial and riverine

Latitude then: 34 degrees South

Latitude now: 34 degrees North

Sea level: +20m (66ft)

Original environment: braided river floodplains

Deposits: bone-rich channel fills and sands

Status: many hillside sites scattered over the Potwar Plateau

Preservation: three-dimensional bones from disarticulated skeletons

• Siwalik Hills

⊕ Earth: *c.* 12MA

• Siwalik Hills

⊕ Fossil site today

First discovered in the 1830s, the Neogene Siwalik formations of northern India and Pakistan are one of the longest and richest sequences of terrestrial vertebrate faunas known. Between 22 and 2 million years ago, sediments from the erosion of the Himalayas were compressed to form the Siwalik Hills, preserving evolving Neogene life that ranges from tiny rodents to early elephants and Asian apes such as *Sivapithecus*.

Between 2 and 5km (1.2–3 miles) thick, the strata now form extensive outcrops over tens of kilometres. Typically, fossils occur as dense aggregates of disarticulated bones, hundreds or even thousands in number, filling small channels within braided river systems.

Plants are very rare, but there are abundant aquatic invertebrates such as snails and clams, bony fish, amphibians, turtles, and crocodiles, along with terrestrial mammals and birds. In a recent analysis of some 40,000 specimens from over 900 sites on the Potwar Plateau, more than 80 per cent are mammals (from 13 separate orders). At any one time there were 50 or more mammal species with some 60 per cent made up of rodents (eg *Eutamius*) and artiodactyls (eg *Hippopotamodon*, *Giraffokeryx*, and *Protragocerus*).

12 million years ago
Serravallian Stage of the Miocene Epoch

Common organisms: rodents and artiodactyls

- ❶ *Hippopotamodon*
- ❷ *Percrocuta*
- ❸ *Platybelodon*
- ❹ *Protragocerus*
- ❺ *Hyainailourus*
- ❻ *Sivapithecus*
- ❼ *Gomphotherium*
- ❽ *Chalicotherium*
- ❾ *Eutamius*
- ❿ *Giraffokeryx*

➲ SIVAPITHECUS JAW (6)
This robust jawbone was built for the attachment of powerful jaw muscles and coping with the stresses imposed by biting on one side. A distinct overbite by the large upper incisors was an adaptation for removing the tough skins of various fruits.

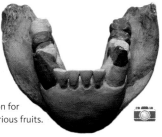

➲ SIVAPITHECUS (6)
With a skull some 15cm (6in) tall and thick enamelled teeth, this pongin hominid was ancestral to the living orang-utans. Like other apes with enamelled teeth, *Sivapithecus* had a fruit-based diet and lived at least partly on the ground – there is little or no evidence for the "suspensory locomotion" seen in orangs and all other living apes.

⬆ GIRAFFOKERYX (10) This okapi-like primitive artiodactyl belonged to the family Giraffidae and ranged across Africa and Eurasia. It is distinguished by two pairs of pointed horn-like ossicones that grew on top of the skull.

OUR EARLIEST HUMAN RELATIVE?
DJURAB DESERT, CHAD

Climate: northern hemisphere subtropical
Biota: terrestrial and lacustrine

Latitude then: 16°N

Latitude now: 16°N

Sea level: +10m (33ft)

Original environment: lakeside gallery forest

Deposits: lakeshore sands and clays

Status: a remote and inaccessible hot desert region with sporadic outcrops covered in shifting sands

Preservation: slightly crushed but otherwise three-dimensional disarticulated bones

⊕ Earth: c.7MA

⊕ Fossil site today

In 2002 the most ancient known hominid, *Sahelanthropus tchadensis*, was named and described. The fossil remains of this metre-sized (40in) creature were found in northern Chad some 2500km (1550 miles) from the East African Rift Valley, where most of our other fossil relatives have been found. Evidently the earliest hominids had a wider distribution than was previously thought.

Sahelanthropus combines human-like features in an otherwise ape-like skull and may be our most ancient unique ancestor, having evolved shortly after the human and ape lineages split (as evidenced by the oldest fossil gorilla, *Chororapithecus*, recently found in 10-million-year-old strata in the Ethiopian section of the Rift Valley).

Around 7 million years ago, Chad's Djurab desert was a lakeside surrounded with gallery forest and savannah grasslands, though the processes of modern desertification had already begun nearby. These environments provided habitats for a rich aquatic biota of fish, crocodiles, and amphibious mammals, alongside terrestrial mammals such as rodents, elephants, equids, bovids, and primates including colobine monkeys and *Sahelanthropus* – some 44 taxa in all.

7 million years ago
Messinian Stage of the Miocene Epoch

Common organisms: fish and bovids

1. *Ictitherium*
2. *Sahelanthropus*
3. *Machairodus*
4. *Anancus*
5. "*Macrotermes*"
6. *Hipparion*
7. *Orycteropus*
8. *Hexaprotodon*
9. *Nyanzachoerus*
10. *Kobus*
11. cercopithecoid
12. *Sivatherium*

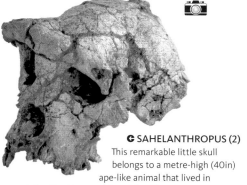

☝ SAHELANTHROPUS (2)
This remarkable little skull belongs to a metre-high (40in) ape-like animal that lived in present-day Chad between 5 and 8 million years ago. The skull combines primitive ape-like features with some more-advanced hominid characters.

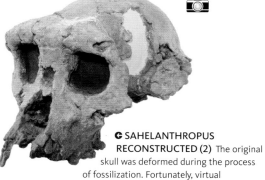

☝ SAHELANTHROPUS RECONSTRUCTED (2) The original skull was deformed during the process of fossilization. Fortunately, virtual reconstruction of the skull can remove the distortion while retaining evidence of hominid characters such as the relatively flat face and very prominent brow ridge.

↻ NYANZACHOERUS (9) The fossil remains of this boar are similar to other close suid relatives from the same evolutionary level from Lothagam in Kenya. Independent dating puts these at 5.2–7.4 million years old, suggesting that they coexisted with *Sahelanthropus*.

BIPEDALISM – A "GIANT STEP" FOR MANKIND
LAETOLI, TANZANIA

Climate: equatorial
Biota: terrestrial

Latitude then: 3°S
Latitude now: 3°S
Sea level: low (-60m/-200ft)
Original environment: dry savannah and woodland
Deposits: volcanic ash
Status: the footprint site is now covered for its protection
Preservation: three-dimensional trace fossils preserved in hardened ash

⊕ Earth: c.3.6MA

⊕ Fossil site today

Up until the 1970s, walking upright on two legs was thought to be a purely modern human attribute. But then came the discovery of some of the most evocative fossils ever found – a set of footprints, some 3.6 million years old, discovered in a layer of hardened ash and mud in a dry riverbed at Laetoli in Tanzania.

The footprints were clearly made by ancient human relatives, capable of walking upright very much as we do today. Yet the "people" who made them were still small, metre-sized (40-in), ape-like beings with small brains. They were probably members of the *Australopithecus afarensis* species, better known as "Lucy's

people" after the famous Ethiopian fossil found a few years previously at Hadar, Ethiopia. The two parallel trackways, apparently in step with one another, have led to all sorts of speculation, such as a male and female walking hand in hand, but this cannot be proven.

The hominid prints are accompanied by those of many other animals, ranging from elephants, rhinoceros, and giraffe, through hyenas, dik-dik, and guinea fowl, down to insects. Freshly fallen ash from a nearby volcano preserved prints from a sample of the local wildlife, and further sediment layers protected them from later erosion.

Common organisms: birds and small mammals

3.6 million years ago
Zanclean Stage of the Pliocene Epoch

● *Madoqua*
● *Numida*
● *Loxodonta*
● *Diceros*
● *Giraffa*
● *Australopithecus afarensis*
● animal tracks
● hominid footprints

↪ AUSTRALOPITHECUS AFARENSIS (6) This fragmentary skeleton is the most complete early hominid found – however, most of the skeleton was found at Hadar – only the jawbone is from Laetoli.

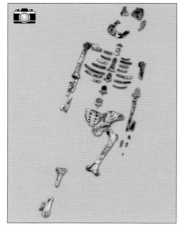

↪ ANIMAL TRACKS (7) The Laetoli footprints not only indicate the presence of hominids but also that of a wide range of animals from elephants, rhinoceros, and giraffes, to hyenas, dik-dik, and guinea fowl, to insects.

↪ FIRST FOOTPRINTS (8) The 70m-long (230ft) fossilized trackway impressed in hardened volcanic ash has been dated radiometrically at 3.6 million years old. The ash preserves the oldest known trace fossil evidence for upright walking by extinct australopithecines – probably *Australopithecus afarensis*.

65.5

THE "CRADLE OF MANKIND"
STERKFONTEIN, SOUTH AFRICA

Climate: hot semi-arid with seasonal rains
Biota: terrestrial vertebrates

Latitude then: 26°S

Latitude now: 26°S

Sea level: +25m (82ft)

Original environment: woodland savannah with little bush cover

Deposits: cave bone deposits

Status: legally protected as a World Heritage Site

Preservation: three-dimensional bones and partial skeletons

⊕ Earth: *c.*2.8MA

⊕ Fossil site today

The discovery in the 1920s of hominid fossils in South Africa provided the first evidence to support Darwin's claim that human ancestry lay in Africa. Before this time fossil finds in the Far East seemed to support an Asian origin for humanity.

The first South African find – a juvenile *Australopithecus africanus* described by Raymond Dart in 1925 – was largely ignored by the international scientific community until Robert Broom discovered more remains in the limestone caves of the Sterkfontein Valley. Broom's adult *A. africanus* skull proved that a group of extinct ape-like beings more primitive than any previously known human relative had evolved in Africa.

Since then, more than 500 hominid fossils, dated at between 4 and 2.8 million years old, have been found in the Sterkfontein caves, with many others in nearby Swartkrans and Kromdraai. They are accompanied by a wide range of large mammals, including plant-eating antelopes and bovids along with carnivores such as hyenas, sabre-tooth cats, lions, and leopards. The bone accumulation in the caves is probably a result of predation and scavenging by cave-dwelling predators – especially hyenas, big cats, and eagles.

2.8 million years ago
Piacenzian Stage of the Pliocene Epoch

Common organisms: bovids

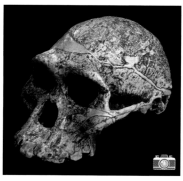

❶ *Protea*
❷ *Megalotragus*
❸ *Panthera pardus*
❹ *Canis mesomelas*
❺ *Australopithecus africanus*
❻ *Stephanoeatus*
❼ *Hippotragus*
❽ *Antidorcas*
❾ *Homotherium*
❿ *Papio*

⤴ AUSTRALOPITHECUS AFRICANUS (5) With its distinctive sloping face, this well-preserved skull was found at Sterkfontein by Robert Broom when he was more than 80 years old. It was nicknamed Mrs Ples, but has since been shown to be male.

⟳ BACKBONES (5)
Robert Broom's discovery of this rare postcranial skeleton, with vertebrae and pelvis intact, proved that despite a very ape-like face, *Australopithecus africanus* was an upright-walking hominid and not an ape. The pelvis is distinctly human-like with wide hips and a bulging abdomen, but overall the proportions of the skeleton are similar to *Australopithecus afarensis*.

⤴ HOMOTHERIUM (9) This extinct big cat, the size of a living male lion, had relatively long front limbs that provided considerable leverage and acceleration for "hot pursuit". Well-developed canines delivered a deadly bite.

TOOLS MAKETH MAN?
OLDUVAI, TANZANIA

Climate: semi-arid tropical cooling into an ice age

Biota: wooded grassland with abundant herbivorous mammals

Latitude then: 3°S

Latitude now: 3°S

Sea level: +25m (82ft)

Original environment: high savannah plains with lakes and rivers

⊕ Earth: c.1.8MA

Deposits: river and lakeshore sediments with volcanic ashes

Status: now within the Ngorongoro conservation area and Serengeti National Park with permits required for fossil collecting

Preservation: mostly subfossil and isolated bone material

⊕ Fossil site today

Around 1.8 million years ago, two small, upright-walking, human-related species lived together in East Africa. One was *Paranthropus boisei*, a specialist plant-eater with muscular jaws. The other was *Homo habilis*, a more omnivorous meat-eater. The highland wooded grasslands, lakes, and rivers of Olduvai provided plenty of food for both hominids, but they in turn were prey to more powerful predators – big cats and packs of hyenas. Many of the early Pleistocene-age animals were recognizably modern, but there were still a few survivors of now-extinct genera, such as the three-toed horse *Hipparion*, and the elephant-like *Deinotherium*.

Fossils were first found at Olduvai in 1913 by the German geologist Hans Reck, but it was not until 1931 that Louis Leakey found crude stone tools in the lowest and oldest Olduvai deposits dated to around 1.8 million years old. The same river and lakeside deposits later produced the beautiful skull of "Nutcracker Man" (*Paranthropus boisei*) in 1959, and in 1960 the first fragments of "Handy Man" (*Homo habilis*). Leakey claimed that despite its small brain, *Homo habilis*, the oldest member of our genus *Homo,* was responsible for the stone tools, and that the East African Rift Valley was the cradle of humankind.

1.8 million years ago
Calabrian Stage of the Pleistocene Epoch

Common organisms: bovids, especially antelopes, and their predators

- ❶ *Paranthropus boisei*
- ❷ *Hipparion*
- ❸ *Panthera leo*
- ❹ *Deinotherium*
- ❺ *Homo habilis*
- ❻ stone tools
- ❼ *Sivatherium*
- ❽ *Pelorovis*
- ❾ *Crocuta*

⤵ **PARANTHROPUS BOISEI (1)** The massive ape-like skull of *Paranthropus boisei* supported a powerful jaw musculature for eating tough fibrous plant material and contained a small 521cc ape-sized brain. Yet these were 1.3m-tall (51in), upright-walking social hominids whose species survived for a million years.

⤴ **HOMO HABILIS (5)** First described from Olduvai, this small (1.3m/51in tall) hominid is sometimes placed in the genus *Australopithecus* because of its small brain. It was probably a gatherer that scavenged kills of other predators.

⤴ **OLDOWAN STONE TOOLS (6)** Found in the lower layers of strata, these primitive pebble tools are now known to be between 1.9 and 1.6 million years old.

EUROPEAN HUMAN ANCESTORS?
GRAN DOLINA, ATAPUERCA, SPAIN

Climate: northern hemisphere warm temperate
Biota: terrestrial

Latitude then: 42°N

Latitude now: 42°N

Sea level: +25m (82ft)

Original environment: karst limestone

Deposits: karst limestone

Status: legally protected World Heritage Site

Preservation: disarticulated but well-preserved three-dimensional bone material

⊕ Earth: *c.*780,000 BP

⊕ Fossil site today

Recent finds from western Europe have complicated our view of human evolution in mid-Pleistocene Ice Age times. Dated at around 780,000 years old, fossils from a limestone fissure at Gran Dolina have provided several fragments of a juvenile skull and bones from the skeletons of several adults. Some of the bones have cut marks made by stone tools, and may indicate that cannibalism was practised by these human relatives.

The Spanish excavators regard them as a new species, *Homo antecessor*, but not all experts agree. The remains resemble the similarly aged skull from Ceprano in Italy, and archaic features of the Spanish skull, such as a prominent double-arched bony brow ridge, connect it to *Homo erectus*. But there are also some more modern features of the teeth that show a link to *Homo heidelbergensis*, ancestor of modern humans. Whatever it is called, the Gran Dolina species may represent a population that was part of our direct ancestry.

What is more, even more recent finds of crude stone tools, dated at around 700,000 years old, from Pakefield in East Anglia, England, reinforce the new view that pre-*heidelbergensis* populations were much more extensive in Europe than previously thought.

780,000 years ago
Ionian Stage of the Pleistocene Epoch

Common organisms: rodents and bovids

1. *Homo antecessor*
2. *"Equus altidens"*
3. *Bison voigtstedtensis*
4. tools
5. *Stephanorhinus etruscus*
6. *Ursus*
7. *Marmota*
8. *Lynx*
9. *Eucladoceros giulii*
10. *Sus scrofa*
11. *Vulpes praeglacialis*
12. *Dama*

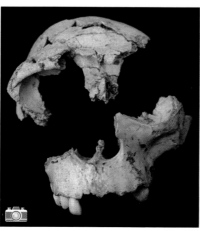

⊙ HOMO ANTECESSOR (1) Spanish investigators claim that this 780,000-year-old partial skull represents a new species that was part of the evolving lineage of *Homo sapiens*, but others claim that it shows features that ally it with another, previously described ancestor of modern humans, *Homo heidelbergensis*.

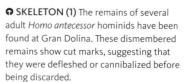

⊙ SKELETON (1) The remains of several adult *Homo antecessor* hominids have been found at Gran Dolina. These dismembered remains show cut marks, suggesting that they were defleshed or cannibalized before being discarded.

⊙ TOOLS (4) Nearly 200 primitive pebble tools have so far been found alongside the remains of *Homo antecessor*. Like the stone tools found at Olduvai, they were made by hitting one pebble with another, producing crude choppers and sharp flakes.

65.5

THE MYSTERY OF ASIAN ORIGINS
ZHOUKOUDIAN, BEIJING, CHINA

Climate: cold dry
Biota: terrestrial vertebrates

Latitude then: 39°N

Latitude now: 39°N

Sea level: +100m (330ft)

Original environment: limestone cave

Deposits: cave floor sediments

Status: designated a World Heritage Site in 1987

Preservation: three-dimensional but disarticulated skeletons

⊕ Earth: c.540,000 BP

⊕ Fossil site today

The discovery of "Peking Man" (originally *Sinanthropus pekinensis* but now redefined as *Homo erectus*) was an international success story of the late 1920s. For nearly a decade, an international team carried out extensive excavations of cave deposits at "Dragon Bone Hill" (Longgushan) above the Zhoukou river. It was the largest excavation of a hominid site yet undertaken, and provided the largest cache of early hominids known. There were primitive stone tools (around 17,000), animal bones, and the remains of some 40 individual hominids – mostly skulls. Despite missing facial bones, these were spectacularly well preserved, and

supported biologist Ernst Haeckel's idea that human origins lay in the Far East.

Tragically, the vast bulk of this fossil treasure was lost en route to America during World War II – but fortunately some well-made casts survived. Further excavations carried out since have provided new finds, dating the site at between 0.46 and 0.23 million years ago.

Modern interpretations see the cave site as providing shelter, access to water, and a vantage point for hunters to watch for prey. But the *Homo erectus* people were by no means always the predators – sometimes they in turn fell victim to big cats and hyenas.

670,000–410,000 years ago
Ionian Stage of the Pleistocene Epoch

Common organisms: sika deer (*Pseudaxis*)

❶ tools
❷ *Homo erectus*
❸ *Pseudaxis*
❹ *Megantereon*
❺ *Megaloceros*
❻ *Macaca*
❼ *Canis*
❽ *Celtis*
❾ *Pachycrocuta*

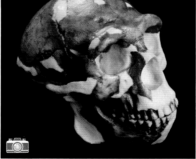

➲ **TOOLS (1)** Stone tools left by successive hominids include flakes, rough choppers, and hammer stones, but no spear heads, or handaxes.

➲ **HOMO ERECTUS (2)** From above, the distinctive visor-like shape of the prominent bony brow ridge, typical of australopithecines and early *Homo*, can be seen. This was thought to provide strengthening to the skull or act as a shading visor, but neither explanation can be proven.

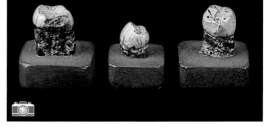

➲ **TEETH (2)** Molar teeth like these were the first hominid remains found at Zhoukoudian – in 1921 by the Austrian paleontologist Otto Zdansky. From the start he recognized that their low, flattened crowns were more human than ape-like. They proved that a hominid had lived at Zhoukoudian alongside the extinct Ice Age animals.

225

THE FIRST HUMAN ART
BLOMBOS CAVE, SOUTH AFRICA

Climate: southern warm temperate

Biota: terrestrial and marine

Latitude then: 34°S

Latitude now: 34°S

Sea level: -75m (-245ft)

Original environment: coastal limestone cliff and cave

Deposits: cave floor sediments

Status: the cave site is on private land

Preservation: three-dimensional artefacts with disarticulated bones

Blombos Cave •

⊕ Earth: *c.*75,000 BP

Blombos Cave •

⊕ Fossil site today

A tiny red ochre crayon from Blombos, just 4cm (1.6in) long and scratched in a geometric pattern, is the oldest known "artwork". Some 75,000 years old, it pre-dates most prehistoric art by 40,000 years, and is the first record of abstract graphic activity by humans. What the pattern represents is unknown, but it is certainly not accidental, and was found alongside ornaments (60 perforated shells of *Nassarius kraussianus*), and bone and stone tools of a type made only by *Homo sapiens*. The 400 small pointed stone tools and flakes suggest that Blombos was a "workshop" using stone from a quarry site 30km (18 miles) away.

Blombos has produced dense shellfish middens (*Donax serra*), along with more than a thousand bones of fish, Cape fur seals, and dolphins. Large hearths and terrestrial animal remains show that a wide range of resources were being exploited, especially dune molerats, hyraxes, eland, and small bovids.

The cave was apparently only occupied for brief periods, during which animal prey was processed and eaten, and red ochre was ground into powder, perhaps for use in personal decoration. Elsewhere in South Africa (Pinnacle Point), ochre use dates back to 164,000 years ago.

75,000 years ago
Upper Stage of the Pleistocene Epoch

Common organisms: marine shellfish and small mammals

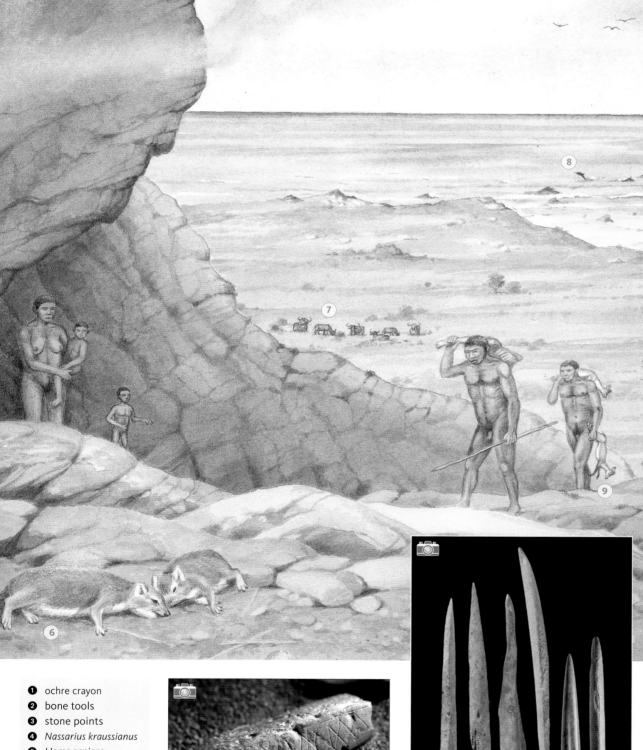

1. ochre crayon
2. bone tools
3. stone points
4. *Nassarius kraussianus*
5. *Homo sapiens*
6. *Procavia*
7. *Syncerus*
8. delphinid
9. *Raphicercus*
10. *Lepus*
11. *Antidorcas*
12. *Taurotragus*
13. *Arctocephalus pusillus*

OCHRE "CRAYON" (1) The surfaces of this 4cm-long (1.6in), crayon-shaped piece of naturally occurring ochre have been ground flat to produce red powder and later scratched with lozenge-shaped patterns – the oldest known consciously created design.

BONE TOOLS (2) Some 20 hand-worked bone tools, many shaped as points, have been found in the hoard of 75,000-year-old artefacts at Blombos. The remains reveal significant behavioural changes and cultural innovations among the humans of the time.

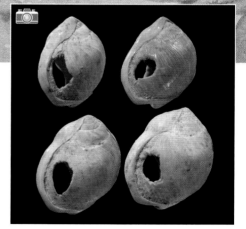

NASSARIUS (4) Some 60 shells of this small marine gastropod were found alongside other human artefacts at the cave. The shells were perforated, and presumably were originally strung together as some kind of personal ornament – the oldest known.

227

65.5

NEANDERTHAL LIFE
FORBES' QUARRY, GIBRALTAR

Climate: Mediterranean
Biota: terrestrial and marine

Latitude then: 36°N

Latitude now: 36°N

Sea level: -75m (245ft)

Original environment: coastal limestone cliff and cave

Deposits: cave floor sediments

Status: some of the caves are only accessible from the sea and are now protected sites

Preservation: three-dimensional disarticulated bones

Forbes' Quarry

⊕ Earth: c.50,000 BP

Forbes' Quarry

⊕ Fossil site today

Remains found in Germany's Neander valley in 1856 sparked scientific recognition of the first known extinct human species – *Homo neanderthalensis*. This led to the reassessment of a female skull found at Forbes' Quarry on Gibraltar in 1848, which also proved to be a remnant of the extinct Neanderthal people.

The Neanderthals existed between about 400,000 and 28,000 years ago, extending across much of western Eurasia, from Wales to the Urals and from Germany to Gibraltar. They overlapped with incoming modern humans for some 10,000 years before becoming extinct. Powerfully built, meat-eating hunters, the

Neanderthals had brains of similar size to *Homo sapiens*, but possibly with a somewhat different internal structure.

In 1926, the skull bones of a Neanderthal child were found alongside animal bones, stone tools, and charcoal, at Devil's Tower in Gibraltar, and dated to around 40,000 years old. Reconstruction of the large-brained skull and dental studies show that the child had suffered a broken jaw and was about 4 years old. General tooth wear and cut marks show that Neanderthals used their teeth as a vice when scraping fat from animals skins, and were predominantly right handed.

23.03   2.59   0

50,000 years ago
Upper Stage of the Pleistocene Epoch

Common organisms: marine shellfish and small mammals

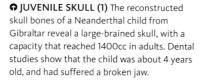

- ❶ *Homo neanderthalensis*
- ❷ stone tools
- ❸ *"Oryctolagus"*
- ❹ *Testudo graeca*
- ❺ *Mytilus*
- ❻ *Capra ibex*
- ❼ *"Haliaeetus"*
- ❽ *Rhinoceros*
- ❾ *Elephas*

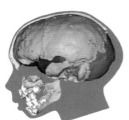
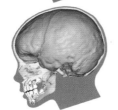

↪ SKULL COMPARISON (1) Skull models of a Neanderthal (*top*) and a modern human child of the same age demonstrate the essential differences in skull and jaw shape.

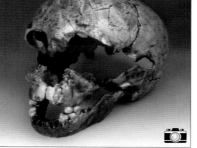

⊙ JUVENILE SKULL (1) The reconstructed skull bones of a Neanderthal child from Gibraltar reveal a large-brained skull, with a capacity that reached 1400cc in adults. Dental studies show that the child was about 4 years old, and had suffered a broken jaw.

⊙ ADULT SKULL (1) Found in 1848 on Gibraltar, the first complete Neanderthal skull to be discovered was not recognized until later. A female, it shows a typically large projecting face, receding cheeks, prominent brow ridge, and large braincase.

⊙ TOOL (2) Stone blades were thought to have been solely the work of modern humans, but it is now clear that Neanderthals also manufactured them.

FIRST AUSTRALIANS
WILLANDRA LAKES, NEW SOUTH WALES, AUSTRALIA

Climate: moist subtropical to arid
Biota: freshwater aquatic and terrestrial vertebrates

Latitude then: 33°S

Latitude now: 33°S

Sea level: -75m (-245ft)

Original environment: lakeside

Deposits: barrier sands

Status: protected within Willandra Lakes World Heritage Site and Mungo National Park

Preservation: three-dimensional articulated human skeleton and cremated remains

☉ Earth: *c.*40,000 BP

☉ Fossil site today

Some 40,000 years ago, a body was carefully buried in a shallow pit beside a lake and covered in red ochre. Not far away, the body of a young woman was burned before her skeletal remains were collected, broken into pieces, interred in another pit, and covered with ash. This oldest known cremation was, until recently, thought to be significantly younger than the ochre burial, but the two interments are now considered contemporaneous.

The modern humans who carried out these ritual burials had already been in Australia for some 6000–10,000 years, according to the dating of stone tools found nearby. The site also demonstrates evidence for climate change at the time, and the interaction of early humans with landscape and biota. As a result, very few of the large and medium-sized animals survived beyond 40,000 years ago.

Genetic and anatomical evidence shows that the ancestors of these early Australians, like those of all modern humans, originated in Africa and first dispersed some 100,000 years ago. Those that got to Australia probably took a coastal route to Southeast Asia and across the Indonesian archipelago when sea levels were significantly lower. Even so, the crossing to Australia would have required boats.

Common organisms: shellfish

- ❶ fish bones
- ❷ *Velesunio*
- ❸ *Homo sapiens*
- ❹ *Macropus*

↩ **FISH BONES (1)** Discarded remains show that the first Australians used the abundant resources of fish and shellfish (such as *Velesunio*) from the Willandra Lakes as food until encroaching sand dunes reduced the lake waters. By about 14,000 years ago, the lakes were entirely dry.

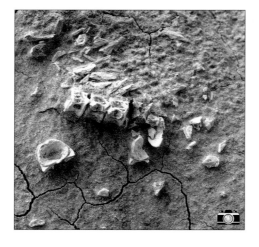

↥ **FIRST AUSTRALIANS (3)** The cranial remains of "Mungo man" reveal a modern "gracile" human, with lightly built thin vault bones, a well-rounded forehead, weak or moderate brow ridge, and relatively small mandibles, teeth, and palate.

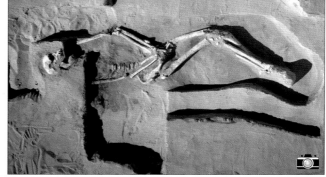

↻ **BURIAL (3)** Found in 1974, the fully articulated skeleton known as Mungo III was carefully buried in a shallow pit beside a lake and covered in red ochre between 43,000 and 41,000 years ago. The burial was discovered when it was exposed again by shifting sand dunes in 1974.

231

65.5

FROZEN IN TIME – RUSSIA'S ICE AGE MEGAFAUNA
MAGADAN, SIBERIA, RUSSIA

Climate: subarctic
Biota: terrestrial vertebrates

Latitude then: 63°N

Latitude now: 63°N

Sea level: -75m (-245ft)

Original environment: tundra permafrost

Deposits: organic muds and silts

Status: the original site has been removed by gold mining activities but there are many similar sites across Siberia that occasionally yield permafrost "mummies"

Preservation: frozen cadaver with soft tissue preservation

⊕ Earth: c.40,000 BP

⊕ Fossil site today

In 1977, gold miners working a placer deposit in the Siberian permafrost discovered the frozen, immaculate remains of an emaciated baby mammoth. Dated to around 40,000 years old, Dima, as it was nicknamed, is one of the best preserved of several frozen "mummies" left over from the Ice Age. The intact body has not been scavenged, or suffered decomposition.

Just over a metre (40in) tall, Dima was a male, and around 7 or 8 months old according to its tooth development. The stomach contained lots of clay, silt, and some fresh-looking plant material, with seed and pollen indicating that it died in summer.

It seems mostly likely that the baby mammoth was tempted to its doom by succulent plants growing around one of the many treacherous pools that appeared as the surface of the permafrost melted in summer. Such pools, like their modern equivalents, would have had vertical and slippery mud sides, making it impossible for a small mammoth to climb out. Although mammoths could swim, the baby would have suffered from hypothermia in the near-freezing water and eventually drowned. The water prevented decay and froze Dima's remains as the harsh Siberian winter set in once again.

40,000 years ago
Upper Stage of the Pleistocene Epoch

Common organisms: large mammalian herbivores

❶ *Coelodonta antiquitatis*
❷ *Equus caballus*
❸ *Canis lupus*
❹ *Mammuthus primigenius*
❺ *Bison priscus*

◆ **"DIMA" (4)** The complete freeze-dried remains of the baby mammoth, nicknamed "Dima" were recovered from terrace deposits of the Kirgiliakh river, a tributary of the Kolyma, north of Magadan in the far northeast of the Russian Republic. The immaculate relic retained traces of hair and stomach contents.

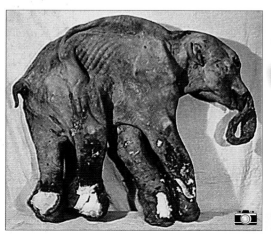

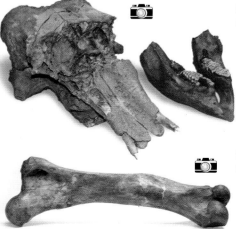

◆ **JUVENILE SKULL (4)** Over many centuries, numerous remains of mammoths and other Ice Age animals such as woolly rhinos and horses have been recovered from the permafrost. Juveniles such as this mammoth were particularly vulnerable to the hazards of the environment.

◆ **LEG BONE (4)** Large, elephant-like leg bones had been found scattered in superficial deposits across Europe and Asia for centuries, and there was much speculation over their origin. Traditionally they were seen as the remains of animals drowned in Noah's flood, but by the mid-19th century it was clear that they belonged to extinct Ice Age mammoths and mastodons.

233

MIRED IN TAR – AMERICA'S ICE AGE WILDLIFE
RANCHO LA BREA, LOS ANGELES, USA

Climate: mid-latitude glacial
Biota: terrestrial with some freshwater aquatics

Latitude then: 34°N

Latitude now: 34°N

Sea level: -75m (-245ft)

Original environment: coastal plain with tar seeps

Deposits: tar

Status: designated a National Natural Landmark since 1963

Preservation: three-dimensional, with many articulated skeletons

⬆ Earth: c.20,000 BP

⬆ Fossil site today

Among the most iconic animals of the Ice Ages are sabre-tooth cats such as *Smilodon*. Some 166,000 bones of this large ambush hunter have been recovered from the La Brea tar seeps of Los Angeles. Fossils from these deposits record the abundance of life between 40,000 and 9,000 years ago, before humans and climate change took their toll.

Plant-eaters became mired in the sticky tar during short-lived periods of warm weather, but on average, it seems, each trapped herbivore attracted and trapped a predatory sabre-tooth, a scavenging coyote, four dire wolves, and numerous raptors.

In total, more than 600 species are known from a million or so bones. Of the 135 North American mammal genera known from this time, 58 are represented, including small rodents, insectivores, bats, hares etc. The missing genera tend to be cold-adapted forms such as moose, musk oxen, caribou, and wolverines.

There are also 138 different birds, 24 reptiles, 6 amphibians, 3 fish, and 56 mollusc species, along with 168 arthropods and 80 plant taxa represented by pollen, seeds, leaves, or wood. These plentiful finds have allowed the life history of many species to be reconstructed for the first time.

20,000 years ago
Upper Stage of the Pleistocene Epoch

Common organisms: rodents, birds, and dire wolves

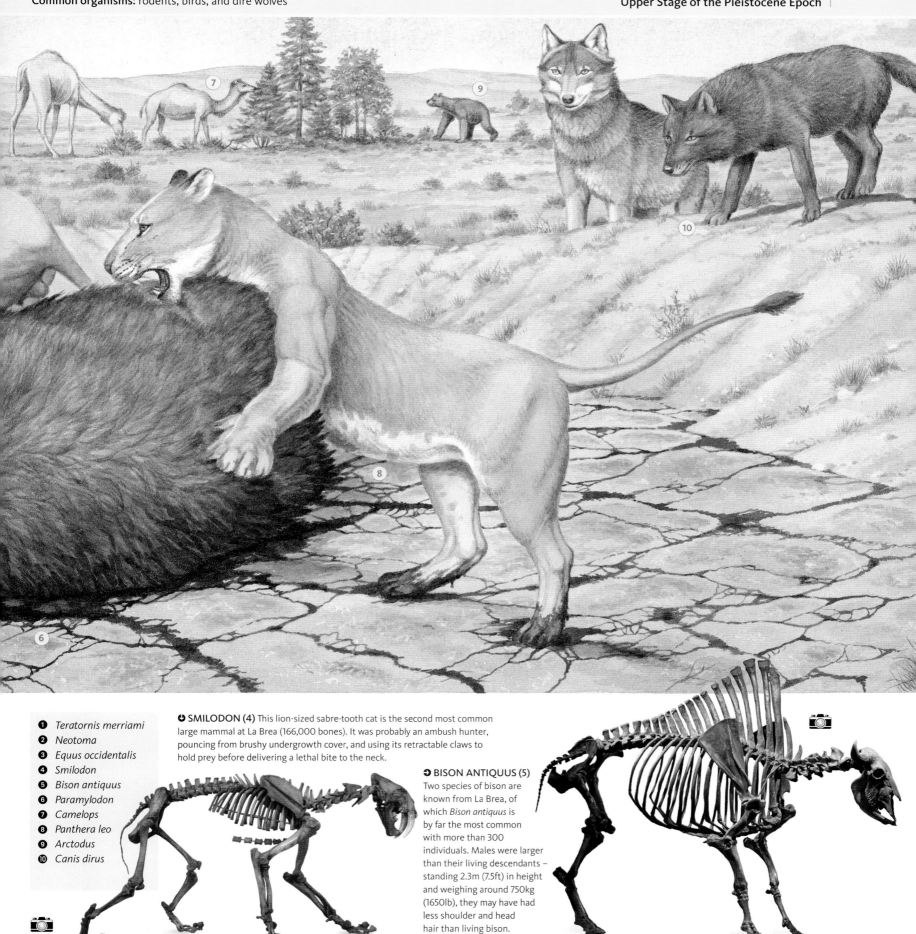

❶ *Teratornis merriami*
❷ *Neotoma*
❸ *Equus occidentalis*
❹ *Smilodon*
❺ *Bison antiquus*
❻ *Paramylodon*
❼ *Camelops*
❽ *Panthera leo*
❾ *Arctodus*
❿ *Canis dirus*

⟳ SMILODON (4) This lion-sized sabre-tooth cat is the second most common large mammal at La Brea (166,000 bones). It was probably an ambush hunter, pouncing from brushy undergrowth cover, and using its retractable claws to hold prey before delivering a lethal bite to the neck.

➲ BISON ANTIQUUS (5)
Two species of bison are known from La Brea, of which *Bison antiquus* is by far the most common with more than 300 individuals. Males were larger than their living descendants – standing 2.3m (7.5ft) in height and weighing around 750kg (1650lb), they may have had less shoulder and head hair than living bison.

PRISTINE CAVE ART FROM THE MEDITERRANEAN
COSQUER CAVE, MARSEILLES, FRANCE

Climate: glacial mid-latitude cool temperate

Biota: coastal marine and terrestrial

Latitude then: 43°N

Latitude now: 43°N

Sea level: -110–130m
(-360–425 ft)

Original environment:
coastal cliff cave with steppe
grasslands

Deposits: limestone

Status: inaccessible and legally
protected submarine cave

Preservation: organic and mineral
pigments almost perfectly
preserved

Cosquer Cave

⊕ Earth: c.18,500 BP

Cosquer Cave

⊕ Fossil site today

In 1985, French diver Henri Cosquer discovered a submerged cave opening on the coast near Marseilles, some 40m (130ft) below present sea level. At the end of a 160m (525ft) tunnel and safely above water lay a cavern with walls covered in a magnificent display of cave art, illustrating the region's late Ice Age fauna.

At the height of the last Ice Age when sea levels were 110–130m (360–425ft) lower, modern human hunters sheltered by the cave entrance and observed the game on the 15km-wide (9 mile) coastal plain below. Occasional brief visits to the cave's deep and dark recesses were made to record the life around them – the animals upon which they depended for food and raw materials. Images range from extinct giant deer and great auk to horses, ibex, and wild cattle, along with human handprints.

Dates gleaned from charcoal lying on the cave floor and from the drawings themselves suggest two periods of activity – an older one around 27,000 years ago, for the handprints and some animal art, and a younger one of 18,500 years ago for most of the animals. We may never know why these images were made, but we are fortunate that they were preserved intact as the Ice Age ended and rising sea levels sealed off the cave entrance.

18,500 years ago
Upper Stage of the Pleistocene Epoch

Common organisms: birds and bovids

❶ cave adornments
❷ *Capra*
❸ *Equus*
❹ *Megaloceros*
❺ *Bison bonasus*
❻ *Sterna paradisaea*
❼ *Homo sapiens*
❽ *Pinguinus impennis*
❾ *Monachus*

➲ **HANDPRINTS (1)** A universal image of Paleolithic times is the human handprint or silhouette, found in rock paintings the world over.

⟳ **IBEX (2)** Related to domestic goats, the metre-tall (40in) ibex was once widespread in rocky regions across Europe and North Africa. The distinctive long curved horns of the male appear in paintings from several locales where the animals are no longer found.

⟳ **GREAT AUK *PINGUINUS* (8)** This extinct 80cm-tall (32in) flightless seabird is one of the few birds illustrated in cave paintings. Most closely related to the living razorbill, it was an excellent swimmer but vulnerable to human predation, finally becoming extinct in the mid-19th century.

65.5

A DWARF HUMAN SPECIES?
FLORES, INDONESIA

Climate: humid tropical
Biota: terrestrial forest dwellers

Latitude then: 8°S

Latitude now: 8°S

Sea level: -110–130m (-360–425 ft)

Original environment: limestone caves

Deposits: cave floor deposits

Status: the cave is still being excavated

Preservation: three-dimensional preservation of bone

⊕ Earth: *c.*16,000 BP

⊕ Fossil site today

One of the hottest debates in human evolution today concerns the status of the metre-high (40in) people who lived in and around the Ling-bua caves on the Indonesian island of Flores – between 18,000 and 15,000 years ago, and perhaps as far back as 95,000 years ago.

Discovered in 2003, and classified as a new species *Homo floresiensis*, the remains seem to be a uniquely dwarfed human relative, but there are claims that they are actually *Homo sapiens* with a genetic or environmental pathology. Dwarfing by isolation on an island is a common phenomenon in other animals – as seen in the Flores *Stegodon*.

Remains of a dozen individuals have been excavated from different levels, alongside primitive stone tools and the butchered remains of extinct animals, some of which have been burned. Analysis of the skull structure indicates similarities to *Homo erectus*, such as the lack of a chin, but the degree of encephalization is closer to australopithecines, as are the shoulder and wrist bones.

If this is a new human species, then the defining brain size for the genus *Homo* (currently 650cc) will have to be lowered, raising the question of how these small-brained beings could have functioned socially.

23.03 2.59 0

18,000–15,000 years ago
Upper Stage of the Pleistocene Epoch

Common organisms: dwarf *Stegodon*, giant rats

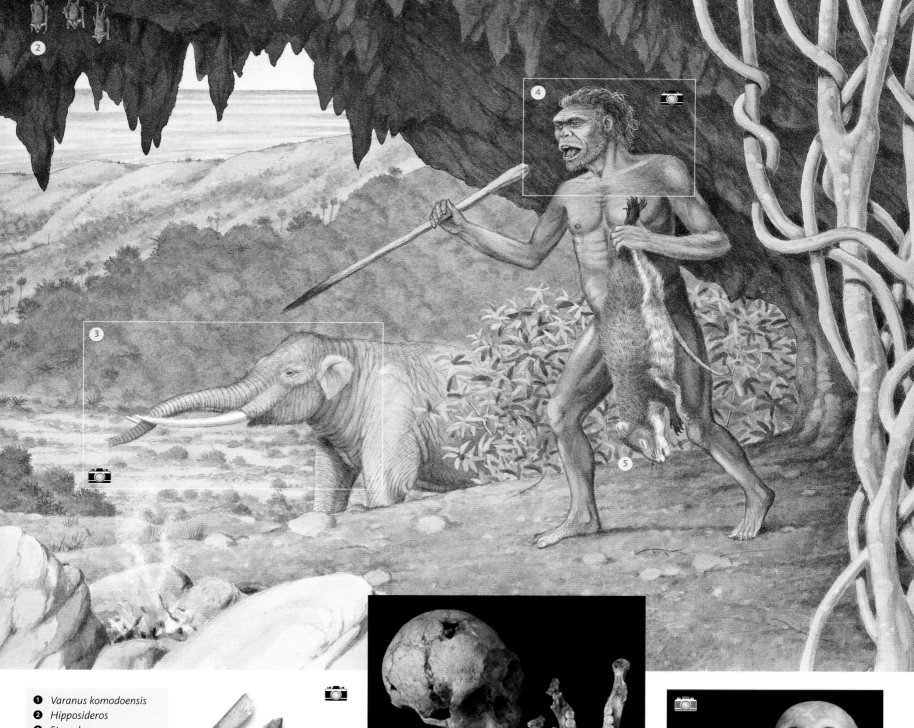

❶ *Varanus komodoensis*
❷ *Hipposideros*
❸ *Stegodon*
❹ *Homo floresiensis*
❺ *Papagomys armandvillei*

➲ **STEGODON (3)** The recovery of teeth from juvenile *Stegodon* within the cave deposits suggests that the *Homo floresiensis* people hunted these extinct pygmy elephants for their meat and skins. However, their stone tools seem to have been used for working fibrous plant material.

⟳ **SKULL AND JAWS (4)** Two similar jawbones have now been recovered from different levels within the cave floor excavation, showing that the population of these diminutive people was large enough to persist for thousands of years. Their survival may have been aided by the island's isolation even in times of low sea level.

⟳ **SKULL COMPARISON (4)** There is a marked contrast between the small skull of a metre-tall (40in) female *Homo floresiensis* whose 380–417cc brain was similar in size to that of a chimp, and a modern human skull.

FIRST AMERICANS
MONTE VERDE, CHILE

Climate: subpolar humid
Biota: terrestrial

Latitude then: 41°S

Latitude now: 41°S

Sea level: -100m (-330ft)

Original environment: riverside

Deposits: fluviatile sands, muds, and gravels

Status: submitted in 2004 for World Heritage Site status and protection

Preservation: some organic materials preserved in waterlogged muds

↻ Earth: c.14,500BP

↻ Fossil site today

Some 14,500 years ago, an open-air settlement on the sandy banks of a stream in southern Chile was home to some of the earliest known Americans. Known as Monte Verde, the site is situated on the wide coastal plain between the mountains of the Andes and the island-studded Pacific Ocean. Here, a group of some 20–30 modern *Homo sapiens* built long, hide-covered, tent-like structures on wooden foundations. Closely associated hearths and pits contain food remains and tools, indicating that Monte Verde was a significant residential site. The big question is when did these "first Americans" arrive?

The only possible land entry is from Siberia via Alaska – a long way from Monte Verde. The present Bering Strait between North America and Asia is shallow, and formed a dry "land bridge" called Beringia during the lower sea levels of glacial periods. Beringia acted as a freeway for movement between continents, but for most of the time it was open the high latitudes of North America were blanketed in ice, which would have barred southward migration until around 12,000 years ago – after Monte Verde. So did the first Americans arrive at the start of the last glacial, before 22,000 years ago, or did they use some other route?

14,500 years ago
Upper Stage of the Pleistocene Epoch

Common organisms: plants, camels, mastodon

❶ cooking hearths
❷ *Homo sapiens*
❸ skin shelters
❹ *Vultur*
❺ tools/weapons
❻ *Lama*
❼ *Mammut americanum*

↻ ☊ **COOKING HEARTHS (1)** A U-shaped structure at the west end of the site had a compacted sand and gravel base. Timber posts were placed every 0.5m (20in) along each side, supporting a skin roof cover. A platform, 3m (10ft) wide and 4m (13ft) long, protruded at the base and turned it into a Y shape overall. The platform's open front was clear apart from evidence of small hearths.

↻ **SKIN SHELTERS (3)** Archeological remains show the presence of two types of wooden structure at Monte Verde. One had a rectangular wood base of stakes and rough-cut planks making a floor some 3–4.5m (10–15ft) long, with vertical posts a metre (40in) apart. Traces of animal fur suggest that skins were used to make the walls, and these were attached by ropes and knotted rushes to wooden pegs.

↻ **TOOLS (5)** Excavation also yielded two bone lance points, similar to projectile tips found in Venezuela, and a wooden lance point. These people were hunter-gatherers who killed mastodon and llamas and gathered a wide variety of plants and marine algae.

241

65.5

THE BIG-GAME HUNTERS OF FOLSOM
FOLSOM, NEW MEXICO, USA

Climate: Younger Dryas millennium-long cold snap
Biota: terrestrial vertebrates

Latitude then: 36°N

Latitude now: 36°N

Sea level: -75m (-245ft)

Original environment:
prairie steppe grasslands

Deposits: bone-rich clays and silts

Status: National Historic Landmark
and New Mexico State Monument

Preservation: three-dimensional
bones with butchered skeletal
remains

⊕ **Earth:** c.10,500 BP

⊕ Fossil site today

Folsom is one of the most important sites in the Americas because it provided the first convincing evidence that native Americans had occupied the continent and coexisted with extinct animal species around the end of the last Ice Age, some 12,600 years ago.

Excavations in the late 1920s uncovered a finely made fluted arrowhead – what is now called a "Folsom point" – lodged between the ribs of an extinct species of bison. This discovery ended one of the most bitter disputes in American archaeology and, although its precise age was not known at the time, proved that American prehistory

began as far back as late Pleistocene times and that the Paleoindians had actively hunted Ice Age animals.

Dozens of other similar kill sites have since been found across the Great Plains, firmly establishing the Folsom and slightly older Clovis cultures. But much of the science of the Folsom site was inadequately known until re-investigation began in the 1970s and again in the late 1990s. It is now understood as a bison kill site where 32 large Ice Age *Bison antiquus* were trapped, killed, and butchered by a small group of mobile hunters, using pointed stone spears and tools.

Common organisms: bison

10,500 years ago
Holocene Epoch

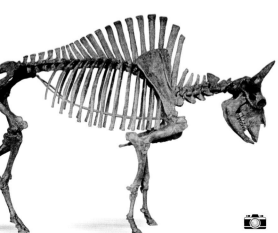

⊃ BISON ANTIQUUS (2)
The extinct late Ice Age *Bison antiquus* was some 20 per cent larger than its surviving descendant, standing over 2m (6.6ft) high and 3.5m (11.5ft) long. At Folsom, 32 carcasses were dismembered into the most valuable and transportable packages – ribs, vertebrae, and upper limbs – leaving just the lower limbs, crania, and jaws (although not until the animals' tongues had been removed).

☾ FOLSOM POINT (3) The 1926 discovery of a fluted chert projectile point lodged between the ribs of an extinct species of bison finally proved that the first Americans were mobile "big-game" hunters who exploited the last of the Ice Age megafauna in the Americas and probably hastened their extinction.

⊃ FOLSOM SPEAR (3) 28 projectile points were finally recovered at the Folsom excavations. Originally, the points were made with their flutes fitted and bound into socketed shafts, with the point and shaft anchored to a wooden spear.

❶ *Homo sapiens*
❷ *Bison antiquus*
❸ folsom point/spear

243

LAST OF THE ICE AGE MEGAFAUNA
PAPATOWAI, OTAGO, NEW ZEALAND

Climate: cool temperate
Biota: terrestrial and marine

Latitude then: 47°S
Latitude now: 47°S
Sea level: as present
Original environment: coastal dunes
Deposits: sand, shell, and bone material
Status: numerous sites, some partly or wholly excavated
Preservation: bone and shell material well preserved

⊕ Earth: c.700 BP

Papatowai

⊕ Fossil site today

When humans arrived on New Zealand's shores around a thousand years ago, they found flourishing wildlife that was totally different from their tropical Pacific island homelands.

New Zealand was part of the Gondwanan "ark" that separated from Australia some 80 million years ago. It had no non-flying terrestrial mammals, and no significant ground predators. The biggest animals were the large flightless birds generally known as moas. They had evolved a diversity of taxa adapted to habitats ranging from forest and open scrub to grassland and included animals that grew up to 2.5m (8ft) tall.

The incoming hunters soon decimated the incumbent wildlife, and brought with them placental intruders – rats and dogs that further reduced the smaller native species.

Archaeological evidence from the estuarine coastal site of Papatowai preserves cultural refuse that reveals the scale of initial hunting of both seals and moas – at least 7000 moa seem to have been taken over several decades, and these resources were soon depleted, so that even intermittent moa hunting had to move inland. With a rate of consumption of some 10 moas per person per year, the last of the megafauna was soon obliterated.

NEOGENE

QUATERNARY

23.03

2.59   0

700 years ago
Holocene Epoch

Common organisms: flightless birds, marine shellfish, and marine mammals

❶ *Petroica*
❷ *Apteryx*
❸ *Dinornis giganteus*
❹ *Euryapteryx*
❺ *Arctocephalus forsteri*
❻ *Homo sapiens*
❼ *Emeus*

↪ MOA HEAD (3) Mummified remains of moas have been found at different sites in New Zealand, including a head of *Megalapteryx didinis* from Cromwell, Otago. This relatively small species had legs only 50cm (20in) long.

↻ MOA SKELETON RECONSTRUCTED (3) The skeleton of *Dinornis giganteus* stands some 2.5m (8ft) tall – when alive, the animal would have weighed some 230kg (510lb). Its fossil distribution indicates a greatest diversity in open forest scrubland and grassland.

↪ FLIGHTLESS BIRD (3) British anatomist Richard Owen examined the first moa bone in 1834 and published his description of the moa as an extinct flightless bird in 1840. He went on to describe many of their different species including this huge *Dinornis giganteus*.

245

65.5

A GEOLOGICAL REVELATION
GRAND CANYON, ARIZONA, USA

Climate: from polar to equatorial
Biota: varied

Latitude then: *c.*30°S

Latitude now: 36°N

Sea level: variable

Original environment: from marine to terrestrial

Deposits: predominantly carbonates and sandstones

Status: National Park (since 1919) and World Heritage Site (since 1979)

Preservation: various, but mostly disarticulated shells

⊕ Earth: present day

⊕ Fossil site today

Clearly visible from space, the Grand Canyon's magnificent breadth, depth, and length are awe inspiring. The gorge is nearly 448km (278 miles) long and has been excavated by the erosive power of the Colorado River over perhaps the past 17 million years as the Colorado Plateau has lifted up. The gorge is up to 1829m (6000ft) deep from rim to floor and between 0.5 and 15km (0.3–9 miles) wide.

More importantly to us, the Canyon reveals a section through some 1800 million years of Earth history, from the metamorphic, late Paleoproterozoic Vishnu rocks, up through a predominantly Paleozoic sedimentary sequence to late Cenozoic times. Like all long sequences of strata, these rocks actually record a complex history of tectonic movement, erosion, and deposition, with more gap than record. Very little of the Precambrian is preserved, there are many breaks in the Paleozoic, and all the Mesozoic and much of the Cenozoic record is missing.

But, even so, the Grand Canyon is one of Earth's most majestic geological sights. As civil war veteran Major John Wesley Powell said when he first navigated the length of the Canyon in 1869, it is like "a Book of Revelations in the rock-leaved Bible of geology".

1.8 billion years ago – present day
Paleoproterozoic Era – Present

Common organisms: from shellfish to modern humans

1. Quaternary surface with sloth fossils
2. Kaibab Limestone (mid Permian)
3. Coconino Sandstone (mid Permian)
4. Supai Group (Pennsylvanian-early Permian)
5. Redwall Limestone (Mississippian)
6. Temple Butte Limestone (Devonian)
7. Tonto Group (Cambrian)
8. Precambrian metamorphics and strata with microfossils

➲ **MICROFOSSILS (8)** Tiny (37–170 microns long) flask-shaped microfossils (left) from the 750-million-year-old Chuar strata are virtually identical to the tests (shells) of living single-celled amoebae (right). Their abundance in late Proterozoic strata shows that life in the oceans was becoming more diverse and complex by this time.

↻ **NOTHROTHERIOPS (1)** Known mostly from cave sites dated at between 112,000 and 15,000 years ago, the Shasta ground sloth was distributed from California to Mexico. Many caves have produced fossils of several individuals including juveniles, suggesting that they were habitual cave dwellers.

↻ **SLOTH CLAW (1)** The 12cm-long (5in) claw core and nail of the Shasta ground sloth was evidently a very effective tool for digging, and was perhaps used as a weapon in territorial conflicts and for keeping predators at bay.

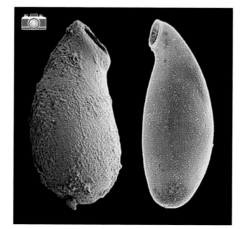

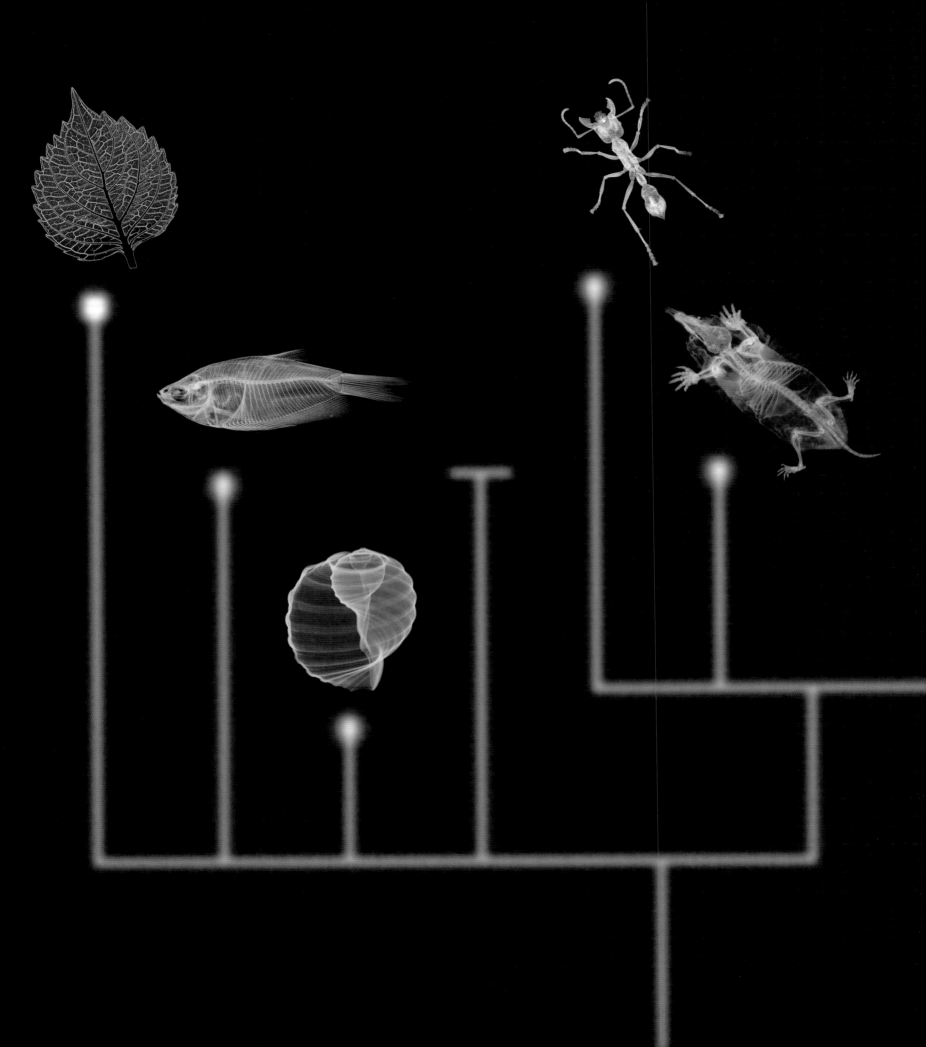

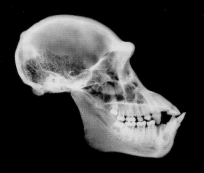

THE TREES OF LIFE

To fully understand the story of evolution, it is necessary to grasp the way in which different species are interrelated. Zoologists and botanists have attempted to put together "trees of life" since before Darwin's time, grouping species into genera, families, orders, classes, and phyla based on their perceived anatomical similarities.

Recent years, however, have seen two interlinked revolutions. The first is cladistics, a more precise system of classification that groups species based on their "phylogeny" or direct evolutionary ancestry, rather than purely anatomical similarities that can be reached through effects such as convergent evolution. The second is molecular phylogenetics – the direct study of DNA and other genetic information from living species in order to ascertain their links to others. The "cladograms" that comprise this section of the book are based on the latest research and current consensus within this fast-developing field, and give a comprehensive picture of the interrelationships between living and extinct organisms.

CLADISTICS – KEY TO THE TREE OF LIFE

THE TAXONOMIC SYSTEM OF CLASSIFICATION, BASED ON SHARED ANATOMICAL FEATURES, WAS THE STANDARD WAY OF CHARTING THE TREE OF LIFE FOR MORE THAN TWO CENTURIES. TODAY, HOWEVER, THE PATTERN OF LIFE IS MAPPED USING CLADISTICS, A SYSTEM BASED ON SHARED ANCESTRY, INFERRED BY MAPPING ANATOMICAL FEATURES ONTO EVOLUTIONARY TREES.

Eighteenth-century botanist Carolus Linnaeus was the first person to apply a consistent approach to classification, using the species, distinguished by the Latin binomial of genus and species name, as a fundamental "unit" of life. He attempted to group like with like, so that more than one species could be included in a genus, similar genera were grouped into orders, and similar orders into classes.

At the time, this hierarchy had no evolutionary subtext – it was simply thought to express a divine order in the natural world. However, its success led to many editions incorporating increasing numbers of living and fossil species, and in the 19th century, it was adapted to reflect proposed evolutionary and "phylogenetic" relationships in a scheme that resembled a genealogical family tree.

However, as scientific understanding of the fossil record and the genetics of living species improved in the 1970s and 1980s, problems emerged with the old Linnaean-based scheme – particularly in the interrelationships of various high-ranking groups. Cladistics offers a radically new approach, putting phylogenetics first in an attempt to rewrite the old order. Above all, it seeks to establish the relationship between monophyletic groups or "clades", each comprised of an ancestor and all of its descendants. It is this scheme that we have used throughout this book.

A cladogram is a hierarchy of species or larger groups of organisms based on their evolutionary history and development. It typically analyses both external features such as those preserved in the fossil record, and genetic information available from living creatures. Using sophisticated analytical techniques, it is possible to recognize the similarities and differences between species, work out the order in which they branched onto separate evolutionary pathways, and even estimate just how long ago these separations happened. By classifying species within nested clades, biologists can piece together the history of life in unprecedented detail.

ACTINISTIA coelacanths

DIPNOI lungfishes, tetrapods

TETRAPODA "Basal tetrapod groups"

Chinlea

Ceratodus

Eusthenopteron

CLADE

A clade is simply a grouping of organisms united by a shared and unique ancestor at some point in their evolutionary history. The clade includes the ancestral species and all of its descendants. Members of a clade are distinguished from non-members by a shared adaptation or "derived character" passed from ancestor to descendants. However, such features are sometimes disguised by later "secondary adaptations" that may even appear to revert to a more primitive form.

ADAPTATION TEXT

These boxes describe the adaptations or characteristics (technically termed "derived characters") that make each clade unique. In this example from the Osteichthyes cladogram (p.268), the Dipnoi or lungfish are marked out from other species by the possession of both gills and functioning lungs.

MULTIPLE LEVELS

One significant advantage of cladistics is that it can work at any level – cladograms can explain the relationships between individual species and whole groups without unnecessary clutter. The Osteichthyes cladogram, for example, highlights the development of several early tetrapod species within the wider scope of the major groupings.

RHIPIDISTIA ▶

◀ SARCOPTERYGII

OSTEICHTHYES ▶

NODE MARKER

These junctions show how two or more individual clades diverge within a larger grouping. All the clades above this "node" are united by a shared ancestor and a newly introduced feature or "derived character" further back in their history.

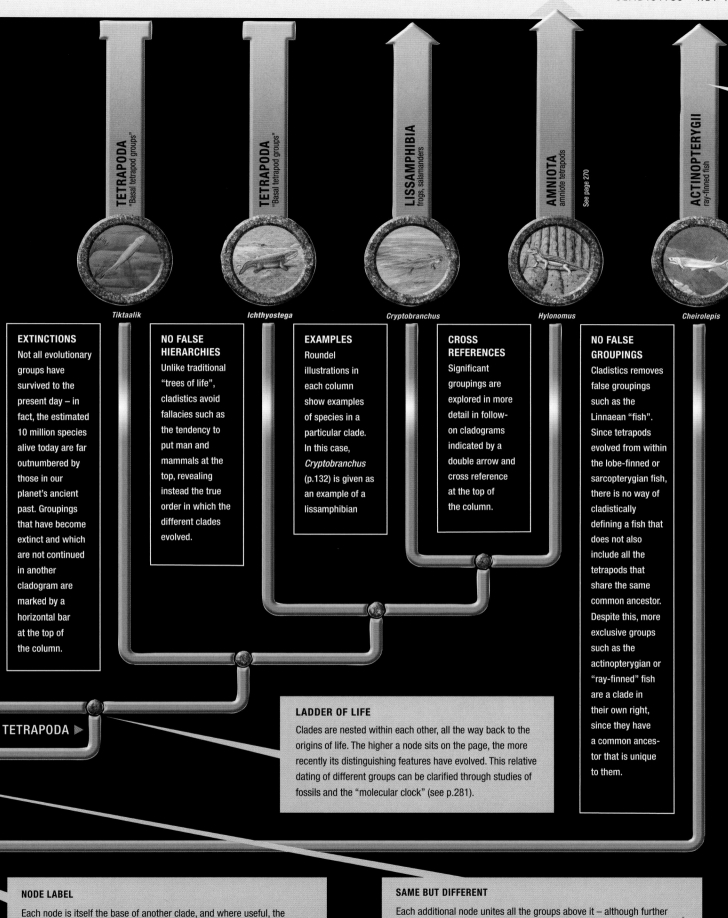

TETRAPODA
"Basal tetrapod groups"

TETRAPODA
"Basal tetrapod groups"

LISSAMPHIBIA
frogs, salamanders

AMNIOTA
amniote tetrapods

See page 270

ACTINOPTERYGII
ray-finned fish

Tiktaalik

Ichthyostega

Cryptobranchus

Hylonomus

Cheirolepis

LIFESPAN

Double arrows at the head of the column indicate a grouping that is continued in another cladogram (see cross reference). A single arrow indicates a grouping that still survives today, while a horizontal bar indicates a clade that is now extinct.

EXTINCTIONS

Not all evolutionary groups have survived to the present day – in fact, the estimated 10 million species alive today are far outnumbered by those in our planet's ancient past. Groupings that have become extinct and which are not continued in another cladogram are marked by a horizontal bar at the top of the column.

NO FALSE HIERARCHIES

Unlike traditional "trees of life", cladistics avoid fallacies such as the tendency to put man and mammals at the top, revealing instead the true order in which the different clades evolved.

EXAMPLES

Roundel illustrations in each column show examples of species in a particular clade. In this case, *Cryptobranchus* (p.132) is given as an example of a lissamphibian

CROSS REFERENCES

Significant groupings are explored in more detail in follow-on cladograms indicated by a double arrow and cross reference at the top of the column.

NO FALSE GROUPINGS

Cladistics removes false groupings such as the Linnaean "fish". Since tetrapods evolved from within the lobe-finned or sarcopterygian fish, there is no way of cladistically defining a fish that does not also include all the tetrapods that share the same common ancestor. Despite this, more exclusive groups such as the actinopterygian or "ray-finned" fish are a clade in their own right, since they have a common ancestor that is unique to them.

BOX FEATURE

On some spreads, the cladograms are supplemented with box features on specific issues including evolutionary landmarks, famous fossil discoveries, and outstanding questions.

TETRAPODA ▷

LADDER OF LIFE

Clades are nested within each other, all the way back to the origins of life. The higher a node sits on the page, the more recently its distinguishing features have evolved. This relative dating of different groups can be clarified through studies of fossils and the "molecular clock" (see p.281).

SCOPE OF THE CLADOGRAMS

Lack of space makes it impossible to include every single species in the book, but the cladograms aim to put all the major groups of life, both living and extinct, in context. Coloured backgrounds link related cladograms – plants, protostome and deuterostome animals, mammals and their extinct ancestors, and the archosaur reptiles including birds.

NODE LABEL

Each node is itself the base of another clade, and where useful, the names of these clades are supplied, along with a description of their unique unifying features.

SAME BUT DIFFERENT

Each additional node unites all the groups above it – although further evolution may do a good job of concealing the original relationship between the different clades.

THE HIERARCHY OF LIFE

HUMANITY'S INTIMATE RELATIONSHIP WITH THE OTHER ORGANISMS THAT SURROUND US HAS MADE THE NAMING OF PLANTS AND ANIMALS ESSENTIAL SINCE PREHISTORIC TIMES. MORE RECENTLY, NATURALISTS HAVE TRIED TO MAKE SENSE OF LIFE'S EXTRAORDINARY ABUNDANCE AND DIVERSITY BY SEEKING OUT COMMON FEATURES TO GROUP LIKE WITH LIKE IN A HIERARCHICAL CLASSIFICATION.

Extant life consists of some two million species described by science, with perhaps four times that number or even substantially more still awaiting discovery and formal description. Living species range from giant redwood conifers weighing up to 2000 tonnes, to microbes that are invisible to the naked eye, and most familiar from their ability to make us ill. All these organisms are united in their basic cellular structure, and the simplest comprise just a single cell. The known fossil record extends the evidence for life and its evolution back across some 3.5 billion years of history to the oceans of early Earth, and through many millions of extinct species both known and unknown.

The unique features of living organisms include the ability to reproduce and increase in number through copying their genetic material (strands of DNA and proteins, called chromosomes) and the information it contains. It is the transmission, mutation, and intermixing of genetic material that drives the changes on which natural selection ultimately acts.

Over recent decades, life has been subdivided at a fundamental level into two groups – the prokaryotes and eukaryotes. The former, with their genetic chromosome material unbounded within the cell, were thought to represent the primitive condition. By contrast, the eukaryotic cell has its genetic material bounded within a separate internal membrane to form a nucleus, and as a result has been widely interpreted as a more advanced or "derived" form of life. But modern investigations of the prokaryote Archaea, long considered to be the least-modified descendants of the first life on Earth, have cast doubt on this previously fundamental hierarchy – the apparent simplicity of the Archaea may in fact be the result of "secondary simplification" from more complex ancestors.

METAZOA Animals *See page 258*

Exellia

FUNGI *See page 254*

Amanita muscaria

PLANTAE Plants *See page 254*

osmundacean

With more than 1.2 million living species, the multicellular metazoans form a major group that ranges from sponges to humans, equating to what was formerly regarded as the "Kingdom Animalia", based on the negative attribute that they did not produce energy by photosynthesis. Instead, metazoans are fundamentally mobile organisms that obtain energy by consuming organic compounds derived from other organisms, which rely in turn on photosynthesis or chemosynthesis to build their tissues.

With some 100,000 living species, the fungi are a major group of eukaryotes. They typically range in size from microscopic to a metre or so (3ft) across – a notable exception is the honey fungus, *Armillaria*, which can cover some 600 hectares (1500 acres), weigh up to 100 tonnes, and live for 1500 years. The fossil record of fungi extends back to at least Devonian times, and their apparent simplicity is probably a secondary development resulting from their parasitic habits.

Although the word "plant" has a long history of common usage, the technical definition of the Plantae is problematic, since experts disagree on whether the green algae should be included. For our purposes, algae are excluded and the Plantae is restricted to land plants or "embryophytes", which form a single monophyletic group. The 270,000 living species range from centimetre-sized (0.5in) to more than 50m (165ft) tall, and their fossil history extends back into Ordovician times.

MULTICELLULAR EUKARYOTES ▶

DNA AND THE GREAT SPLIT

The recent upset in the prevailing view of eukaryotes as more complex descendants from the apparently simpler prokaryotes owes much to molecular analysis of bacterial DNA. This has revealed a gulf within the prokaryotes, with Archaea and Eubacteria separated by just as large a genetic difference as the one that separates both from the eukaryotes.

MULTICELLULAR EUKARYOTES

This outmoded but still commonly used grouping, which included all the plants, animals, and fungi along with the red and brown seaweeds, has been overtaken by molecular studies.

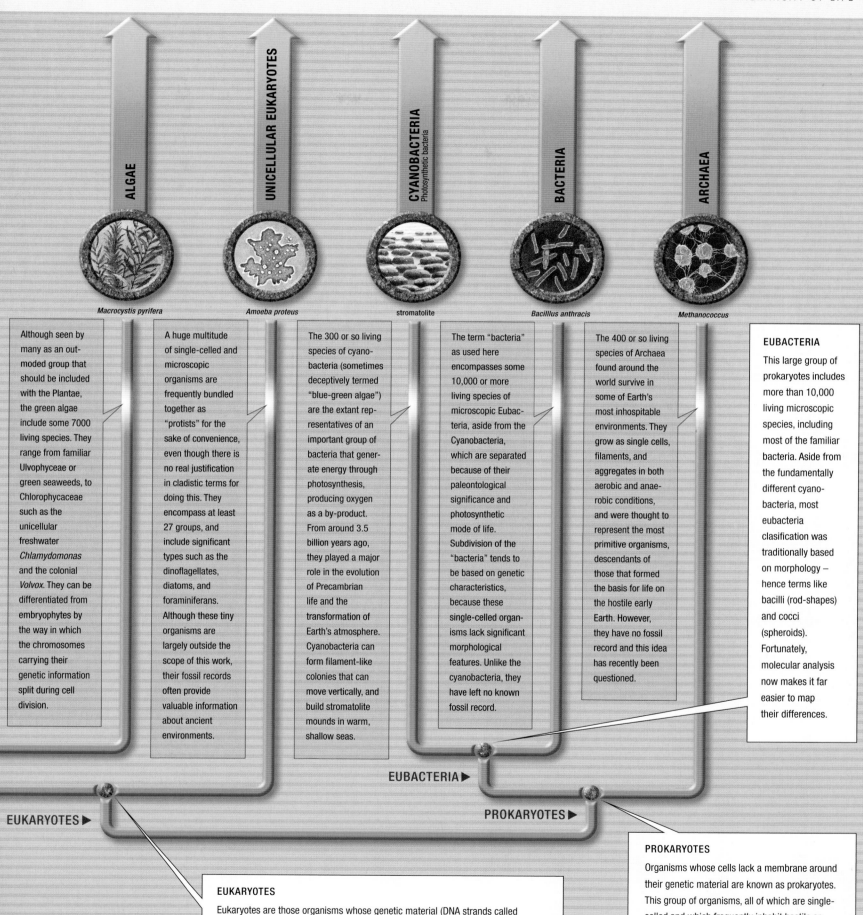

ALGAE

Macrocystis pyrifera

UNICELLULAR EUKARYOTES

Amoeba proteus

CYANOBACTERIA
Photosynthetic bacteria

stromatolite

BACTERIA

Bacilllus anthracis

ARCHAEA

Methanococcus

Although seen by many as an out-moded group that should be included with the Plantae, the green algae include some 7000 living species. They range from familiar Ulvophyceae or green seaweeds, to Chlorophycaceae such as the unicellular freshwater *Chlamydomonas* and the colonial *Volvox*. They can be differentiated from embryophytes by the way in which the chromosomes carrying their genetic information split during cell division.

A huge multitude of single-celled and microscopic organisms are frequently bundled together as "protists" for the sake of convenience, even though there is no real justification in cladistic terms for doing this. They encompass at least 27 groups, and include significant types such as the dinoflagellates, diatoms, and foraminiferans. Although these tiny organisms are largely outside the scope of this work, their fossil records often provide valuable information about ancient environments.

The 300 or so living species of cyano-bacteria (sometimes deceptively termed "blue-green algae") are the extant rep-resentatives of an important group of bacteria that gener-ate energy through photosynthesis, producing oxygen as a by-product. From around 3.5 billion years ago, they played a major role in the evolution of Precambrian life and the transformation of Earth's atmosphere. Cyanobacteria can form filament-like colonies that can move vertically, and build stromatolite mounds in warm, shallow seas.

The term "bacteria" as used here encompasses some 10,000 or more living species of microscopic Eubac-teria, aside from the Cyanobacteria, which are separated because of their paleontological significance and photosynthetic mode of life. Subdivision of the "bacteria" tends to be based on genetic characteristics, because these single-celled organ-isms lack significant morphological features. Unlike the cyanobacteria, they have left no known fossil record.

The 400 or so living species of Archaea found around the world survive in some of Earth's most inhospitable environments. They grow as single cells, filaments, and aggregates in both aerobic and anae-robic conditions, and were thought to represent the most primitive organisms, descendants of those that formed the basis for life on the hostile early Earth. However, they have no fossil record and this idea has recently been questioned.

EUBACTERIA

This large group of prokaryotes includes more than 10,000 living microscopic species, including most of the familiar bacteria. Aside from the fundamentally different cyano-bacteria, most eubacteria clasification was traditionally based on morphology — hence terms like bacilli (rod-shapes) and cocci (spheroids). Fortunately, molecular analysis now makes it far easier to map their differences.

EUBACTERIA ▶

EUKARYOTES ▶

PROKARYOTES ▶

EUKARYOTES

Eukaryotes are those organisms whose genetic material (DNA strands called chromosomes) is confined in a semipermeable membrane to form a distinct nucleus to each cell. They include both unicellular and multicellular organisms.

PROKARYOTES

Organisms whose cells lack a membrane around their genetic material are known as prokaryotes. This group of organisms, all of which are single-celled and which frequently inhabit hostile or extreme environents, can be divided into Eubacteria and Archaea based on DNA evidence.

EMBRYOPHYTA

THE COLONIZATION OF EARTH'S LANDSCAPE HAS BEEN ALMOST ENTIRELY DUE TO THE SUCCESS OF THE LAND PLANTS OR EMBRYOPHYTES. ALL LAND-LIVING TETRAPODS DEPEND ON FOOD CHAINS UNDERPINNED BY THE PHOTOSYNTHETIC EMBRYOPHYTES, AND THE RECORD OF PLANTS ON LAND DATES BACK TO FOSSILIZED SPORES FROM ORDOVICIAN TIMES.

The Embryophyta includes 270,000 living species, uniting all plants that have colonized land and freshwater, and a few that have returned to the sea. Plants are found in all but the most inhospitable land environments, from high latitudes that endure long periods without sunlight, to arid tropics where annual rainfall is minimal. Over the 450 million years or more since they first colonized land, they have become extraordinarily diverse and successful.

Conquering dry land required a number of evolutionary adaptations to the inhospitable conditions of life out of water (see box opposite). These were only achieved in stages, as successive groups of plants evolved over long periods. At first, land plants were restricted in height and confined to wet environments to ensure their mobile sperm could swim and fertilize the egg cells of other plants. As they grew larger and stronger, the first lowland forests were established around 390 million years ago, but modern vegetation, dominated by the flowering plants, did not become widespread until the Cenozoic, around 60 million years ago.

MARCHANTIOPHYTA — Liverworts

Marchantiia polymorpha

ANTHOCEROPHYTA + BRYOPHYTA — Hornworts and mosses

Anthoceros levis

LYCOPSIDA — Clubmosses

Stigmaria

SPHENOPSIDA — Horsetails

Calamites

The 9000 or so living species of liverworts are small, primitive land plants that lack roots or vascular conduction and support systems, and so cannot grow upright against gravity to any great extent. While they lack true "stomata", they do have openings in their leaf-like tissues that allow an exchange of gases with the environment for photosynthesis. They grow in damp conditions on rocks, in soil, and sometimes on other plants as "epiphytes". Liverwort-like spores have been found in Silurian rocks, but their fossil record is more secure from Late Devonian times.

The hornworts comprise some 300 living species, and their fossil record extends back to late Silurian times. Lacking vascular support and root systems, they only grow a centimetre or so (0.5in) from the ground, but some tropical species are epiphytes. By contrast, mosses have at least 15,000 living species and a fossil record extending back to the Early Devonian. Growing up to 8cm (3in) high, they pioneered life in moist places on land, and as they died and decayed, they formed humus layers essential to the development of soils.

This small group of primitive land plants has some 1200 living species, none more than a few centimetres high. However, early lycopsids that evolved in Devonian times came to dominate extensive forests and swamps during the Carboniferous, with tree-sized forms growing up to 40m (130ft) high. Typically, they have blade-shaped leaves arranged in spirals, and live in wet environments because their reproduction utilizes mobile sperm that must swim in order to fertilize the egg.

Today, the horsetails are restricted to just 20 living species, none growing above a metre (3ft) tall. However, they evolved in Devonian times, and sub-sequently developed tree-sized Carbon-iferous forms up to 20m (66ft) high. Their jointed upright stems grow from rhizome-like structures beneath the ground, while simple microphyll leaves develop in whorls from the stem joints. Like the lycopsids, their means of reproduction confines them to wet environments.

STOMAPHYTA ▶

EMBRYOPHYTA ▶

The mountain pine (*Pinus mugo*) is a living example of the ancient Coniferopsida, which first evolved during the Carboniferous Period more than 300 million years ago.

EMBRYOPHYTA

Modern analysis supports a single and unique origin for the embryophytes. However, some experts still argue that they are polyphyletic and arose from different ancestral groups.

FILICOPSIDA
Ferns

GINKGOALES
Ginkgos

CONIFEROPSIDA
Conifers

CYCADOPHYTA
Cycads

GNETALES

ANGIOSPERMAE
Flowering plants

See page 256

Phlebopteris

Ginkgoites

Taxodium

Taeniopteris

ephedroid sp.

Platanus

Ferns are the most successful survivors of the ancient embryophytes, with some 9500 living species, mostly about a metre (3ft) high, and typically herbaceous or shrubby. Tree-sized forms were more common in the past. Evolving in Devonian times, they reproduce with spores and a sexual "gametophyte" stage – which frees them from watery environments.

The ancient ginkgos are represented by just one surviving species, the maidenhair tree, *Ginkgo biloba*. This living fossil has separate sexes, characteristic veined and lobed leaves, and can grow to 40m (130ft), with a superficial resemblance to deciduous angiosperm trees. The group was at its most abundant and diverse in Mesozoic times when it was food for many verte-brate plant-eaters.

The ancient group of conifers evolved during late Carboniferous times, rose to domi-nance through the Mesozoic, and is represented today by some 600 living species, often tolerant of high latitudes and some-times growing up to 80m (260ft) tall. The conifers reproduce by means of seeds borne in single-sex cones, which many distribute using the wind.

The warmth-loving cycads evolved in Permian times, diversified globally during the Mesozoic to reach even high latitudes, and today include about 300 living species. With palm-frond-like leaves, some have short, squat trunks, while others have thick, unbranched stems growing up to 10m (33ft) high. Sexes are separate and they reproduce by seeds grouped into cones.

With just 91 living tropical species, these obscure gymnosperms may not form a natural group but certainly share some unique features including a "flower" that has developed independently from angiosperm flowers. Their fossil record extends back to the Permian, and there are some features that link them with the extinct Bennetti-tales group, and to the angiosperms.

Angiosperms (plants with flowers and fruit) are the dominant plants on Earth today, with more than 233,000 living species and enormous diversity from small grasses to giant trees and aquatic species. Their fossil record extends back to the Cretaceous, and they have provided Earth's landscapes with a kaleidoscope of coloured blooms, especially since Cenozoic times.

◀ SPERMATOPHYTA

STOMAPHYTA

This major grouping of the embryophytes, which contains over 260,000 living species, is based on the possession of specialized stomata used for the regulation of gas exchange during photosyn-thesis. Stomata are formed from two semicircu-lar guard cells whose opening and closing allows the passage of gas between the atmosphere and the plant tissue.

SPERMATOPHYTA

The spermatophyta includes some 234,000 living species and combines the surviving gymnosperms (seedplants) with the angiosperms (flowering plants), but their phylogeny is somewhat problematic. Molecular data supports the idea of living gymnosperms as a monophyletic group, but there is fossil evidence from the extinct seed ferns to suggest that the cycads, extinct bennettitales, and angiosperms are derived from within the gymnosperms.

GREENING THE LAND

The oldest known land plants are tiny, centimetre-long (0.5in) carbonized stems that are a millimetre or so wide, flattened, and preserved in Silurian mudstones from the Welsh Borders. Microscopic examination shows that these vascular plants, called *Cooksonia*, had already evolved stomata for exchanging gases with the atmosphere, and that the tube-shaped "tracheid" cells within their stems were already strengthened by the organic protein lignin. These cells are essential for carrying fluids and helping the stem grow upright against gravity. No root structures have yet been found, but the ends of the forked branches carry club-shaped spore capsules that were used for reproduction, and *Cooksonia* would have required wet conditions for transport of their gametes and spores. As a result, all these early vascular plants are thought to have lived very close to water.

ANGIOSPERMAE

WITHOUT THE ANGIOSPERMS, OUR MODERN WAY OF LIFE COULD NOT BE SUSTAINED. THIS LARGE AND DIVERSE GROUP OF FLOWERING PLANTS INCLUDES STAPLE GRAIN CROPS, SUCH AS RICE AND WHEAT, ALONG WITH THE FRUIT-BEARING PLANTS AND VEGETABLES, UPON WHICH HUMANS AND MANY OTHER ANIMALS SUBSIST.

Since the beginning of Cenozoic times some 65 million years ago, almost all terrestrial habitats have been colonized by the flowering plants, of which there are now some 234,000 known species. From polar latitudes to the tropics and from alpine heights to the sunlit waters of lakes and seas, flowering plants have come to dominate most terrestrial habitats. In the 135 million years since they arose in the Early Cretaceous, they have adapted their habits to include vines, herbs, shrubs, and trees. Common angiosperm features include a central female part with ovaries enclosing ovules (potential seeds) and a ring of male pollen-producing stamens. A central stalk emerging from the ovaries ends in a pollen-receiving surface called the stigma, so that a flower is essentially a bisexual organ with both male and female parts.

Traditionally the number of cotyledons or seed leaves were used to subdivide angiosperms into monocotyledons and dicotyledons, but only the former group is truly monophyletic and descended from a single ancestor. Despite this discovery, the terms are still widely used.

NYMPHAEACEAE
water lilies

nymphaealean

ARACEAE
arum lilies

arum lily

ARECACEAE
Seasquirts

palms

TYPHACEAE
cattails

Typha

Commonly called water lilies, the Nymphaceae are the principal family among the Nymphaeales, and contain some 70 or more living species that are common in both temperate and tropical fresh-waters. While the leaves and flowers of these plants typically float on the surface of the water, roots extend down into the sediment below, and their length restricts water lilies to relatively shallow waters. Modern analysis suggests that Nymphaceae form a sister group to other flowering plants, evolving independently from their early angiosperm ancestors.

This large group contains some 3700 living species of monocotyledon angiosperms, and is also referred to as the aroids. They are widely distributed, but are particularly common and diverse in the tropics of the Americas. Characteristically, the flowers of the Araceae are borne on an inflorescence known as a spadix, and partially enclosed within a hood-like sheath. Recent research has shown that the Lemnaceae or duckweeds also belong within this group.

Mostly restricted to the tropics and warm temperate climates, this large group of some 2600 living species includes the palms, some of which grow to 50m (165ft) tall. Palms are mostly distinguished by large evergreen leaves clustered in spirals at the top of an unbranched stem. However, there are many divergences from this arrangement, and they are found throughout a wide range of habitats from deserts to rainforests.

Today the cattails are a small group — some 12 species of monocotyledonous perennial flowering plants, mostly found in marshy habitats of the northern hemisphere. Typically growing as large herbs up to 7m (23ft) tall, they have alternating leaves at the base, and a jointless stem rising to a brownish spike of compact unisexual flowers. The male organs are reduced to a pair of stamens at the top, and pollination and seed distribution relies on the wind.

LILIOPSIDA ▶

ANGIOSPERMAE ▶

Water lilies (Nymphaeales) are modern representatives of the Nymphaeaceae, thought to be the most primitive form of flowering plants.

ANGIOSPERMAE

The most widespread modern land plants are also the most recently evolved major group, originating in late Mesozoic times. Along with the living gymnosperms, they are the only surviving seedplants.

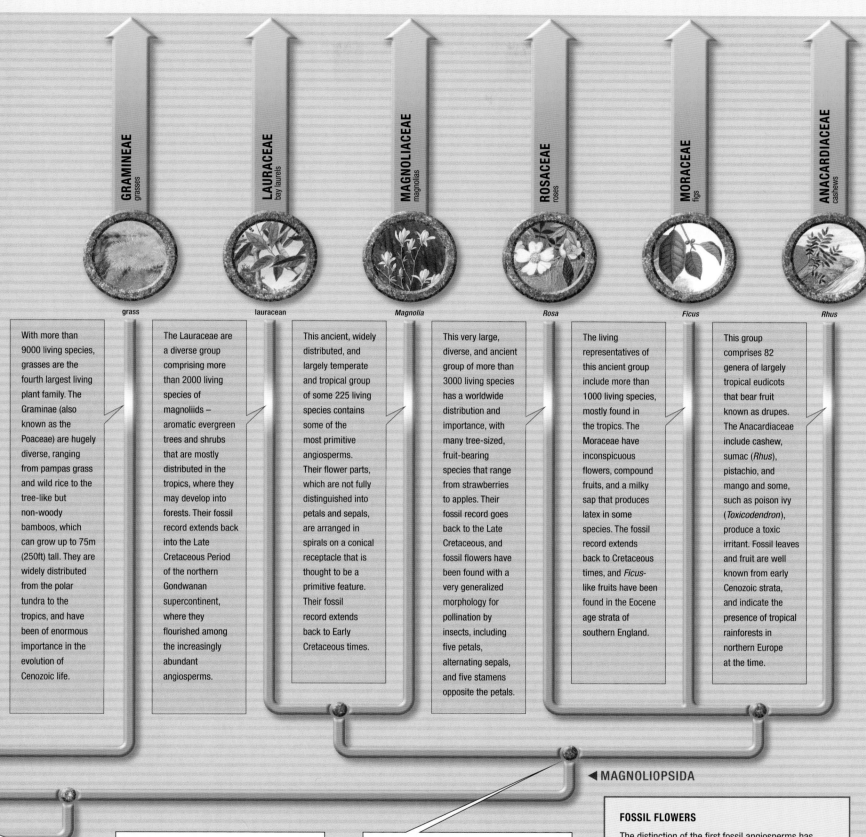

GRAMINEAE
grasses

grass

LAURACEAE
bay laurels

lauracean

MAGNOLIACEAE
magnolias

Magnolia

ROSACEAE
roses

Rosa

MORACEAE
figs

Ficus

ANACARDIACEAE
cashews

Rhus

With more than 9000 living species, grasses are the fourth largest living plant family. The Graminae (also known as the Poaceae) are hugely diverse, ranging from pampas grass and wild rice to the tree-like but non-woody bamboos, which can grow up to 75m (250ft) tall. They are widely distributed from the polar tundra to the tropics, and have been of enormous importance in the evolution of Cenozoic life.

The Lauraceae are a diverse group comprising more than 2000 living species of magnoliids – aromatic evergreen trees and shrubs that are mostly distributed in the tropics, where they may develop into forests. Their fossil record extends back into the Late Cretaceous Period of the northern Gondwanan supercontinent, where they flourished among the increasingly abundant angiosperms.

This ancient, widely distributed, and largely temperate and tropical group of some 225 living species contains some of the most primitive angiosperms. Their flower parts, which are not fully distinguished into petals and sepals, are arranged in spirals on a conical receptacle that is thought to be a primitive feature. Their fossil record extends back to Early Cretaceous times.

This very large, diverse, and ancient group of more than 3000 living species has a worldwide distribution and importance, with many tree-sized, fruit-bearing species that range from strawberries to apples. Their fossil record goes back to the Late Cretaceous, and fossil flowers have been found with a very generalized morphology for pollination by insects, including five petals, alternating sepals, and five stamens opposite the petals.

The living representatives of this ancient group include more than 1000 living species, mostly found in the tropics. The Moraceae have inconspicuous flowers, compound fruits, and a milky sap that produces latex in some species. The fossil record extends back to Cretaceous times, and *Ficus*-like fruits have been found in the Eocene age strata of southern England.

This group comprises 82 genera of largely tropical eudicots that bear fruit known as drupes. The Anacardiaceae include cashew, sumac (*Rhus*), pistachio, and mango and some, such as poison ivy (*Toxicodendron*), produce a toxic irritant. Fossil leaves and fruit are well known from early Cenozoic strata, and indicate the presence of tropical rainforests in northern Europe at the time.

◄ MAGNOLIOPSIDA

LILIOPSIDA

This subdivision of angiosperms includes the majority of monocotyledons, in which the embryo has only one initially formed "cotyledonous" leaf, and parts of the flower are usually arranged in threes.

MAGNOLIOPSIDA

This subdivision of the angiosperms equates to the traditional dicotyledon group, in which the developing plant embryos have two first-formed cotyledonous leaves, and the parts of the flowers are arranged in twos and fives.

FOSSIL FLOWERS

The distinction of the first fossil angiosperms has proved highly elusive, but angiosperm-like pollen has been recovered from Early Cretaceous strata, some 135 million years old. Some three-dimensional, millimetre-sized flowers, preserved in charcoal from Sweden's Late Cretaceous, show some defining angiosperm features including the ring of scale-like sepals that surround the petals.

METAZOA

ALL EUKARYOTE ANIMALS, FROM SPONGES TO HUMANS, ARE METAZOANS, WITH BODIES COMPRISING MANY CELLS THAT COLLABORATE TO GIVE THE BODY FORM AND FUNCTION AND ALLOW GROWTH TO A SUBSTANTIAL SIZE. MOST METAZOANS ARE CAPABLE OF INDEPENDENT MOVEMENT – A NECESSITY FOR A LIFE FEEDING OFF OTHER ORGANISMS.

Scientists have long recognized a number of major "grade changes" in the organization of life forms throughout the history of evolution. One of the most significant was that between single-celled (protistan) and multicelled (metazoan) organisms. A number of further grade changes within the metazoans were traditionally portrayed as simple increases in complexity, but this is no longer accepted. The sponges or "Porifera", for example, were thought to form a natural group but are now recognized as para-phyletic, with different ancestries. Nevertheless, the sponges, with their simple cell structures, grade into eumetazoans whose cells form distinct and specialized tissues and organs.

Other interesting questions remain unanswered. For instance, the coelom, a fluid-filled cavity, was thought to have developed as a single adaptation, but its appearance in both protostomes and deuterostomes raises the question of whether it arose independently on two separate occasions. Similar issues surround the segmentation associated with both groups.

PLACOZOA
"flat animals"

Trichoplax adhaerens

Only a single living species of this millimetre-sized marine metazoan group is known, and that was discovered by chance more than 120 years ago in a saltwater aquarium in Austria. The placozoans have no known fossil representatives, but are thought to be the most primitive of all metazoans. Their bodies consist of a few thousand cells arranged in a flattened, roughly discoid shape. An upper and lower layer of cells equipped with hair-like "cilia" are used to move across surfaces and change the placozoan's body shape.

DEMOSPONGIAE
common sponges

Vauxia

With over 8000 living species and a fossil record going back to the Cambrian, the Demospongiae are very diverse in form and habit. Typically 30–40cm (12–16in) high and plant-like in form, their bodies are supported by a skeletal mesh of silica spicules, held together by a protein called spongin. They are found in fresh and salt water from the poles to the tropics, and at all depths. As in all sponges, water is drawn in through pores and filtered for food by the motion of whip-like flagellae.

The ancestry of this group is uncertain.

HEXACTINELLIDA
glass sponges

Euplectella aspergillum

There are some 1000 living species of these "glass sponges", with a fossil record extending back into Cambrian times. Purely marine, they typically live in deep waters, where they build extraordinarily intricate skeletons, up to 50cm (20in) tall, out of interlocking silica spicules with a six-rayed structure. Remarkably, these skeletons are somehow constructed by collaboration between the spicule-forming cells, yet they form only very simple cell layers and have no specialized tissues or body organs as such.

CALCAREA
calcareous sponges

Clathrina

The calcareous marine sponges form a group of around 1000 living species, and have a fossil record reaching back into early Cambrian times. As their name suggests, they have skeletons made of calcium carbonate spicules. Typically, they live in relatively shallow waters on continental shelves, and require a hard substrate onto which their larvae can attach. The initial part of their skeleton often grows to a large size before forming its spicules.

METAZOA ▶

Modern tropical reefs are typically home to a variety of metazoans, ranging from sponges and corals to protostome invertebrates and deuterostome echinoderms and fish.

METAZOA

The metazoans are defined as multicellular and independently motile eukaryotes whose body plan is determined as they develop, although some undergo a change of form (metamorphosis) later in life.

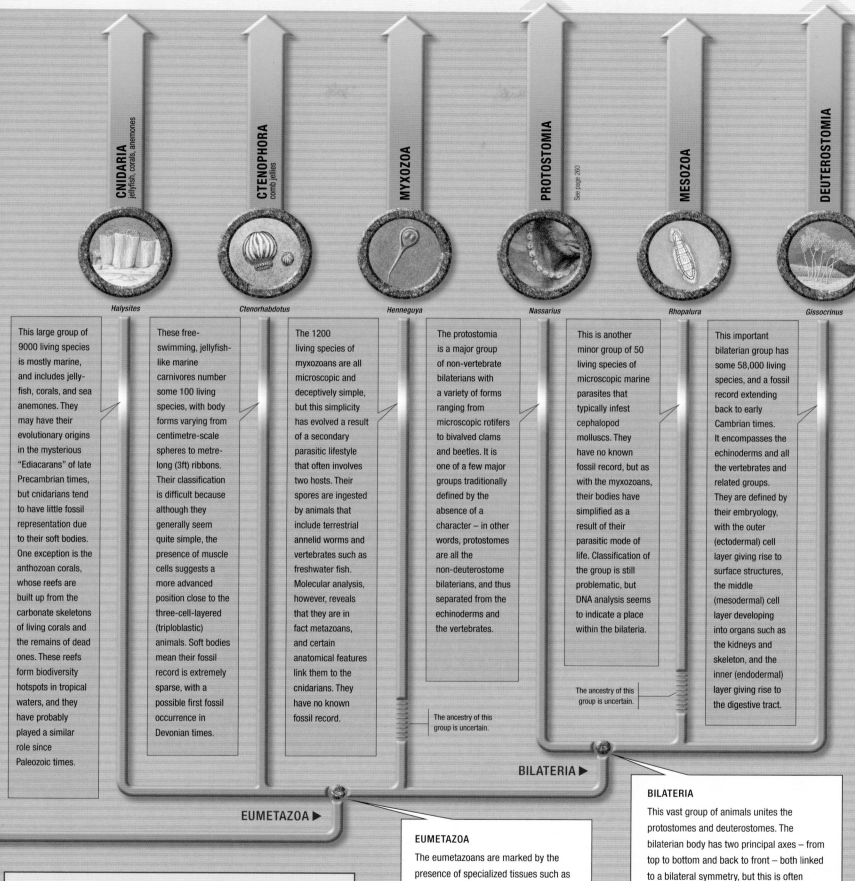

CNIDARIA
jellyfish, corals, anemones

CTENOPHORA
comb jellies

MYXOZOA

PROTOSTOMIA
See page 260

MESOZOA

DEUTEROSTOMIA
See page 266

Halysites

Ctenorhabdotus

Henneguya

Nassarius

Rhopalura

Gissocrinus

This large group of 9000 living species is mostly marine, and includes jellyfish, corals, and sea anemones. They may have their evolutionary origins in the mysterious "Ediacarans" of late Precambrian times, but cnidarians tend to have little fossil representation due to their soft bodies. One exception is the anthozoan corals, whose reefs are built up from the carbonate skeletons of living corals and the remains of dead ones. These reefs form biodiversity hotspots in tropical waters, and they have probably played a similar role since Paleozoic times.

These free-swimming, jellyfish-like marine carnivores number some 100 living species, with body forms varying from centimetre-scale spheres to metre-long (3ft) ribbons. Their classification is difficult because although they generally seem quite simple, the presence of muscle cells suggests a more advanced position close to the three-cell-layered (triploblastic) animals. Soft bodies mean their fossil record is extremely sparse, with a possible first fossil occurrence in Devonian times.

The 1200 living species of myxozoans are all microscopic and deceptively simple, but this simplicity has evolved a result of a secondary parasitic lifestyle that often involves two hosts. Their spores are ingested by animals that include terrestrial annelid worms and vertebrates such as freshwater fish. Molecular analysis, however, reveals that they are in fact metazoans, and certain anatomical features link them to the cnidarians. They have no known fossil record.

The protostomia is a major group of non-vertebrate bilaterians with a variety of forms ranging from microscopic rotifers to bivalved clams and beetles. It is one of a few major groups traditionally defined by the absence of a character – in other words, protostomes are all the non-deuterostome bilaterians, and thus separated from the echinoderms and the vertebrates.

This is another minor group of 50 living species of microscopic marine parasites that typically infest cephalopod molluscs. They have no known fossil record, but as with the myxozoans, their bodies have simplified as a result of their parasitic mode of life. Classification of the group is still problematic, but DNA analysis seems to indicate a place within the bilateria.

This important bilaterian group has some 58,000 living species, and a fossil record extending back to early Cambrian times. It encompasses the echinoderms and all the vertebrates and related groups. They are defined by their embryology, with the outer (ectodermal) cell layer giving rise to surface structures, the middle (mesodermal) cell layer developing into organs such as the kidneys and skeleton, and the inner (endodermal) layer giving rise to the digestive tract.

The ancestry of this group is uncertain.

The ancestry of this group is uncertain.

BILATERIA ▶

EUMETAZOA ▶

EUMETAZOA

The eumetazoans are marked by the presence of specialized tissues such as nerve, muscle, and sensory cells from cell layers that form in the early stages of embryonic development. The mouth forms the entrance to a differentiated gut.

BILATERIA

This vast group of animals unites the protostomes and deuterostomes. The bilaterian body has two principal axes – from top to bottom and back to front – both linked to a bilateral symmetry, but this is often masked in the adult form. The position of the mouth and a concentration of sense organs at the front leads to increasing development of a distinct head ("cephalization").

TRIPLOBLASTIC ANIMALS

Bilaterians are sometimes referred to as triploblastic animals, from the way that the blastula, the ball of cells formed as the embryo develops, differentiates rapidly into three layers of endodermal, mesodermal, and ectodermal cells.

PROTOSTOMIA

THE BULK OF THE PROTOSTOMIAN FOSSIL RECORD IS MADE UP OF PRESERVED SHELLS. BUT THIS BIAS GREATLY UNDERESTIMATES THE SHEER DIVERSITY AND NUMBERS OF PROTOSTOME SPECIES. THE MAIN REASON FOR THIS IS THAT EUARTHROPODS SUCH AS INSECTS FORM A MAJOR ELEMENT WITHIN THE GROUP, BUT DO NOT NORMALLY FOSSILIZE.

Protostomians make up a majority of living animals – some 1,140,000 known species ranging from microscopic rotifers to giant squids. They are distinguished from other bilaterians by the way their embryos develop – the first opening to form, known as the blastopore, eventually becomes the mouth (protostome means "mouth first").

Many protostomians lack a skeleton, but where one is present it is an exoskeleton. The nervous system, meanwhile, runs below the gut (apart from the ganglia associated with the head region). Traditionally, the animals in this group were referred to as "invertebrates", based on the absence of a backbone, but this term would also include some deuterostomes.

The protostomians have adapted to all environments from polar wastes to dry deserts and ocean depths. Their fossil record stretches back to at least the early Cambrian but, depending on interpretation, may extend into the Precambrian. "Molecular clock" estimates, meanwhile, suggest the group originated around 800 million years ago.

A wide variety of flatworms (Platyhelminthes) make bizarre but beautiful inhabitants of underwater environments such as reefs.

ROTIFERA
rotifers

Philodina

With some 1800 living species and little fossil record, the rotiferans are a group of microscopic protostomes that are typically spherical or tubular and less than 2mm in size. Important constituents of freshwater ecosystems, their ability to resist desiccation also allows survival in moist terrestrial habitats, especially soils. The rear of the animal has cilia for swimming, while the front has a foot-like structure for attaching to a substrate. There are no respiratory or circulatory systems.

PLATYHELMINTHES
flatworms

Pseudoceros dimidiatus

These "flatworms" number some 14,000 living species, mostly aquatic with many parasitic forms, but have very little fossil record. They are characterized by a complex gut with a single opening that serves as both mouth and anus. The body is flattened, but there is usually a distinctive head except in some parasitic forms. Although they lack respiratory and circulatory systems, some species – particularly parasites – can reach lengths of several metres.

MOLLUSCA
molluscs

See page 262

Orthoceras

The 120,000 or so living species of molluscs comprise a diverse and important group of protostomes that have colonized many aquatic and terrestrial environments. Many of them secrete a protective calcium carbonate shell (rarely several metres long) that can be preserved by fossilization, and mollusc shells form a significant proportion of the fossil record back to early Cambrian times.

ANNELIDA
annelid worms

Canadia

The annelid worms have soft, elongated, cylindrical bodies that are segmented, and occasionally grow to a few metres long. While they lack skeletons, some are armed with hard mandibles that can fossilize, and many live in burrows that can also form trace fossils. Occupying a range of aquatic and terrestrial habitats since the early Cambrian, today they number 14,000 living species, including earthworms.

◄ EUTROCHOZOA

EUTROCHOZOA
The eutrochozoans possess a band of hair-like cilia around the mouth that are used for locomotion. They include both segmented forms such as the annelid worms, and unsegmented animals such as molluscs.

LOPHOTROCHOZOA
The lophotrochozoans are a major group of protostomians that possess a larval type known as a trochophore, a ring of tentacles around the mouth, and particular developmental genes.

◄ LOPHOTROCHOZOA

PROTOSTOMIA ►

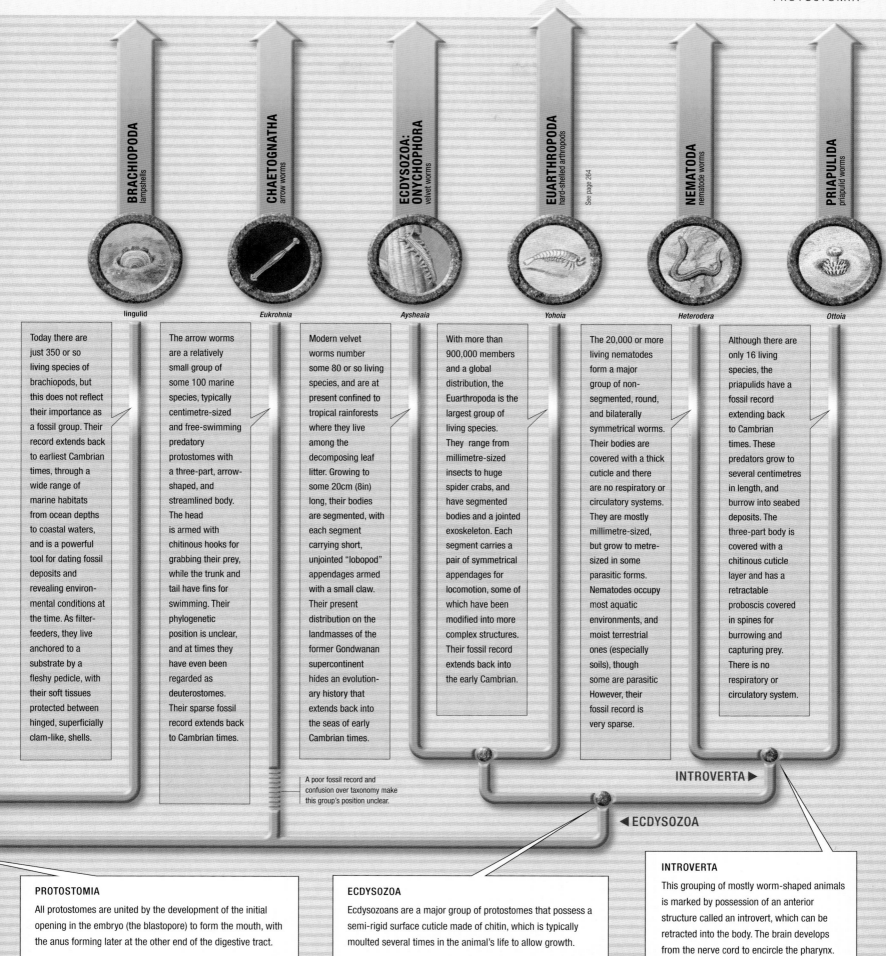

BRACHIOPODA
lampshells

lingulid

Today there are just 350 or so living species of brachiopods, but this does not reflect their importance as a fossil group. Their record extends back to earliest Cambrian times, through a wide range of marine habitats from ocean depths to coastal waters, and is a powerful tool for dating fossil deposits and revealing environmental conditions at the time. As filter-feeders, they live anchored to a substrate by a fleshy pedicle, with their soft tissues protected between hinged, superficially clam-like, shells.

CHAETOGNATHA
arrow worms

Eukrohnia

The arrow worms are a relatively small group of some 100 marine species, typically centimetre-sized and free-swimming predatory protostomes with a three-part, arrow-shaped, and streamlined body. The head is armed with chitinous hooks for grabbing their prey, while the trunk and tail have fins for swimming. Their phylogenetic position is unclear, and at times they have even been regarded as deuterostomes. Their sparse fossil record extends back to Cambrian times.

A poor fossil record and confusion over taxonomy make this group's position unclear.

ECDYSOZOA: ONYCHOPHORA
velvet worms

Aysheaia

Modern velvet worms number some 80 or so living species, and are at present confined to tropical rainforests where they live among the decomposing leaf litter. Growing to some 20cm (8in) long, their bodies are segmented, with each segment carrying short, unjointed "lobopod" appendages armed with a small claw. Their present distribution on the landmasses of the former Gondwanan supercontinent hides an evolutionary history that extends back into the seas of early Cambrian times.

EUARTHROPODA
hard-shelled arthropods

See page 264

Yohoia

With more than 900,000 members and a global distribution, the Euarthropoda is the largest group of living species. They range from millimetre-sized insects to huge spider crabs, and have segmented bodies and a jointed exoskeleton. Each segment carries a pair of symmetrical appendages for locomotion, some of which have been modified into more complex structures. Their fossil record extends back into the early Cambrian.

NEMATODA
nematode worms

Heterodera

The 20,000 or more living nematodes form a major group of non-segmented, round, and bilaterally symmetrical worms. Their bodies are covered with a thick cuticle and there are no respiratory or circulatory systems. They are mostly millimetre-sized, but grow to metre-sized in some parasitic forms. Nematodes occupy most aquatic environments, and moist terrestrial ones (especially soils), though some are parasitic However, their fossil record is very sparse.

PRIAPULIDA
priapulid worms

Ottoia

Although there are only 16 living species, the priapulids have a fossil record extending back to Cambrian times. These predators grow to several centimetres in length, and burrow into seabed deposits. The three-part body is covered with a chitinous cuticle layer and has a retractable proboscis covered in spines for burrowing and capturing prey. There is no respiratory or circulatory system.

INTROVERTA ▶

◀ ECDYSOZOA

PROTOSTOMIA
All protostomes are united by the development of the initial opening in the embryo (the blastopore) to form the mouth, with the anus forming later at the other end of the digestive tract.

ECDYSOZOA
Ecdysozoans are a major group of protostomes that possess a semi-rigid surface cuticle made of chitin, which is typically moulted several times in the animal's life to allow growth.

INTROVERTA
This grouping of mostly worm-shaped animals is marked by possession of an anterior structure called an introvert, which can be retracted into the body. The brain develops from the nerve cord to encircle the pharynx.

MOLLUSCA

WITH MORE THAN 117,000 LIVING SPECIES, THE MOLLUSCS ARE A REMARKABLY DIVERSE GROUP OF ANIMALS, WITHIN WHICH A FURTHER EIGHT GROUPS ARE RECOGNIZED. THESE RANGE FROM UNFAMILIAR CENTIMETRE-SIZED AND WORM-SHAPED FORMS COVERED WITH CALCAREOUS SPINES, TO BETTER KNOWN GASTROPODS AND CEPHALOPODS, INCLUDING GIANT SQUIDS AT LEAST 10M (33FT) LONG.

In the basic mollusc form, a flat muscular "foot" on the underside is used for locomotion. There is a distinct head and tail, and a "visceral hump" on top of the body, containing the organs and covered by a protective mantle that folds over the gills and commonly secretes a calcium carbonate shell, into which the animal can retract its body.

This body form has proved remarkably adaptable, with the muscular foot taking on many different functions that range from creeping across substrates to burrowing in sediment, swimming, and capturing prey. The head and associated sensory organs can be highly sophisticated, as seen in the cephalopods with their well-developed eyes and learning ability, but other groups, such as the bivalves and scaphopods, show reduction and loss of the head and have a body that is almost entirely enclosed within the shell.

The origin of the group is, as yet, unknown – it must lie in the late Precambrian, but the fossil record starts with early Cambrian millimetre-sized and single-shelled forms with a variety of conical shapes. These probably gave rise to the major cephalopod, gastropod, and bivalve groups. Subsequently, molluscs adapted to most marine environments, invading freshwaters during the Devonian and the land in Carboniferous times. There are also slug-like molluscs in both aquatic and terrestrial environments that have lost their shells and do not leave any significant fossil remains. Molluscs are only absent from the coldest and driest environments.

CAUDOFOVEATA

Chaetoderma intermedium

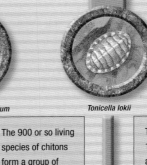

POLYPLACOPHORA
chitons

Tonicella lokii

MONOPLACOPHORA

Neopilina

The Caudofoveata are one of two low-diversity, soft-bodied, worm-like groups of marine molluscs whose evolutionary position is uncertain. They habitually burrow into seabed sediments, and have no known fossil record, but include some 100 living species growing to around 5cm (2in) long. The other group, the Solenogastres, with 350 living species, has a similar habit and a single known fossil from the late Carboniferous of Illinois, USA. Both are covered with a cuticle bearing small calcareous scales and spicules.

The 900 or so living species of chitons form a group of marine grazers that use a tough organic "radula" to rasp algae from rock surfaces. Their flattened oval bodies grow to 30cm (12in) long, with a distinct bilateral symmetry about a head-to-tail axis. There is a flat ventral muscular foot covered by a mantle, gills, and a carbonate shell consisting of several articulated plates, which may reflect a primitive molluscan condition. Their fossil record extends back to late Cambrian times.

This small group of 15 marine species are classic "living fossils". They were known from remains dating back to early Cambrian times long before their first modern representatives were discovered in 1952. Their body form may be close to the primitive mollusc condition, with a single cap-shaped shell covering the viscera, and a flat muscular foot. Paired muscles were once thought to show the presence of basic segmentation, but this is now doubted.

EUMOLLUSCA ▶

MOLLUSCA ▶

Modern cuttlefish are advanced cephalopods with highly developed brains and sense organs suited to their predatory lifestyle.

MOLLUSCA

Within the protostomes, development defines a major division between the Lophotrochozoa (groups such as worms, molluscs, and brachiopods) and the Cuticulata (including arthropods and priapulids).

GASTROPODA
snails

CEPHALOPODA
squids, octopus, nautilus

BIVALVIA
clams

SCAPHOPODA
tusk shells

Sinuites

nautilid

Aviculopecten

Antalis entalis

FOSSILS AND GEOLOGICAL DATING

Many mollusc groups secrete calcium carbonate shells to protect their soft tissues, and these preserved mineralized shells form a very important component of the fossil record. However, since their shells are usually made from the relatively unstable mineral aragonite, many fossil molluscs are either preserved by sediment moulds or have shells replaced by calcite.

Cephalopods make a particularly significant contribution to our understanding of the geological past. Their diversity and rapid evolution has made their coiled shells common fossils in the marine deposits of Paleozoic and Mesozoic seas. In the late 19th century, German paleontologist Albert Oppel recognized the potential of cephalopod ammonite fossils for subdividing and correlating Jurassic strata. Identification of rapidly evolving ammonite species allowed the fine subdivision of marine strata into successive "biozones" that could be correlated wherever the fossils were found.

Since Oppel's pioneering work, the development of similar "biozonal" schemes, using the fossils of a variety of life forms ranging from Cambrian marine trilobites to Cenozoic land-living rodents, has allowed strata of all ages and environments to be subdivided and correlated around the world.

With more than 103,000 living species and a fossil record extending back to the late Cambrian, the gastropods are one of the most successful and diverse molluscan groups. Growing up to 50cm (20in) long, with viscera that are twisted through 180 degrees, mostly protected by a coiled shell into which they can retreat. Gastropods have been particularly successful in spreading from marine to fresh water, and onto land.

Although today they number just 700 living marine species, including squid, cuttlefish, octopus, and the nautilus, the Cephalopoda were far more abundant in the geological past. In particular, the nautiloids and ammonoids with their coiled shells and prehensile tentacles were highly successful. Their fossil record extends back to late Cambrian times.

With a pair of carbonate shells up to 1m (40in) long to protect their bodies, bivalves have successfully diversified from early Cambrian seas to occupy a wide range of fresh-waters since the Devonian. Today they number some 12,000 species and have successfully adapted their shells to different habits such as burrowing, attachment to substrates, and free swimming to escape predators.

There are some 400 living species of these burrowing marine molluscs. Beneath an elongated, conical shell a few centimetres long, the body is highly modified with a greatly reduced head, a rudimentary respiratory system that lacks gills, and a cylindrical ventral foot for burrowing. Their fossil record extends back to Late Ordovician times.

◀ VISCEROCONCHA

LOBOCONCHA ▶

◀ CONCHIFERA

LOBOCONCHA

The Loboconcha are those conchiferans with enlarged shells that have developed to cover the entire body, with just the foot emerging. The group unites the bivalves with the scaphopods, and makes a marked contrast with the Visceroconcha (gastropods and cepalopods), in which the head can emerge for some distance out of the shell.

CONCHIFERA

Several important molluscan groups, totalling some 116,000 living species, are conchiferans. They are characterized by the possession of a single carbonate shell in the larval stage, which is then modified into different forms ranging from single cones to bivalves. In some conchiferan species, the shell is later absorbed into the body or lost entirely. The earliest fossil representatives, monoplacophorans and bivalves, appear in early Cambrian times.

EUMOLLUSCA

The large eumolluscan group of more than 117,000 living species is defined by the emergence of the muscular foot, gills, and dorsal mantle covering, with paired muscles and the ability to secrete protective structures made from calcium carbonate.

EUARTHROPODA

IT IS HARD TO EXAGGERATE THE IMPORTANCE OF THE EUARTHROPODA. HOWEVER THEY ARE MEASURED – WHETHER BY NUMBERS OF INDIVIDUALS, NUMBER OF SPECIES, DIVERSITY, OR BIOMASS – THIS GROUP OF PROTOSTOMES ECLIPSES ALL OTHER ANIMALS BY A LONG WAY, AND IS VITAL TO THE ECOSYSTEMS OF OUR PLANET.

With marine origins and a fossil record going back to the early Cambrian, the euarthropods were among the first animals to exploit a wide variety of niches. However, a number of extinct groups with unclear lineages raise the possibility that modern euarthropods may not share a unique common ancestor, and may not be a truly "monophyletic" group.

A tough, waterproof exoskeleton and paired, jointed appendages meant the euarthropods were preadapted for their invasion of the land from freshwater during the Ordovician. Further diversification on land has been integrally linked to the evolution of plants and consequently much of terrestrial life.

The hexapods (mostly insects) outnumber all other euarthropods. They have a fossil record that extends back to the Devonian, although preservation requires special conditions and fossils are not common. Nevertheless, now these conditions are better understood, increasing numbers of ancient insects are being recovered.

The Colorado beetle (*Leptinotarsa decemlineata*) is a representative of the wildly successful and diverse Coleoptera, part of the hexapod clade that, in terms of species, outnumbers all other life on Earth.

PYCNOGONIDA
sea spiders

Palaeoisopus

MEROSTOMATA
horseshoe crabs and eurypterids

Mesolimulus

ARACHNIDA
scorpions, spiders

Pulmonoscorpius

MYRIAPODA
centipedes, millipedes

millipede

The 1000 or so living species of carnivorous sea spiders are found at all depths of marine waters, where they typically feed on corals, sponges, and organic debris. They are mostly small, centimetre-sized euarthropods with small bodies, long paired and jointed legs, and a superficial resemblance to terrestrial spiders. However, their unique identifying features include a proboscis used to suck tissue directly from their prey. Although their fossil record is sparse, it extends back to the late Cambrian.

Today, just five species of mero-stomes survive. All are limulids, king crabs or horseshoe crabs up to 40cm (16in) long – a group whose fossil record extends back to mid-Cambrian times. They have a horseshoe-shaped carapace, a pair of chelicerae, five pairs of walking legs, and six pairs of legs on the opisthosoma that are modified as gills. The extinct, scorpion-like eurypterids, which grew to 2m (6.6ft) long, are also merostomes, with a fossil record from Ordovician to Permian times.

Most of the 75,000 or so species of living arachnids are centimetre-sized terrestrial spiders, but the group also includes ticks, mites, scorpions, uropygids (whip-scorpions), amblypygids and others. The prosoma of these animals is partly covered by a shield, and the appendages of the opisthosoma have been reduced or lost. Respiration is through the trachea (windpipe) and/or lungs. Their diversity results from a long fossil record extending back to the Silurian.

The myriapods are characterized by a body that is not divided into a thorax and abdomen, but instead has a distinct head and a large number of body segments, each of which carries a pair of legs. In the diplopods (millipedes) pairs of segments fuse and carry two pairs of legs. The 12,000 or so living species of centimetre-sized myriapods are primarily terrestrial, plant-eating detritivores that live in soil and leaf litter, but a fossil record stretching back to Silurian times reveals much larger extinct species.

CHELICERATA ▶

◀ CHELICERIFORMES

EUARTHROPODA

This grouping includes all arthropod groups with a hardened exoskeleton, and distinguishes them from other animals that are also sometimes classed as "arthropods".

◀ EUARTHROPODA

CHELICERATA

Chelicerata are characterized by the division of their bodies into a frontal prosoma with a carapace-like shield, a central opisthoma, and a telson (tail).

REMIPEDIA

CEPHALOCARIDA

MAXILLOPODA

BRANCHIOPODA

MALACOSTRACA
crabs, shrimps

HEXAPODA
insects

Lasionectes entrichoma

Lepidurus packardi

Valdiviella insignis

Lepidocaris

Crangospis

Notocupoides

The 11 living species of centimetre-sized remipedes were only recently discovered living in the submarine caves of the Gulf of Mexico. They have a distinct head carrying branched antennae, and numerous identical body segments carrying appendages for swimming. Their apparently primitive body plan suggests that they may be basal pancrustaceans, but contradictory evidence links them to the more advanced malacostracans. A sparse fossil record may extend back to the Carboniferous.

Nine species of millimetre-sized cephalocarids are all that survive from a much larger group whose fossil record extends back to Early Devonian times and possibly earlier. The living representatives of these small, bottom-dwelling aquatic animals have only recently been discovered. Their bodies have a three-fold division into a shielded head, a thorax with nine segments and paired appendages, and an abdomen that lacks appendages.

The 15,000 or more species included in the Maxillopoda are predominantly aquatic, millimetre- to centimetre-sized pancrustaceans, ranging from cirrepedes (barnacles), copepods, and ostracods, to the parasitic branchiurans. The head carapace is reduced, the thorax has a maximum of six segments, and the abdomen has no appendages. Their diversity and a fossil record extending back to Cambrian times may indicate that the group is in fact paraphyletic.

The branchiopods are millimetre- to centimetre-sized, mostly freshwater animals, but many of the 1000 or so living species can cope with more saline waters, and some have eggs that can survive prolonged dessication. These include the shrimp-like anostracans, notostracans, and conchostracans, and the cladocerans (water fleas). Typically the abdominal appendages have been lost. Their fossil record extends back to Devonian times.

This large group of some 23,000 living species has varied body forms, and lifestyles that range from swimming to burrowing. It includes familiar "eumalacostracans" such as the crabs, lobsters, and shrimps (decapods), sand hoppers (amphipods) and woodlice (isopods), as well as the phyllocarids. They first diversified in early Cambrian seas, and have since conquered most aquatic habitats, and some terrestrial ones.

With 900,000 or so living species, the hexapods are characterized by three pairs of walking legs. They include a remarkable diversity of body forms ranging from millimetre-sized beetles to butterflies with 20cm (8in) wing-spans, and lifestyles from the parasitic to the highly social. But all share several adaptations, such as a tracheal respiratory system, for a predominantly terrestrial mode of life. Their fossil record extends back to Early Devonian times.

PANCRUSTACEA ▶

MANDIBULATA (or ANTENNATA) ▶

MANDIBULATA

This grouping is characterized by paired mandibles appended to the head and used for feeding. These mandibles are toughened in groups such as the locusts and crabs.

PANCRUSTACEA

This group is marked by the possession of a particular larval form called the nauplius, with three pairs of swimming appendages attached to the head, and a telson. The monophyly of the group is supported by several molecular studies.

TAKING TO THE AIR

Flight was a major evolutionary adaptation first acquired by insects. The earliest fossils of flying hexapods are dragonfly-like animals from the Carboniferous, and there is debate over whether the first wings evolved from movable gills (such as those still found in mayfly larvae) or as extended lobes on segments of the thorax. However they evolved, in terms of numbers and versatility, insects are still by far the most successful fliers.

265

DEUTEROSTOMIA

TODAY'S LARGE ANIMALS ARE ALMOST EXCLUSIVELY VERTEBRATES,
A GROUP WHOSE ORIGINS LIE IN THE DEUTEROSTOMIAN GROUP
OF EARLY "BILATERIAN" ANIMALS. THE MAJOR DIVISION BETWEEN
PROTOSTOMES AND DEUTEROSTOMES WAS FIRST RECOGNIZED ON
EMBRYOLOGICAL EVIDENCE IN THE 19TH CENTURY, BUT THE RISE
OF MOLECULAR BIOLOGY HAS LED TO SIGNIFICANT REVISION OF
LONG-HELD BELIEFS ABOUT THEIR CLASSIFICATION.

The deuterostomes are defined by features of their embryological development and, in particular, by the fact that it is the anus, rather than the mouth, that first develops from the blastopore of the growing embryo (in contrast to the protostomes the name deuterostome can be translated as "mouth second"). Other significant differences include the fact that the embryo's outer layer or ectoderm gives rise to the nervous system and skin epidermis.

The group has an extraordinary diversity that extends from starfish and seasquirts to humans. The inclusion of the former groups might seem strange when their adult forms are considered, but the embryological and larval development of these echinoderm and urochordate groups reveals their deuterostome connections. Echinoderms undergo a radical change (metamorphosis) from a bilaterally symmetrical larva to produce a radially symmetrical adult form. The acquisition of the "chordate" condition, (as seen primitively in larval seasquirts) marks an important innovation on the evolutionary path towards true vertebrates.

Tracking the earliest fossil record of the deuterostomes depends on the interpretation of a few small, soft-bodied animals from the Cambrian deposits of Chengjiang in China and Canada's Burgess Shale. Further finds of early deuterostomes with their soft tissues preserved are sure to bring further insights into the group's origins.

ECHINODERMATA Starfish, sea urchins
Salteraster

HEMICHORDATA Graptolites
Monograptus

UROCHORDATA Seasquirts
Corella parallelogramma

The 6000 or so living species of echinoderms grow up to a few tens of centimetres across, and include the starfish (Aster-oidea), brittlestars (Ophiuroidea), sea urchins (Echinoidea), sea lilies (Crinoidea), and the less familiar Holothuroidea or sea cucumbers. All have a unique calcium carbonate endoskeleton made from numerous plates or spicules, and as a result they have left a good fossil record extending back to early Paleozoic times. Exclusively marine, they have bilaterally symmetrical and free-swimming "pelagic" larvae that typically develop into adults with five-fold symmetry.

The 85 living species of marine hemichordates include the acorn worms, and colonial forms that live in linear arrangements within tubular organic skeletons, growing to a maximum size of a few tens of centimetres. The body is divided into three parts, with a distinctive proboscis. Their fossil record goes back to Cambrian times and includes a large number of extinct groups. Because many of these evolved rapidly, they are useful for the relative dating of early Paleozoic marine strata.

The urochordates consist of 1300 or so living species of marine sea-squirts (which live a static or "sessile" life on the sea-floor), and free-swimming salps (some of which are colonial) and appen-dicularians. They grow to some 10cm (4in) and show little resemblance between adult and larval forms. The larva swims freely, with a tail and notochord that are lost during meta-morphosis into the adult form. Their fossil record is poor but extends back to Cambrian times.

DEUTEROSTOMIA ▶

Modern stingrays (Dasyatidae) are living examples of the ancient chondricthyan line of cartilaginous fish – the first vertebrates to evolve jaws.

CHORDATA

More than 52,000 living species, from seasquirts to humans, comprise this grouping, united by the possession of a notochord (flexible support rod) at some stage in development

CEPHALOCHORDATA
Lancelets

MYXINOIDEA
Hagfish

PETROMYZONTIFORMES
Lampreys

CHONDRICHTHYES
Sharks, rays

SARCOPTERYGII
Lobe-finned fish
See page 268

ACTINOPTERYGII
Ray-finned fish
See page 268

Pikaia

hagfish

Lampetra

Hybodus

Onychodus

Dastilbe

The living lancelets are small chordates, up to 8cm (3in) long. About a dozen species are known, all of which live in tropical seas as shallow burrowing filter-feeders that suck water through their mouths and through branchial slits (primitive gills) for feeding and respiration. The stiff but flexible dorsal notochord is flanked by segmented muscles that allow the animal to swim. *Pikaia*, from Cambrian rocks in Canada, may be a fossil cephalochordate.

There are some 20 living species of hagfish – eel-like marine craniates that grow to about 60cm (24in) and burrow into seabed sediment. They lack jaws, but the mouth is surrounded by tooth plates that form a rasping apparatus for biting and sucking flesh from their prey. The arrangement of the gills differs from that in vertebrates, and their fossil record extends back into Carboniferous times.

The 40 or so living species of jawless lampreys have freshwater larvae and mostly marine adults that follow a parasitic lifestyle using their sucker-like mouth and horny teeth. The elongated, eel-shaped body, up to 1m (40in) long, has a specialized cartilaginous skeleton without paired fins. Their fossil record may extend back to the Cambrian.

Today, cartilaginous fish such as sharks and rays grow up to several metres long and are widespread but purely marine. There are 850 living species, characterized by special copulatory organs in the males. A rich fossil record from Silurian times onward includes both marine and freshwater forms.

The 26,000 or more living species in this group range from lungfish to humans, with a fossil record extending back to Silurian times. Defining features include paired fins attached to the pelvic and pectoral girdles by a single articulation.

The 24,000 or so living species of this group comprise the ray-finned fish (ie excluding cartilaginous and lobe-finned fish), and have a fossil record extending back to Silurian times.

◀ OSTEICHTHYES

◀ GNATHOSTOMATA

◀ VERTEBRATA

◀ CRANIATA

◀ CHORDATA

CRANIATA
The development of a braincase (skull) around the sense organs at the front of the body characterizes the craniate chordates. This large group has a fossil record extending back to Cambrian times.

VERTEBRATA
The possession of backbones (vertebrae) surrounding and replacing the notochord defines this major group of more than 50,000 living species. The inner ear has two semicircular canals, and there are extrinsic eye muscles and lateral sensory lines. The fossil record of the vertebrates extends back to Cambrian times.

GNATHOSTOMATA
This large group of vertebrates is defined by the possession of biting jaws that develop from the foremost gill arches. The gills are located outside the branchial skeleton and the inner ear has developed a third semicircular canal. The extinct acanthodians, which originated in Late Ordovician times, preserve a primitive stage in the evolution of the jaws.

CLASSIFICATION PROBLEMS
There are several unsolved problems in the inter-relationships between deuterostomes. Questions have been raised over the status of the hemichordates, and particularly whether they are monophyletic or not (since the organization of several features such as the gill slits seems to be different from the chordates). There is also a curious group of extinct asymmetrical organisms, with a skeleton of echinoderm-like calcareous plates, called the calcichordates. Their relationship to both the echinoderms and the chordates is still a subject of debate, as are the relationships of hagfish and lampreys to a number of extinct jawless fish-like groups.

OSTEICHTHYES

VERTEBRATE EVOLUTION HAS PASSED THROUGH A SERIES OF STAGES
MARKED BY THE DEVELOPMENT OF SIGNIFICANT NEW FEATURES. THE
DEVELOPMENT OF JAWS DEFINED THE GNATHOSTOMES, A GROUP
THAT UNITES THE CARTILAGINOUS CHONDRICHTHYANS WITH THE
OSTEICHTHYANS OR BONY FISH. THIS LATTER GROUP IS SUBDIVIDED
INTO ACTINOPTERYGIANS (RAY-FINNED FISH), AND LOBE-FINNED
SARCOPTERYGIANS WITH THEIR TETRAPOD DESCENDANTS.

A common feature of osteichthyans is their internal bony skeleton,
modified from the cartilaginous skeleton of other vertebrates. With
more than 50,000 living species, they incorporate most of what are
commonly considered "fish" (excluding the chondrichthyan
sharks and rays), but also, in cladistic terms, include the hugely
diverse tetrapods, ranging from salamanders, snakes, and birds to
humans. Both osteichthyans and chondrichthyans share a
common gnathostome ancestor, probably somewhere in the
Silurian, but this important distinction demonstrates why old
terms such as "fish" (the Linnaean Class Pisces) are obsolete in
scientific terms, since they are not truly monophyletic groupings.

There are two types of osteichthyan bone – endochondral and
dermal. The latter is a more ancient tissue whose origins lie in the
extinct agnathans as a covering of bony plates protecting head and
thorax. Endochondral bone, meanwhile, replaces cartilage in the
internal endoskeleton. In addition to vertebrae and a pelvic girdle,
this skeleton has a series of "branchial arches" at the front, with
the foremost modified to form biting jaws armed with bony teeth.

Another important innovation is the possession of air sacs that
are mostly used to regulate buoyancy, but were originally lungs.
Their presence in emerging Devonian fish groups was an
important aid to survival in tropical freshwaters with low oxygen.

The angelfish (Pomacanthidae) are typical examples of the bony actinopterygian fish, with
fins given rigidity by unjointed "rays" of dermal bone.

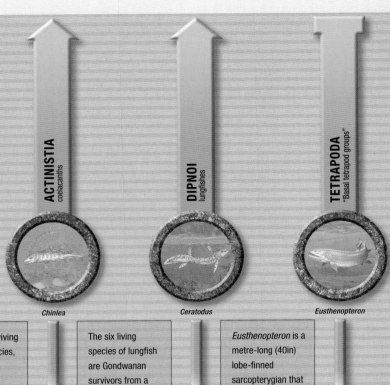

ACTINISTIA coelacanths

Chinlea

DIPNOI lungfishes

Ceratodus

TETRAPODA "Basal tetrapod groups"

Eusthenopteron

The single surviving
actinistian species,
the coelacanth
*Latimeria
chalumnae*, is a
remarkable "living
fossil" – sole
survivor of a
much larger and
diverse group of
coelacanths with
paired lobefins,
growing to
2m (6.6ft) long.
They originated in
Devonian times and
were thought to
have died out in the
Cretaceous.
Coelacanths have a
bizarre internal fin
skelton to their anal
and front dorsal
fins, resembling the
skeletons of their
pectoral and pelvic
fins. The tail fin is
symmetrical about
a small lobe that
is an extension of
the notochord.

The six living
species of lungfish
are Gondwanan
survivors from a
much larger group
of sarcopterygians
that evolved during
Devonian times
in tropical marine
and fresh waters.
Growing to some
2m (6.6ft) long, the
dipnoans have both
gills and functioning
lungs that allow
them to obtain
oxygen directly
from the
atmosphere, and
thus survive in
ephemeral rivers
and lakes. Fossil
lungfish can be
identified by the
distinctive forms
of their teeth.

Eusthenopteron is a
metre-long (40in)
lobe-finned
sarcopterygian that
shares some unique
features with early
tetrapods, including
the arrangement of
bones in its paired
limbs, and the
structure of skull,
teeth, and internal
nostrils. More than
2000 specimens
have been
recovered from the
Late Devonian
strata of Quebec,
Canada, including
juveniles that
(unlike later
tetrapods) show no
sign of a larval
stage followed by
metamorphosis.

RHIPIDISTIA ▶

◀ SARCOPTERYGII

OSTEICHTHYES ▶

SARCOPTERYGII

The 26,000 or so living species of sarcopterygian are defined by the possession of
muscular lobed fins, with narrow bases attached to the skeleton's pelvic girdles by a
single articulation.

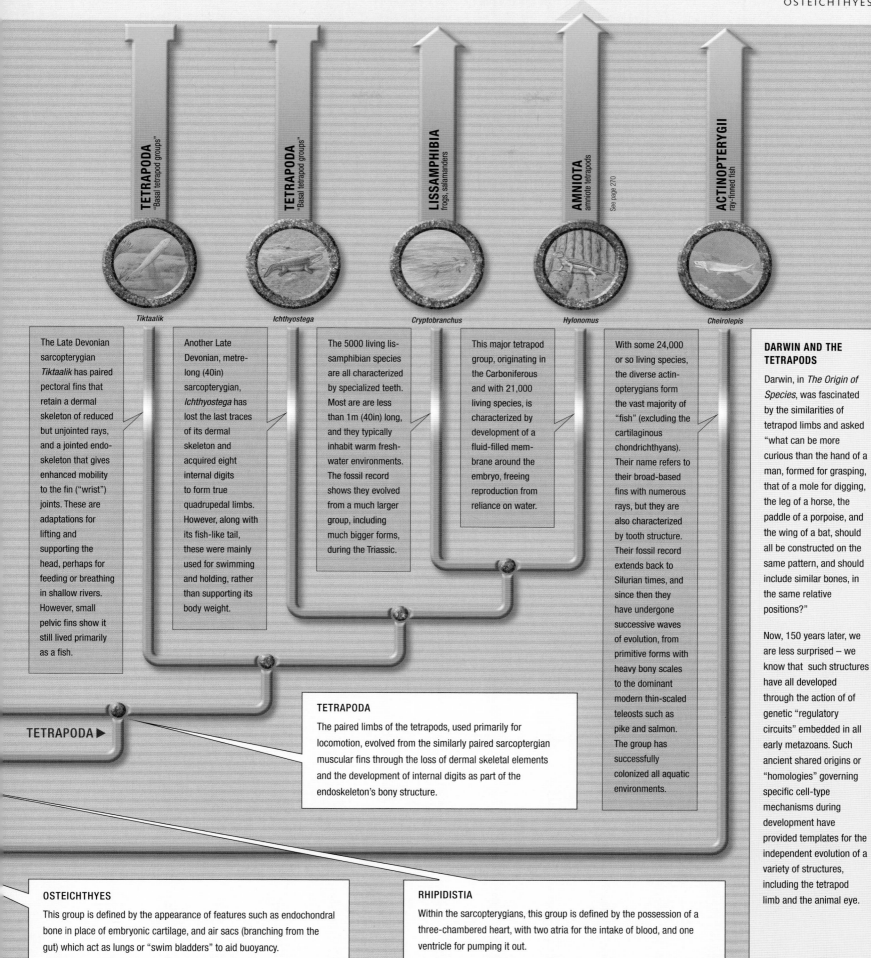

TETRAPODA
"Basal tetrapod groups"

TETRAPODA
"Basal tetrapod groups"

LISSAMPHIBIA
frogs, salamanders

AMNIOTA
amniote tetrapods

See page 270

ACTINOPTERYGII
ray-finned fish

Tiktaalik

Ichthyostega

Cryptobranchus

Hylonomus

Cheirolepis

The Late Devonian sarcopterygian *Tiktaalik* has paired pectoral fins that retain a dermal skeleton of reduced but unjointed rays, and a jointed endo-skeleton that gives enhanced mobility to the fin ("wrist") joints. These are adaptations for lifting and supporting the head, perhaps for feeding or breathing in shallow rivers. However, small pelvic fins show it still lived primarily as a fish.

Another Late Devonian, metre-long (40in) sarcopterygian, *Ichthyostega* has lost the last traces of its dermal skeleton and acquired eight internal digits to form true quadrupedal limbs. However, along with its fish-like tail, these were mainly used for swimming and holding, rather than supporting its body weight.

The 5000 living lis-samphibian species are all characterized by specialized teeth. Most are are less than 1m (40in) long, and they typically inhabit warm fresh-water environments. The fossil record shows they evolved from a much larger group, including much bigger forms, during the Triassic.

This major tetrapod group, originating in the Carboniferous and with 21,000 living species, is characterized by development of a fluid-filled mem-brane around the embryo, freeing reproduction from reliance on water.

With some 24,000 or so living species, the diverse actin-opterygians form the vast majority of "fish" (excluding the cartilaginous chondrichthyans). Their name refers to their broad-based fins with numerous rays, but they are also characterized by tooth structure. Their fossil record extends back to Silurian times, and since then they have undergone successive waves of evolution, from primitive forms with heavy bony scales to the dominant modern thin-scaled teleosts such as pike and salmon. The group has successfully colonized all aquatic environments.

DARWIN AND THE TETRAPODS

Darwin, in *The Origin of Species*, was fascinated by the similarities of tetrapod limbs and asked "what can be more curious than the hand of a man, formed for grasping, that of a mole for digging, the leg of a horse, the paddle of a porpoise, and the wing of a bat, should all be constructed on the same pattern, and should include similar bones, in the same relative positions?"

Now, 150 years later, we are less surprised – we know that such structures have all developed through the action of of genetic "regulatory circuits" embedded in all early metazoans. Such ancient shared origins or "homologies" governing specific cell-type mechanisms during development have provided templates for the independent evolution of a variety of structures, including the tetrapod limb and the animal eye.

TETRAPODA ▶

TETRAPODA

The paired limbs of the tetrapods, used primarily for locomotion, evolved from the similarly paired sarcoptergian muscular fins through the loss of dermal skeletal elements and the development of internal digits as part of the endoskeleton's bony structure.

OSTEICHTHYES

This group is defined by the appearance of features such as endochondral bone in place of embryonic cartilage, and air sacs (branching from the gut) which act as lungs or "swim bladders" to aid buoyancy.

RHIPIDISTIA

Within the sarcopterygians, this group is defined by the possession of a three-chambered heart, with two atria for the intake of blood, and one ventricle for pumping it out.

TETRAPODA

TRADITIONAL IDEAS ABOUT THE TETRAPODS – FOUR-LIMBED
VERTEBRATES SUCH AS AMPHIBIANS, MAMMALS, REPTILES, AND
BIRDS – HAVE LATELY UNDERGONE A RADICAL TRANSFORMATION.
RECENT DISCOVERIES AND NEW TECHNIQUES OF ANALYSIS FOR
BOTH LIVING AND FOSSIL REPRESENTATIVES OF THIS LARGE AND
DIVERSE GROUP HAVE BROUGHT AN IMPROVED UNDERSTANDING
OF BOTH THEIR ORIGINS AND THEIR INTER-RELATIONSHIPS.

The 26,000 or more living species of tetrapod range from frogs and
salamanders through mammals, turtles, and lizards, to birds and
crocodiles. They are characterized by four limbs ending in digits,
although these may be secondarily reduced and even lost in groups
such as the gymniophone lissamphibians, squamate snakes, and
some cetacean mammals. Nevertheless, a fossil record stretching
back to the Devonian shows that the primitive tetrapod limb pairs
are derived from the paired fins of extinct Devonian
sarcopterygian fish, and proves that they have subsequently been
adapted for a variety of different modes of locomotion in different
groups: the underlying bony structures in the wings of a bird and a
bat, the swimming flippers of a seal, and the arm of a chimp are
all homologous.

Other notable tetrapod innovations include the appearance of a
neck, with the pectoral girdle separated from the head and the
specialization of the first neck vertebrae. The nostril and nasal sac
becomes linked to the eye via the lacrimal duct (a condition
seen in extinct sarcopterygian "tetrapodomorph" groups, such as
the panderichthyids and osteolepiformes). Additionally, one of the
jawbones, the hyomandibular, is transferred to the auditory system,
becoming the stapes bone of the inner ear.

The Galapagos tortoise (*Geochelone niger*) is the largest living representative of the Chelonia, a group
that first appeared in Triassic times, more than 200 million years ago.

GYMNIOPHONA caecilians

Boulengerula boulengeri

URODELA salamanders

Chunerpeton

ANURA frogs

Callobatrachus

Some 160 living
species of
"caecilians" –
worm-shaped and
legless amphibians
growing up to 1m
(40in) long – live
mostly in moist
terrestrial habitats.
The reduction
and loss of their
limbs along with
their girdles is
a secondary
adaptation for a
burrowing mode
of life, whose
development is
recorded through
fossil forms as far
back as Jurassic
times. The
Gymniophona have
also developed
a unique retractile
sensory "tentacle"
that can be
extended from
the animal's mouth.

The 400 or so
species of living
salamanders have
lizard-shaped
bodies and grow
to nearly 2m (80in)
long. They are
mostly terrestrial,
but depend
upon water for
reproduction. Some
salamanders are
"neotenous",
reaching sexual
maturity while
retaining juvenile
features. In general,
though, they
show no unique
anatomical features.
However, their
monophyly is
supported by
genetic analysis,
and their fossil
record extends back
to Jurassic times.

Frogs and toads
have adapted to life
in a wide range of
terrestrial habitats
since they first
evolved in Jurassic
times. They are
characterized by the
unique structure of
their pelvis, with an
elongated ilium
bone that is linked
to their hopping
mode of locomotion.
Today, there are
around 4300 living
species, making
them by far the
most successful
lissamphibians,
although many are
endangered.

BATRACHIA ▶

◀ LISSAMPHIBIA

TETRAPODA ▶

LISSAMPHIBIA

The 5000 living amphibian species
are grouped as lissamphibians and
characterized by features such as
their tooth structure, short or absent
ribs, and unique eye musculature. The
first fossils date to the Early Triassic.

TETRAPODA

The tetrapods are distinguished by
the development of paired limbs
with internal bony digits.

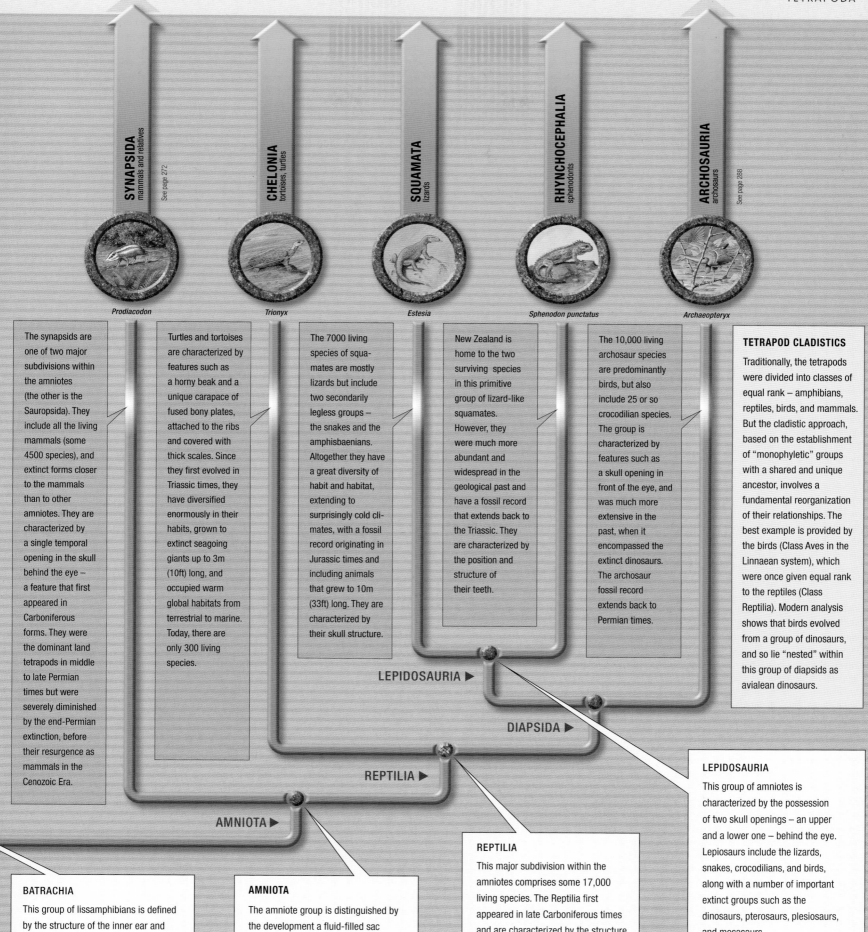

SYNAPSIDA
mammals and relatives
See page 272

CHELONIA
tortoises, turtles

SQUAMATA
lizards

RHYNCHOCEPHALIA
sphenodonts

ARCHOSAURIA
archosaurs
See page 288

Prodiacodon

Trionyx

Estesia

Sphenodon punctatus

Archaeopteryx

The synapsids are one of two major subdivisions within the amniotes (the other is the Sauropsida). They include all the living mammals (some 4500 species), and extinct forms closer to the mammals than to other amniotes. They are characterized by a single temporal opening in the skull behind the eye – a feature that first appeared in Carboniferous forms. They were the dominant land tetrapods in middle to late Permian times but were severely diminished by the end-Permian extinction, before their resurgence as mammals in the Cenozoic Era.

Turtles and tortoises are characterized by features such as a horny beak and a unique carapace of fused bony plates, attached to the ribs and covered with thick scales. Since they first evolved in Triassic times, they have diversified enormously in their habits, grown to extinct seagoing giants up to 3m (10ft) long, and occupied warm global habitats from terrestrial to marine. Today, there are only 300 living species.

The 7000 living species of squamates are mostly lizards but include two secondarily legless groups – the snakes and the amphisbaenians. Altogether they have a great diversity of habit and habitat, extending to surprisingly cold climates, with a fossil record originating in Jurassic times and including animals that grew to 10m (33ft) long. They are characterized by their skull structure.

New Zealand is home to the two surviving species in this primitive group of lizard-like squamates. However, they were much more abundant and widespread in the geological past and have a fossil record that extends back to the Triassic. They are characterized by the position and structure of their teeth.

The 10,000 living archosaur species are predominantly birds, but also include 25 or so crocodilian species. The group is characterized by features such as a skull opening in front of the eye, and was much more extensive in the past, when it encompassed the extinct dinosaurs. The archosaur fossil record extends back to Permian times.

TETRAPOD CLADISTICS

Traditionally, the tetrapods were divided into classes of equal rank – amphibians, reptiles, birds, and mammals. But the cladistic approach, based on the establishment of "monophyletic" groups with a shared and unique ancestor, involves a fundamental reorganization of their relationships. The best example is provided by the birds (Class Aves in the Linnaean system), which were once given equal rank to the reptiles (Class Reptilia). Modern analysis shows that birds evolved from a group of dinosaurs, and so lie "nested" within this group of diapsids as avialean dinosaurs.

LEPIDOSAURIA ▶

DIAPSIDA ▶

REPTILIA ▶

AMNIOTA ▶

BATRACHIA

This group of lissamphibians is defined by the structure of the inner ear and connections of the outer and inner nostrils.

AMNIOTA

The amniote group is distinguished by the development a fluid-filled sac around the developing embryo.

REPTILIA

This major subdivision within the amniotes comprises some 17,000 living species. The Reptilia first appeared in late Carboniferous times and are characterized by the structure of their vertebrae.

LEPIDOSAURIA

This group of amniotes is characterized by the possession of two skull openings – an upper and a lower one – behind the eye. Lepiosaurs include the lizards, snakes, crocodilians, and birds, along with a number of important extinct groups such as the dinosaurs, pterosaurs, plesiosaurs, and mosasaurs.

SYNAPSIDA

THE SYNAPSIDS INCLUDE ALL THE LIVING MAMMALS AND
THEIR EXTINCT RELATIVES EXTENDING BACK TO THE LATE
CARBONIFEROUS – GROUPS THAT HAVE IN THE PAST
BEEN GIVEN THE MISLEADING NAME OF "MAMMAL-LIKE
REPTILES". DESPITE THEIR RANGE AND DIVERSITY, THESE
EXTINCT FORMS ARE STILL RELATIVELY UNKNOWN
OUTSIDE THE WORLD OF PALEONTOLOGY.

The appearance of the synapsids was a major
event in amniote evolution, and can be
reconstructed through the fossil record of its
various stages. The group was originally defined
by the presence of a lower temporal opening
behind the eye, which provided space for the
passage of enlarged jaw muscles. This
in turn was connected to improvements in
processing food through tooth specialization,
which in turn required restructuring of the lower
jaw and its articulation. The snout region was
extended in the carnivorous forms, but took on a
very different shape in some of the plant-eaters,
where the front teeth were lost and replaced with
a horny, turtle-like beak for shearing through
tough plant material.

Synapsids arose in equatorial regions in late
Carboniferous times, and were initially lizard-like
in appearance. The Permian and Triassic saw the
origination and rapid expansion of successively
more advanced groups, many of which soon died
out to be replaced by new forms.

CASEASAURIA

Cotylorhyncus

One of two main
groups of early
synapsids, the
caseasaurians are
only known from
Permian times.
They include two
famiies – the
large plant-eating
caseids, which
grew up to
4m (13ft) long,
and the smaller
carnivorous
eothyridids, which
look superficially
different but are
united by the
structure of the
snout and nostrils.
They were a
successful group
in early Permian
times, but by the
middle of the
period were
challenged by the
rapidly evolving
therapsids. They
ultimately died out
in the end-Permian
extinction.

VARANOPSEIDAE

Varanops brevirostris

As their name
suggests, these
small- to medium-
sized synapsids,
which grew up to
1.5m (5ft) long,
resemble varanid
(monitor) lizards
with slender bodies
and long limbs.
The slender, lightly
built skull with
large eyes, a long
narrow snout,
and jaws with
specialized sharp,
recurved teeth, was
adapted for hunting
down small but
active prey. The
varanopseids were
probably the most
agile predators of
the time. They
evolved during the
late Carboniferous
and died out in late
Permian times.

OPHIACODONTIDAE

Archaeothyris

This group evolved
in the late Carbon-
iferous and died out
in mid-Permian
times. It includes
a number of long-
snouted forms up to
3m (10ft) long, and
the earliest known
pelycosaur – the
50cm (20in)
Archaeothyris,
which was
discovered in the
bole of a fossil
lycopod tree,
preserved by late
Carboniferous "coal
measure" strata at
Joggins in Nova
Scotia. Its slender
snout is armed with
sharply pointed
teeth, one of which
shows a developed
canine-like form.
The group includes
aquatic, semi-
aquatic, and fully
terrestrial animals.

EDAPHOSAURIDAE

Edaphosaurus

This group of
specialized plant-
eaters, up to 3m
(10ft) in length, are
among the earliest
known herbivorous
amniotes. They bear
a superficial
resemblance to the
caseids, but differ
in skull details. The
head was very
small in relation to a
bulky body, but the
jaw had powerful
muscles and teeth
adapted for
cropping vegetation.
Elongated neural
spines supported a
large dorsal "sail".
They evolved in late
Carboniferous times
and became
extinct in the
middle Permian.

SYNAPSIDA ▶

Inostrancevia from the Permian of Russia is a
member of the short-lived gorgonopsian group
of synapsids.

SYNAPSIDA

This very large group is characterized by several
features of the skull and teeth, such as the presence
of a lower temporal opening and the development of
canine-like "caniniforms".

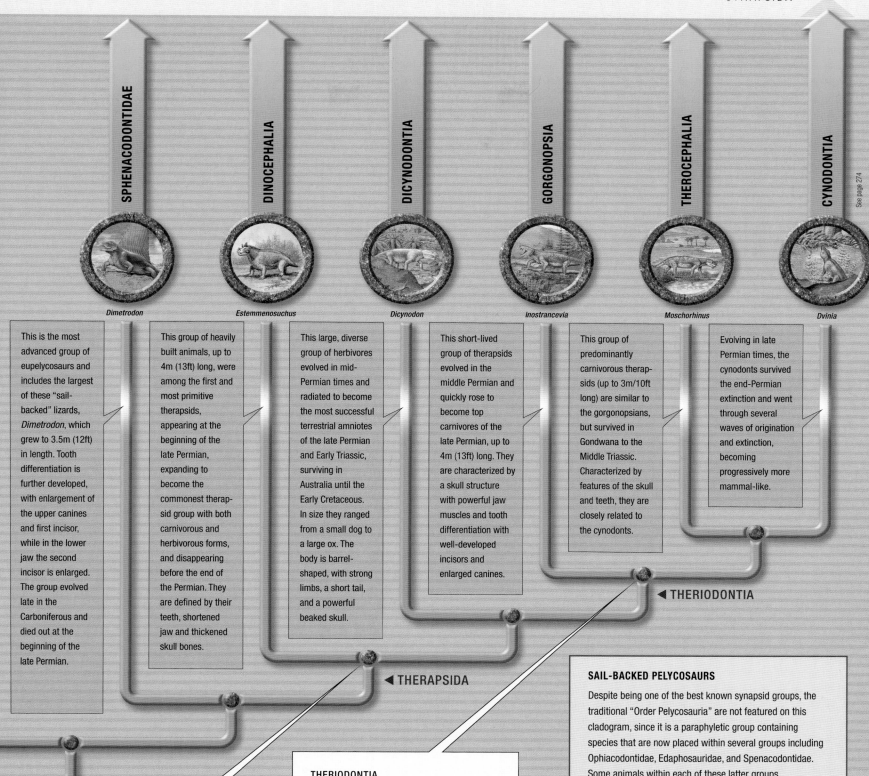

SPHENACODONTIDAE

Dimetrodon

DINOCEPHALIA

Estemmenosuchus

DICYNODONTIA

Dicynodon

GORGONOPSIA

Inostrancevia

THEROCEPHALIA

Moschorhinus

CYNODONTIA

Dvinia

See page 274

This is the most advanced group of eupelycosaurs and includes the largest of these "sail-backed" lizards, *Dimetrodon*, which grew to 3.5m (12ft) in length. Tooth differentiation is further developed, with enlargement of the upper canines and first incisor, while in the lower jaw the second incisor is enlarged. The group evolved late in the Carboniferous and died out at the beginning of the late Permian.

This group of heavily built animals, up to 4m (13ft) long, were among the first and most primitive therapsids, appearing at the beginning of the late Permian, expanding to become the commonest therapsid group with both carnivorous and herbivorous forms, and disappearing before the end of the Permian. They are defined by their teeth, shortened jaw and thickened skull bones.

This large, diverse group of herbivores evolved in mid-Permian times and radiated to become the most successful terrestrial amniotes of the late Permian and Early Triassic, surviving in Australia until the Early Cretaceous. In size they ranged from a small dog to a large ox. The body is barrel-shaped, with strong limbs, a short tail, and a powerful beaked skull.

This short-lived group of therapsids evolved in the middle Permian and quickly rose to become top carnivores of the late Permian, up to 4m (13ft) long. They are characterized by a skull structure with powerful jaw muscles and tooth differentiation with well-developed incisors and enlarged canines.

This group of predominantly carnivorous therapsids (up to 3m/10ft long) are similar to the gorgonopsians, but survived in Gondwana to the Middle Triassic. Characterized by features of the skull and teeth, they are closely related to the cynodonts.

Evolving in late Permian times, the cynodonts survived the end-Permian extinction and went through several waves of origination and extinction, becoming progressively more mammal-like.

◄ **THERIODONTIA**

◄ **THERAPSIDA**

THERAPSIDA

This grouping includes the biarmosuchians, dinocephalians, dicynodonts, and theriodonts. It is characterized by enlarged fenestrae (the typical synapsid holes in the skull), further specialization of the teeth into incisors, canines, and molars, and a more upright posture.

THERIODONTIA

This group is characterized by a number of skull features such as laterally flaring cheek bones, more specialized and mammal-like teeth, and an enlarged jawbone with improved surfaces for muscle attachment. The theriodonts include the gorgonopsians, therocephalians, cynodonts, and all their mammalian descendants. Apart from cynodonts, the last theriodonts became extinct in the Early Triassic.

SAIL-BACKED PELYCOSAURS

Despite being one of the best known synapsid groups, the traditional "Order Pelycosauria" are not featured on this cladogram, since it is a paraphyletic group containing species that are now placed within several groups including Ophiacodontidae, Edaphosauridae, and Spenacodontidae. Some animals within each of these latter groups independently evolved a strange dorsal "sail" or "fin" structure on their backs – a web of skin supported by extreme elongations of the neural spines from the vertebrae. Probably the best known of these "sail-backed" species are the carnivorous sphenacodontid *Dimetrodon* and the herbivorous *Edaphosaurus*. The function of the sail structure has generated a great deal of argument, but the two most popular suggestions are that it may have been used in mating displays, or for heat-exchange (allowing sunlight to warm the animal in cold conditions, and helping the animal to cool down when hot).

CYNODONTIA

SYNAPSID EVOLUTION WENT THROUGH A SERIES OF STAGES FROM THE EARLIEST LIZARD-LIKE FORMS, WITH THE DEVELOPMENT OF INCREASINGLY MAMMAL-LIKE FEATURES UNTIL THE PERMIAN CYNODONTS WERE BECOMING VERY CLOSE TO THE TRUE MAMMAL CONDITION. THE TEETH OF BOTH HERBIVORES AND CARNIVORES SHOW INCREASING DIFFERENTIATION, ALLOWING MORE EFFECTIVE PROCESSING OF THE FOOD BEFORE DIGESTION.

Alongside these improvements came an enlargment of the dentary bone of the jaw and modification of its articulation to cope with more powerful jaw musculature. The smaller bones of the lower jaw are displaced and become part of an improved auditory system. A further advance is seen in the formation of the secondary palate, which first appears in the therocephalian synapsids. This "roof" to the mouth allows a separation of breathing and feeding, so that both can occur together.

Another progressive change is seen in the stance and gait, with the adoption of a more upright position. The limbs are brought in from their sideways-splayed primitive stance, until they are positioned beneath the body, with their articulation to the pelvic and pectoral girdles duly modified. As a result, the feet can be positioned close to the mid-line of the body, which allows for more efficient locomotion – especially in running. Such changes impact upon the movement of the backbone and the coordination of running and breathing. The spine's motion changes from sideways to up-and-down, and this in turn requires modifications to the vertebrae and their articulations with one another and the skull.

The back of the skull, meanwhile, is enlarged to accommodate a growing brain and elaborated sense organs of sight, smell, and hearing – improved senses offered unquestionable benefits for many of these otherwise defenceless, small, or moderate sized animals.

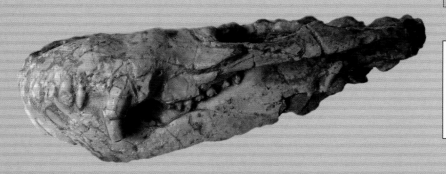

The long skull of *Procynosuchus* displays the beginnings of specialized cheek teeth and other developments that would ultimately develop until the cynodonts became true mammals.

PROCYNOSUCHIDAE

Procynosuchus

The first cynodonts from the very late Permian show a number of mammalian features in the skull and lower jaw, providing increased efficiency in pre-processing food for digestion. For instance, there is a well-marked crest on the skull for the attachment of jaw muscles, and the zygomatic arch (cheek bone) is widely curved to accommodate these muscles. The dentary bone grows to comprise most of the lower jaw, cheek teeth become increasingly elaborate, and the secondary palate, separating off the nasal passage from the mouth (in order to eat and breath at the same time) is almost complete.

THRINAXODONTIDAE

Thrinaxodon

The thrinaxodonts show further changes such as modified bones at the base of the skull for better articulation with the first vertebrae of the backbone. By contrast, reptiles and more primitive synapsids have only a single condyle. The back bones are further modified into specific thoracic and lumbar forms – the latter with short, fused ribs. The tail is growing more elongated and slender, while the hind limbs and their articulation with the pelvic girdle are modified for a more upright posture, with the limbs pulled under the body.

CYNOGNATHIDAE

Cynognathus

This carnivorous eucynodont group is typified by *Cynognathus*, a large predator whose 30cm (12in) skull displays powerful jaw muscles and a dentary bone comprising more than 90 per cent of the lower jaw. Cynognathids also show further development in the jaw towards mammal-like jaw articulation. They were dog-sized terrestrial predators of Early and Middle Triassic age, but their group's position within the eucynodonts is a matter of dispute.

◀ CYNODONTIA

CYNODONTIA

This diverse group of theriodonts includes the living mammals and their extinct relatives. They range in age from the late Permian to the middle Cretaceous (and to the present when the mammals are taken into account).

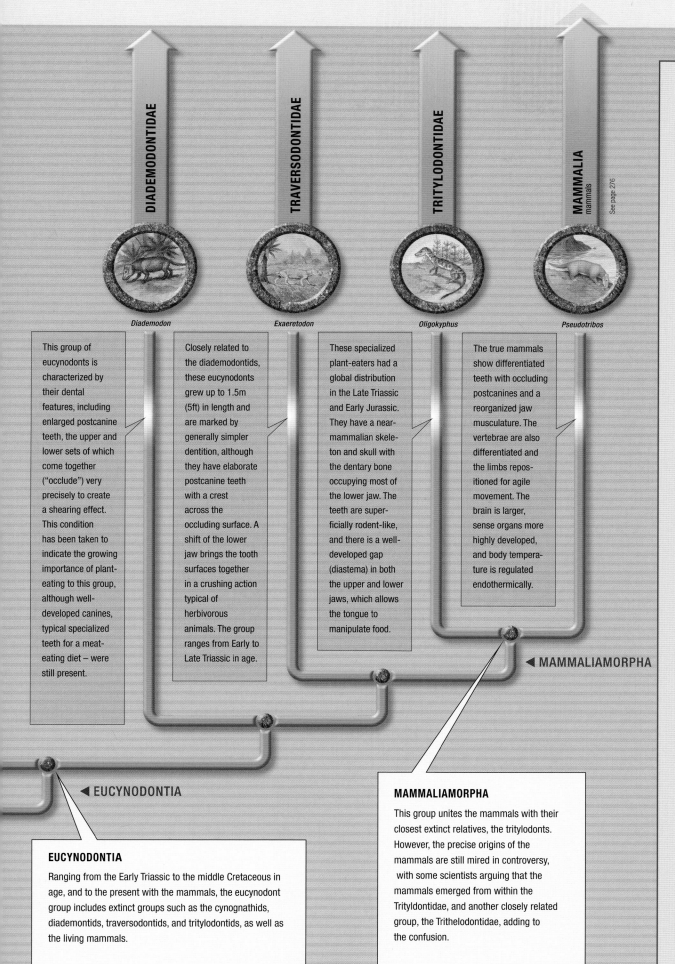

DIADEMODONTIDAE

Diademodon

TRAVERSODONTIDAE

Exaeretodon

TRITYLODONTIDAE

Oligokyphus

MAMMALIA
mammals

See page 276

Pseudotribos

This group of eucynodonts is characterized by their dental features, including enlarged postcanine teeth, the upper and lower sets of which come together ("occlude") very precisely to create a shearing effect. This condition has been taken to indicate the growing importance of plant-eating to this group, although well-developed canines, typical specialized teeth for a meat-eating diet – were still present.

Closely related to the diademodontids, these eucynodonts grew up to 1.5m (5ft) in length and are marked by generally simpler dentition, although they have elaborate postcanine teeth with a crest across the occluding surface. A shift of the lower jaw brings the tooth surfaces together in a crushing action typical of herbivorous animals. The group ranges from Early to Late Triassic in age.

These specialized plant-eaters had a global distribution in the Late Triassic and Early Jurassic. They have a near-mammalian skeleton and skull with the dentary bone occupying most of the lower jaw. The teeth are superficially rodent-like, and there is a well-developed gap (diastema) in both the upper and lower jaws, which allows the tongue to manipulate food.

The true mammals show differentiated teeth with occluding postcanines and a reorganized jaw musculature. The vertebrae are also differentiated and the limbs repositioned for agile movement. The brain is larger, sense organs more highly developed, and body temperature is regulated endothermically.

◄ MAMMALIAMORPHA

◄ EUCYNODONTIA

EUCYNODONTIA

Ranging from the Early Triassic to the middle Cretaceous in age, and to the present with the mammals, the eucynodont group includes extinct groups such as the cynognathids, diademontids, traversodontids, and tritylodontids, as well as the living mammals.

MAMMALIAMORPHA

This group unites the mammals with their closest extinct relatives, the tritylodonts. However, the precise origins of the mammals are still mired in controversy, with some scientists arguing that the mammals emerged from within the Tritylodontidae, and another closely related group, the Trithelodontidae, adding to the confusion.

THE EVOLUTION OF ENDOTHERMY

So-called "warm-bloodedness" is a key mammalian feature. More correctly known as endothermy, this ability to regulate internal body temperature and maintain it independently of warm or cold surroundings gives mammals some significant evolutionary advantages compared to "cold-blooded" animals such as reptiles.

Endothermy allows mammals to maintain a high metabolism and remain active at all times, regardless of environmental conditions. But it also exacts a biological cost, since it requires them to eat more at frequent intervals in order to generate the necessary energy. Mammals today regulate their bodies at a range of different temperatures. Monotremes generally have lower body temperatures than other mammals, which may reflect their early divergence. However, all mammals maintain an almost constant temperature that is significantly higher than their environment.

Although it is impossible to directly measure the metabolism of an extinct species, there is ample fossil evidence that the synapsids, and especially the cynodonts, were evolving towards endothermy. One key feature is the reduction and loss of abdominal ribs, allowing the developing diaphragm to pump the lungs and increase their efficiency. Another clue lies in their increasingly upright posture, which may have allowed them to breathe while running over short distances (which is physiologically impossible for an animal with a sprawling gait).

One of the major problems that cynodonts would have faced, however, was the reduction in size that accompanied their development towards mammals. Smaller animals have a larger surface area to volume ratio than larger ones, and so tend to lose heat more rapidly. This in turn means that a smaller endothermic animal must eat more food for its size compared to a larger one, simply in order to maintain its body temperature. One possible solution would have been for the early warm-blooded mammals to go into a form of temporary hibernation when temperatures fell too low.

However it developed, endothermy ultimately allowed mammals to occupy hot and cold environments across the planet, and exploit evolutionary opportunities made available with the disappearance of non-avian dinosaurs at the end of the Mesozoic.

MAMMALIA

NUMEROUS GROUPS OF MOSTLY SMALL, PRIMITIVE MAMMALS EVOLVED AND DIED OUT WITHIN THE MESOZOIC ERA, COEXISTING AND EVEN COMPETING WITH DINOSAURS, THE DOMINANT AMNIOTES OF THE TIME. MOST ARE REPRESENTED BY LITTLE MORE THAN THE FOSSILS OF THEIR TEETH AND JAWBONES, WHICH HAS MADE THE ESTABLISHMENT OF THEIR EVOLUTIONARY RELATIONSHIPS VERY DIFFICULT.

Over the last 150 years and more, paleontologists have been discovering tiny isolated teeth, bits of jawbone, and other occasional bony fragments with distinctly mammalian features in rocks of Mesozoic age. The earliest such find to be recognized was made in the 1820s by William Buckland, digging in the Middle Jurassic Stonesfield 'Slate' of Oxfordshire, England. Thus, well before the Darwin–Wallace theory of evolution was published, there was fossil evidence that small mammals had coexisted with the dinosaurs and other extinct Mesozoic amniotes.

Since then, many more mammalian fossil remains have been recovered from these times. However, a major problem has been the lack of skulls and complete skeletons discovered until recent years. As a result, paleontologists struggled to make sense of the relationships between these extinct species and the surviving mammal groups – the monotremes, marsupials, and placentals. Perhaps the most surprising thing is that so many fossil species can be distinguished from differences in their teeth and jawbones alone. This is probably because these structures were so important to the group's Mesozoic survival and success that they evolved rapidly, diversifying as mammal populations spread worldwide, became isolated, and generated new species.

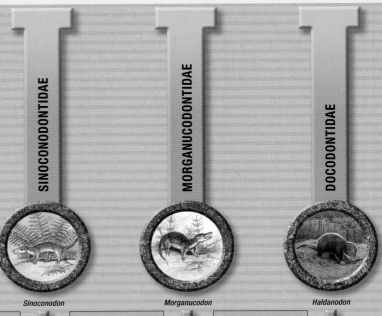

SINOCONODONTIDAE
Sinoconodon

MORGANUCODONTIDAE
Morganucodon

DOCODONTIDAE
Haldanodon

A complete skull of this Early Jurassic animal from China preserves a number of features that suggest it lies close to the base of the mammals – but otherwise it shows a mosaic of primitive and more derived features. The teeth, especially the post-canines, have a relatively simple, primitive structure with little evidence of true occlusion between the upper and lower sets. Furthermore, the way in which they were replaced seems more cynodont than mammal-like. The skull grew to 6cm (3.5in) long, but the post-cranial skeleton is not yet known.

Originally described from isolated Late Triassic teeth found in southwest England, this group survived into the Middle Jurassic. At first, the multi-cusped teeth were seen as the oldest true mammal remains. Since then, the discovery of many more finds including post-cranial skeletal remains, has shown that the morganucodonts were small, shrew-like animals with skulls 2–3cm (1in) long. There is evidence of an enhanced hearing system, and the jaw has an unusual double joint. The total body length is some 10–15cm (4–6in).

This group from the Middle to Late Jurassic is characterized by molar teeth whose crowns have complex occluding surfaces comparable to those of the modern mammal groups. Primitive sprawling limbs, as seen in *Haldanodon*, may be a secondary adaptation for burrowing or perhaps swimming. Cladistic analysis of their combination of features suggests they are more derived than the morganucodonts, but still less advanced than all other mammals.

HOLOTHERIA ▶

MAMMALIA ▶

Eomaia was a small and probably arboreal Cretaceous mammal, with dental features that put it in the Theria, an ancestor of the placental mammals.

MAMMALIA

Typical mammal features include an enlarged brain and sense organs, erect gait, modified teeth, and endothermic temperature regulation.

MONOREMATA + AUSKITRIBOSPHENIDA
Monotremes

EUTRICONODONTIA

MULTITUBERCULATA

"SYMMETRODONTIA"

DRYOLESTDAE

THERIA
Marsupial and placental mammals

See page 278

Bishops

Gobiconodon

Ptilodus

Zhangheotherium

Henkelotherium

Sinodelphys

The monotremes are today represented by just three genera – the Australasian platypus and two echidna genera. They share several features, such as a hindlimb "spur", that suggest a monophyletic group, but their relationship to other mammals is still uncertain. Recent discoveries, such as the Mesozoic fossil *Bishops*, suggest that monotremes were part of a fossil group called the Ausktribosphenida.

Although there is some doubt about their monophyly, the Middle Jurassic to Late Cretaceous eutriconodonts (growing up to 1m/40in long) are characterized by uniquely interlocking lower molars. Recent fossil discoveries of postcranial skeletons show this was a diverse group, widely distributed from Europe to China and including the badger-sized predator *Repenomamus*.

This specialized and abundant group of extinct rodent-like mammals (Middle Jurassic to Paleogene in age and up to 30cm/12in long) is clearly defined by features of the skull, teeth, and postcranial skeleton. The teeth are unique, with the upper incisors reduced to a maximum of three, and the enlarged second incisor working against a single enlarged lower incisor.

Mostly known from isolated teeth and incomplete jaws (Late Triassic to Early Cretaceous in age if *Kuenotherium* is included in the group, otherwise Late Jurassic to Early Cretaceous), these extinct mammals share features with the holotherians but probably do not form a monophyletic group.

This widespread group (Late Jurassic to Paleocene in age) is characterized by the presence of an angular "process" on the dentary jaw-bone. The upper and lower molars form a series of interlocking triangles, as in the symmetrodonts but with further elaborations.

All living mammals except the surviving monotremes are therians – some 4500 species. They all give birth to live young, but their two groups – marsupials and placentals – are otherwise distinct.

◀ EUPANTOTHERIA

THERIIMORPHA ▶

THERIIMORPHA

The Theriimorpha is a grouping that unites all the modern mammals including both monotremes and the extinct closer relatives of the marsupial and placental mammals, while excluding some earlier forms.

EUPANTOTHERIA

This group of extinct Mesozoic mammals, in which the dryolestids are one of three groups, are placed between the symmetrodonts and more advanced therian mammals with tribosphenic (three-pointed) molars.

THE NEW MAMMAL REVOLUTION

In recent years, many ideas about the nature and diversity of extinct Mesozoic mammals have been overturned by the discovery of more and more complete skeletons from terrestrial deposits in Mongolia and China – including sites such as Daohuguo (p.132). These finds help to give a much better idea of how these animals lived alongside the dominant amniotes of the day, but also reveal that they were not all the small, shrew-sized creatures of popular imagination – some grew into badger-sized predators, some developed semi-aquatic, beaver-like lifestyles, and others took to the trees and developed gliding techniques similar to modern flying squirrels.

HOLOTHERIA

This large group of Mesozoic mammals is distinguished by cheek teeth with a clear triangulation of the main cusps associated with a shearing action between upper and lower teeth.

THERIA

THE THERIAN MAMMALS ARE ALL THOSE THAT GIVE BIRTH TO LIVE
YOUNG WITHOUT THE PROTECTION OF A SHELL, AND INCLUDE
THE MARSUPIALS AND EUTHERIAN "TRUE" PLACENTAL MAMMALS.
WITHIN THE THERIA, THE MARSUPIALS AND THEIR FOSSIL RELATIVES
ARE UNITED IN A GROUP KNOWN AS THE METATHERIA. LIKE
EUTHERIANS, THEY GIVE BIRTH TO LIVE YOUNG, BUT THEIR
REPRODUCTIVE BIOLOGY IS DISTINCT IN MANY OTHER WAYS.

The living marsupials consist of some 270 species, most of which
are found in Australia, with about 70 species in South America and
the remaining handful in Central and North America. This
fundamental division in their distribution is the basis for the
recognition of the Ameridelphia and Australidelphia, but these
groups are also supported by anatomical differences. Their largely
southern-hemisphere distribution would seem to relate to a
heritage in the Gondwanan supercontinent, but once the fossil
record is taken into account it seems marsupials were once far
more widespread, and extended into Europe and Asia.

The marsupials can easily be distinguished from the placental
mammals by their means of reproduction, in which a short uterine
development is followed by an early birth. The still-immature
offspring is then nourished by its mother's milk, generally within a
pouch where it firmly attaches to the teat for prolonged periods.

Because the soft tissues of the pouch are not generally
preserved, identification of fossil marsupials relies on other
information from skeletal anatomy, and in particular the teeth.
Their early record is very fragmentary, but extends back to Early
Cretaceous times and the remarkable little 15cm (6in) fossil of
Sinodelphys from 125-million-year-old Chinese deposits.

The cat-sized Virginia opossum (*Didelphis virginiae*) is the largest species in the Ameridelphia,
the American branch of the marsupial group. It is also the only marsupial found
in North America north of the Rio Grande.

DELTATHEROIDA

Deltatheridium

AMERIDELPHIA
Opossums

Herpetotherium

AUSTRALIDELPHIA
Australian marsupials

Macropus

This extinct group
of mammals has
a fossil record that
mostly comes from
the Late Cretaceous
of Mongolia, with
a few earlier fossils
from North America.
Apart from the
usual teeth and jaw
fragments, there
are some skulls
and skeletons,
showing animals
that grew to some
16cm (6.5in) long
and were typically
carnivorous, with
sharply cusped
cheek teeth. The
arrangement of
teeth corresponds
more closely to that
of the marsupials
than placental
mammals, and
some experts place
them within the
Metatheria.

This group consists
of all the American
marsupials, today
numbering nearly
100 living species
of opossum. In
Paleocene times,
South America
was the source of
a major radiation
of marsupials
thoughout the
Americas, including
carnivorous,
insectivorous, and
omnivorous animals
that were only
displaced when
placental mammals
arrived in later
Cenozoic times.
Their early fossil
record is problem-
atic, but probably
includes extinct
forms from the Late
Cretaceous that
may have originated
in North America.

Three quarters of all
marsupials fall
within this group,
which is dominated
by diverse
Australian forms of
varying sizes, and
which formed
almost the entire
pre-human
mammal fauna of
Australia. They
diversified in the
isolation of the
Australasian
continent, but
unfortunately their
early Paleocene
fossil record is
poor, and it is not
until late Oligocene
times that it
improves with
spectacular sites
such as Riversleigh
(see p.208).

◀ MARSUPIALIA

METATHERIA ▶

THERIA ▶

MARSUPIALIA

The 270 living species of
marsupial share unique features
in their skull, teeth, and mode
of reproduction.

THERIA

As well as a variety of anatomical
features, therians are distinguished
by their ability to bear live young.

See page 280

AFROTHERIA

Phiomia

On the basis of DNA analysis, 75 living afrotherian species are united. They represent an old group that originated in Africa during the Paleocene and subsequently spread into Asia. This major group of eutherians includes animals as diverse as the aardvark and elephant, but the living species show only a small part of their earlier variety – extinct members included embrithopods such as *Arsinotherium* (see p.202), and a group of extinct marine mammals called the desmostylians.

PILOSA
Sloths and anteaters

Eurotamandua

This group of 10 or so living species includes the surviving sloths (Folivora) and anteaters (Vermilingua), along with fossil relatives whose record extends back to the Paleocene. Spectacular giant sloths first evolved from small, tree-dwelling forms in Oligocene times, with some adapting to life on the ground and growing to lengths of 6m (20ft). They became extinct in late Pleistocene times. The fossil record for anteaters is much poorer, but may extend back to the Eocene.

CINGULATA
Armadillos

armadillo

With only 20 or so living species, these strange placental mammals of the Americas have an armoured leathery shell and are less than 1.5m (5ft) long. Expert burrowers, they use their powerful arms and claws to dig dens and search for their insect food. They have a fossil record back to the Paleocene, and their extinct relatives the glyptodonts thrived in the Americas from the late Miocene to the Pleistocene.

EUARCHONTOGLIRES
Rodents, lagomorphs, and primates

See page 282

Eutamius

The 2300 living species of this group include diverse forms that range from tree shrews and flying lemurs to primates (including ourselves) and rodents. The group has a fossil record extending back to Paleocene times and its members are primarily identified through molecular analysis. Certain features of the placenta are unique to the group, but since these do not fossilize, extinct euarchontoglires are mostly identified through skeletal and dental features.

LAURASIATHERIA

See page 286

Gazella

This large group contains some 1800 living species that range from hedgehogs and bats to hippos and whales. From origins during Cretaceous times, they have diversified enormously, with some members becoming adapted to a life in water and others specialized in flight. With such diversity, it is hard to believe this group shares a single unique ancestor, and indeed their identification is based primarily on genetic analysis.

BOREOEUTHERIA

This grouping of over 4100 living species, with a fossil record that extends back into Cretaceous times, is supported by genetic studies. While Xenarthra, Afrotheria, and Boreoeutheria are all recognized as monophyletic groups, the relationships between them are still obscure – some studies suggest the Afrotheria split first (as shown here), while others propose that the Xenarthra and Afrotheria are joined in a common clade. However, it seems likely that all the major mammal groups split from each other in a relatively short period of time.

XENARTHRA ▶

BOREOEUTHERIA ▶

EUTHERIA ▶

EUTHERIA NODE

As well as their method of reproduction, placental mammals are distinguished by skeletal structure of the feet and ankles, along with features of the jaws and teeth. The oldest known is the Cretaceous age *Eomaia*.

XENARTHRA

The 30 or so species of living xenarthrans include the armadillos (Cingulata), along with the anteaters (Vermilingua) and sloths (Folivora). There is also a spectacular fossil record from Paleocene times of large forms up to 4m (13ft) long. The name means "extraneous joint" and refers to the articulation between the body and tail. They also have a unique pelvis and much-reduced teeth.

PARALLEL EVOLUTION?

The marsupial method of reproduction has been a subject of much debate – it appears to be a primitive feature, transitional to the eutherian placentals with their more prolonged gestation period, but some argue that it might be an independent evolution of viviparity (live birth).

AFROTHERIA

THE AFROTHERIA ARE ONE OF TWO MAJOR GROUPINGS OF PLACENTAL MAMMALS (THE OTHER BEING THE BOREOEUTHERIA), BOTH BASED PRIMARILY ON GENETIC EVIDENCE. THE EARLY FOSSIL RECORD OF THESE ANIMALS IS FRAGMENTARY AND POOR IN COMPARISON WITH THAT OF THE BOREOEUTHERIA, BUT THEY PROBABLY ORIGINATED IN THE LATE CRETACEOUS, AROUND 100 MILLION YEARS AGO.

The association of the elephants, dugongs, hyraxes, aardvarks, elephant shrews, tenrecs, and golden moles might at first seem to be a fairly random collection of placental mammals, with no outward signs of being an inter-related group with a common ancestry. But this group of 75 or so living species, collectively known as the afrotherians, is well supported by molecular and distributional data. This cuts across traditional Linnaean classification and has had a particularly severe impact on the old order-level grouping of the Insectivora, whose members are now broken up and assigned to other groups.

Despite their name and present distribution, representatives of the Afrotheria have extended well beyond Africa at various times in their history. Most notably, the proboscideans (elephants and their relatives) were widespread throughout Eurasia and North America right through until the recent extinction of the ice-age megafauna. The extinct Afrotherian embrithopods, such as *Arsinotherium* from Fayum (p. 202), also appear in Europe's fossil record. In general, however, the early fossil record of many afrotherian groups is very poor, because African fossil-bearing deposits of early Paleogene times are restricted to a few sites in North Africa. However, it does seem that the afrotherians were the major mammal fauna in Africa until Miocene times – when Laurasiatherian groups began to make an impact.

The eastern rock elephant shrew (*Elephantulus myurus*) is a widespread member of the Macroscelidea group found across southern Africa.

TUBULIDENTATA
aardvarks

Orycteropus

AFROSORICIDA
tenrecs

tenrec

MACROSCELIDEA
elephant shrews

Rhynchocynon

The sub-Saharan aardvark is the sole surviving species of this afrotherian group – a nocturnal and long-snouted specialist termite-eater, growing to 70cm (28in) long. A poor fossil record extends back to Oligocene times in Europe. The aardvark's body shows some convergent features in common with the termite-eating pangolins (pholidotans). The group is defined by its dental features, with continuously growing peg-like teeth lacking in enamel. The muscular forelegs, meanwhile, are armed with tough claws on all four digits.

The 45 living species of tenrecs and golden moles have a distribution limited to sub-Saharan Africa, Madagascar, and offshore islands in the Indian Ocean. Traditionally these small and commonly fossorial (burrowing) animals, up to 30cm (12in) long), were grouped with insectivores such as the hedghogs and shrews, but molecular studies have shown that they are a unique group in their own right. Their fossil record is very poor and fragmentary and only extends back to Miocene times.

The insectivorous "elephant shrews" (macroscelids) are widely distributed across Africa, grow to around 30cm (12in) long, and number some 15 species. Because of their body form, they were thought to be related to the shrews, but can be distinguished through their teeth. Fossils from Paleocene times appear to show similarities with the ancestral paenungulates, but this may be due to convergent evolution – genetic evidence firmly places them closer to the Afrosoricida.

◄ AFROINSECTIPHILIA

AFROTHERIA ▶

AFROINSECTIPHILIA

Based on genetic evidence, this proposed grouping unites afrotherian groups that were previously considered part of the now-obsolete Insectivora, with the Tubilidentata. As yet, however, there is little anatomical evidence to back this up.

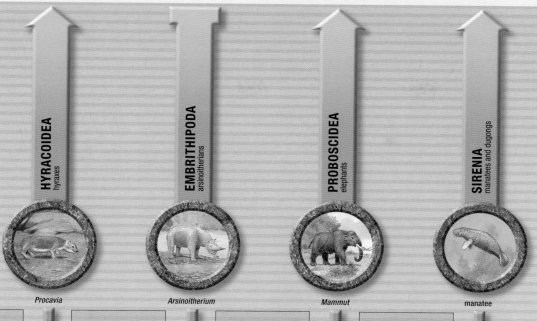

HYRACOIDEA
hyraxes

EMBRITHIPODA
arsinoitherians

PROBOSCIDEA
elephants

SIRENIA
manatees and dugongs

Procavia

Arsinoitherium

Mammut

manatee

DNA LINKS

The discovery of the Afrotheria is just one of many advances made in recent years thanks to the growing use of "molecular phylogenies" – maps of the relationship between organisms based on genetic evidence from the analysis of DNA rather than comparative anatomy.

Genetic studies use several techniques, but mitochondrial DNA (mtDNA) offers a good example. Found in the energy-producing mitochondrion of all eukaryotic cells, mtDNA is unique because it is inherited solely from an organism's mother – unlike most DNA it is not rewritten with every generation by the random combination of genes from two parents.

But despite this, mtDNA is not immune to change. With each generation it is subject to random "genetic drift" – chance mutations that accumulate over the millennia until the mtDNA is quite different from that of the maternal ancestor. Thanks to the recently developed ability to replicate and "sequence" DNA on a large scale, it is now possible to analyse the mtDNA of different organisms and estimate the number of generations since they diverged from their common ancestor. This can lead to surprising discoveries such as the fact that elephant shrews are more closely related to elephants than they are to other shrews.

It is sadly impossible to extract DNA from most fossilized animals, so it might seem that the genetic revolution has rendered fossils obsolete as a means of mapping evolutionary relationships. But they still have a role to play – they provide evidence to calibrate this "molecular clock" and do something that DNA alone can never do – reveal the strange forms that life has developed in adapting to its environment.

The six extant species of hyrax are all terrestrial, and widespread in a range of African environments. Although mostly plant-eaters, they occasionally eat insects. They are all less than 70cm (28in) long, and look superficially like guinea pigs – but in fact they are unrelated, and retain some primitive placental features. They have a fossil record that extends back to the early Eocene, and although they have a flat-footed gait, pads on the soles of the feet and hoof-like nails on the toes indicate ancestors that walked on their toes in an "ungulate" fashion.

This extinct Paleogene group of large terrestrial herbivores had a superficially rhinoceros-like appearance, including nose horns in some forms, such as the 3m (10ft) *Arsinotherium*, but they are structurally different. They formed an important part of the widespread paenungulate radiation in Eocene times and are thought to be more closely related to the probos-cideans than the sirenians, though since they died out in the Oligocene, there is no genetic material to study.

There are only two surviving species of elephants, the remains of a successful and widespread group of terrestrial herbivores whose fossil record extends back to the Eocene. Elephants are the largest living land animals and are characterized by the loss of lower canine teeth and the transformation of the third upper incisors into tusks. The origin of such adaptations can be seen in extinct Eocene forms.

The five living sirenian species are superficially seal-like animals with hands adapted into five-fingered paddles. As in the whales, the legs have been lost, though a vestigial pelvis remains, and the tail is unwhale-like and constructed from a fold of skin. Sirenians are thought to have become aquatic from a tethytherian ancestor more than 40 million years ago. Growing to some 4m (13ft) long, they have a global distribution in tropical coastal waters.

TETHYTHERIA ▶

PAENUNGULATA ▶

TETHYTHERIA

This grouping of proboscideans, sirenians, and their extinct relatives is united by the position of the eye socket (orbit) in the skull and originated in Paleocene times. The name Tethytheria comes from their supposed origins along the shores of the ancient Tethys Ocean.

PAENUNGULATA

Uniting the hyracoids, proboscideans, and sirenians, this group has genetic characteristics and features of the skull in common. The name means "almost ungulates", referring to the superficial similarity of some features to laurasiatherian hoofed mammals.

AFROTHERIA NODE EVENT

Based on genetic evidence, the Afrotheria are thought to have diverged from the rest of the placental mammals around 105 million years ago.

EUARCHONTOGLIRES

ONE OF THE MAJOR SURPRISES PRODUCED BY THE NEW GENETIC CLASSIFICATIONS OF THE LATE 1990S WAS THE DISCOVERY THAT PRIMATES ARE CLOSELY RELATED TO THE RABBITS AND RODENTS, IN A GROUPING KNOWN AS THE EUARCHONTAGLIRES. THEIR UNGAINLY NAME IS AN AMALGAMATION OF TWO LONG-IDENTIFIED ASSOCIATIONS – THE GLIRES (RODENTS AND RABBITS) AND THE EUARCHONTA (PRIMATES, SCANDENTIANS, AND DERMOPTERANS).

The association of rabbits, mice, and primates like ourselves in a single major evolutionary group might seem unlikely. Certainly, traditional anatomical studies of both living and fossil species had never previously made such an association between animals that are so diverse in body form and adaptation, but there is now extensive molecular data supporting this grouping. And on closer inspection, the group does have some distinguishing anatomical features that unite all its members, such as the structure of the placenta.

At least the fossil record has long supported an association between primates, tree shrews (scandentians) and colugos (dermopterans) in a group that was traditionally united as the Archonta, and extends back to Paleocene times. The link between the lagomorphs (rabbits, hares, and their relatives) and the rodents to form a group known as the Glires is equally clear from fossil and anatomical studies. Within this group, the rodents are the most successful group with a remarkable radiation into more than 2000 species since Paleocene times. In contrast, there are only some 200 species of primates alive today, although they had a much greater diversity in early Neogene times.

The European rabbit (*Oryctolagus cuniculus*) is one of the most successful species of the lagomorph group, but despite wide success as an introduced species beyond its native Iberian peninsula, it is in decline in its original habitat.

DERMOPTERA
colugos

colugo

There are just two living species of colugos or "flying lemurs", with a Southeast Asian distribution in tropical forests. Tree-dwellers that grow to 40cm (16in) long, they are not true fliers but well-adapted gliders, using flaps of skin stretched between their legs. Nor are they actually lemurs – they are in fact an independent group of euarchontans with a fossil record extending back to Paleocene times. They are characterized by their distinctive comb-shaped incisors, which are used for cutting off their plant food of leaves, buds, and flowers.

SCANDENTIA
tree shrews

tree shrew

The 20 living species of tree shrews are small, forest-dwelling, superficially squirrel-like euarchontans, with slender bodies and long tails. Found today across Southeast Asia, their fossil record is very poor but extends back to Eocene times. Traditionally they were classified as insectivores and regarded as possible ancestral primates. However, molecular data place them as a group on the same cladistic level as primates and dermopterans.

PRIMATES
lemurs, monkeys, apes etc.

See page 284

Cantius

Humans aside, the 200 or so living species of primate have a global distribution in tropical regions except for Australasia, and a fossil record extending back to Paleocene times. The group includes apes, monkeys, tarsiers, galagos, lorises, lemurs, and the aye-aye – animals with a wide range of terrestrial and arboreal lifestyles, large brains, an advanced visual sense, and often complex social interactions. They range in height from a few centimetres up to 2m (6.6ft), and in weight from a few grams to 225kg (500lb).

◄ EUARCHONTA

EUARCHONTA

This grouping of some 200 living species of flying lemurs, tree shrews, and primates is united by molecular data and has a fossil record extending back to Paleocene times. The name means "true ancestors".

EUARCHONTOGLIRES ►

LEPORIDAE
rabbits, hares

OCHOTONIDAE
pikas

SCIUROMORPHA
squirrels, dormice

CASTORIMORPHA
beavers

MYOMORPHA
rats, mice

HYSTRICHOMORPHA
porcupines, capybaras

leporid

Ochonta

Marmota

Castor

Papagomys

cavy

Rabbits and hares are small terrestrial mammals, up to 70cm (28in) long, largely nocturnal, and well adapted for rapid running in short bursts. They have elongated hind legs (clearly less exagerrated in fossil forms), four clawed toes on each foot, and hairy soles for improved traction against the ground. There are around 50 living species, and the fossil record extends back to late Eocene times when they first spread throughout Eurasia and North America.

The 30 or so living species of pika are small, superficially hamster-like animals with a short tail, short limbs, and rounded ears. They grow to around 20cm (8in) long, and are cold-adapted terrestrial herbivores distributed throughout North America and Eurasia. Feeding on grasses and other low-lying plants, they tend to be diurnal or active at dawn and dusk, in contrast to the often-nocturnal hares and rabbits. Their fossil record extends back to Oligocene times.

There are more than 300 living species of squirrels, chipmunks, marmots, and flying squirrels with a wide distribution through the Americas, Eurasia, and Africa. Their fossil record stretches back to Eocene times and the group has at various times been taken to include other groups such as the mountain beavers and dormice, based on the routing of the masseter muscle in the skull. However, recent molecular studies only support the association of the squirrels and dormice.

This group of rodents includes the semi-aquatic beavers, kangaroo rats, and gophers, plus a number of extinct forms that originated in the early Eocene. They have a mostly American distribution, with the notable exception of beavers, which were once widespread through Eurasia. The extinct North American giant beaver grew to 2.5m (8ft) in length and only became extinct at the end of the last ice age, some 10,000 years ago.

With some 1140 living species of mice, rats, gerbils, hamsters, voles, and lemmings, the myomorphs are the most successful rodent group, achieving an almost global distribution thanks in part to a close association with humans who have introduced them (often accidentally) to many new locations. Usually nocturnal seed-eaters, some have adapted to many different habitats and foods. They are united by features of the jawbones, jaw muscles, and teeth.

This diverse group of some 200 living species includes the porcupines, chinchillas, cavies, and capybaras (which can grow to 1.2m/48in long). The grouping is supported by molecular data and anatomical features. The fossil record suggests that they originated in South America in early Oligocene times and spread northwards when the land bridge between the two continents formed around 3 million years ago — an event known as the Great American Interchange.

LAGOMORPHA ▶

RODENTIA ▶

GLIRES ▶

RODENTIA

Originating in Paleocene times and with well over 2000 living species, the rodents have spread almost globally. Their success and rapid evolution makes even fragmentary fossils useful for the study of global climate fluctuations and "biozonal" dating.

GLIRES

There are some 2100 living species in this widely distributed grouping of the lagomorphs and rodents, united by features such as specialized incisor teeth.

LAGOMORPHA

The 80 living species of lagomorphs include hares, rabbits, and pikas with an almost global distribution and a fossil record extending back to Paleocene times.

EUARCHONTOGLIRES

This major group of therians includes over 2300 living species, mostly rodents and lagomorphs, and is united by molecular studies.

PRIMATES

FROM EARLY CENOZOIC BEGINNINGS AS RELATIVELY SMALL, TREE-DWELLING ANIMALS, OUR CLOSEST RELATIVES DIVERSIFIED AND ACHIEVED A WIDE DISTRIBUTION BY MIOCENE TIMES. THE GREATEST EVOLUTIONARY CHANGE HAS TAKEN PLACE IN THE LAST 10 MILLION YEARS, AS ONE GROUP, THE HOMININI, HAVE SPREAD AROUND THE WORLD.

The primates were formally recognized in 1758 by Linnaeus. At first, he called the group the "Anthropomorpha" and included our own species, *Homo sapiens*, within it. Linnaeus was attacked for presuming to group humans with chimps, but defended himself by challenging his detractors to show any significant anatomical differences between the two.

Even in Darwin's day, the issue of human links to the other primates was so problematic that he avoided the issue in *The Origin of Species*. Some of the first fossil primates had appeared in the early 1800s, but it was not until the end of the century that the antiquity of primates and their connection to modern humans was generally accepted.

Since the 1920s and the first finds of fossil australopithecines, some 20 extinct species of Hominini have been discovered, mostly within Africa. Interesting questions remain about the earlier evolution of the group, which had a much wider distribution beyond Africa.

Squirrel monkeys (*Saimiri* sp.) are highly social, ominivorous primates of the "New World" platyrrhine group, distinguished from the "Old World" monkeys by flattened noses with sideways-facing nostrils.

LORISIFORMES
lorises

Loris tardigradus

Today, there are 10 living species of lorises and galagos (bush babies), widely distributed from sub-Saharan Africa to India and Southeast Asia. They also have a sparse fossil record that extends back to Eocene times in North Africa. Growing to no more than 40cm (16in) in total length, these forest tree-dwellers are nocturnal omnivores and are distinguished by the structure of their inner ear. The strange nocturnal aye-aye of Madagascar may be the sole living representative of a group ancestral to both lemurs and lorisiformes.

LEMURIFORMES
lemurs

Lemur catta

Although today's 20 or more living species of arboreal lemurs are confined to Madagascar, the fossil record of their extinct relatives extends back to Eocene times and is widely distributed across Europe and into North America. The lemurs include some of the smallest primates, and others that grow to more than a metre (40in) in total length, but much larger forms survived until relatively recently. They include plant-eaters, omnivores, and specialist insect-eaters, and the group is characterized by the structure of the inner ear.

TARSIIFORMES
tarsiers

Tarsius

The three living species of tarsiers are confined to Southeast Asia but have an ancient fossil record that extends back into Paleocene times, with a distribution across Asia, Europe, the Americas, and into North Africa. These tiny, arboreal, forest-dwelling primates, with body lengths of 10cm (4in) or so, are distinguished by their round heads, large, forward-facing eyes, the structure of their eye sockets, and, once again, the inner ear. Their long hind legs are adapted for leaping, and their forward-faciing eyes provide stereoscopic vision that helps them judge distance.

PLATYRRHINI
New World monkeys

Ateles paniscus

As their common name of "New World monkeys" suggests, the 50 or so living species of this group are confined to the Americas – as is their sparse fossil record, which extends back to Oligocene times. They are essentially arboreal forest, dwellers (although some also exploit grassy gaps in the forest), with diets that vary from insects to leaves and fruit. In size, they range from tiny marmosets to large spider monkeys with long arms and tails and a total body length of a metre (40in) or so. The group is defined by the structure of the nostrils and ear.

◀ STREPSIRRHINI   HAPLORRHINI ▶

STREPSIRRHINI

This grouping is distinguished from other primates by the comb-like structure of their front teeth.

PRIMATES ▶

PRIMATES

As a group, primates are characterized by features including an enlarged brain, enhanced vision, and opposable thumbs.

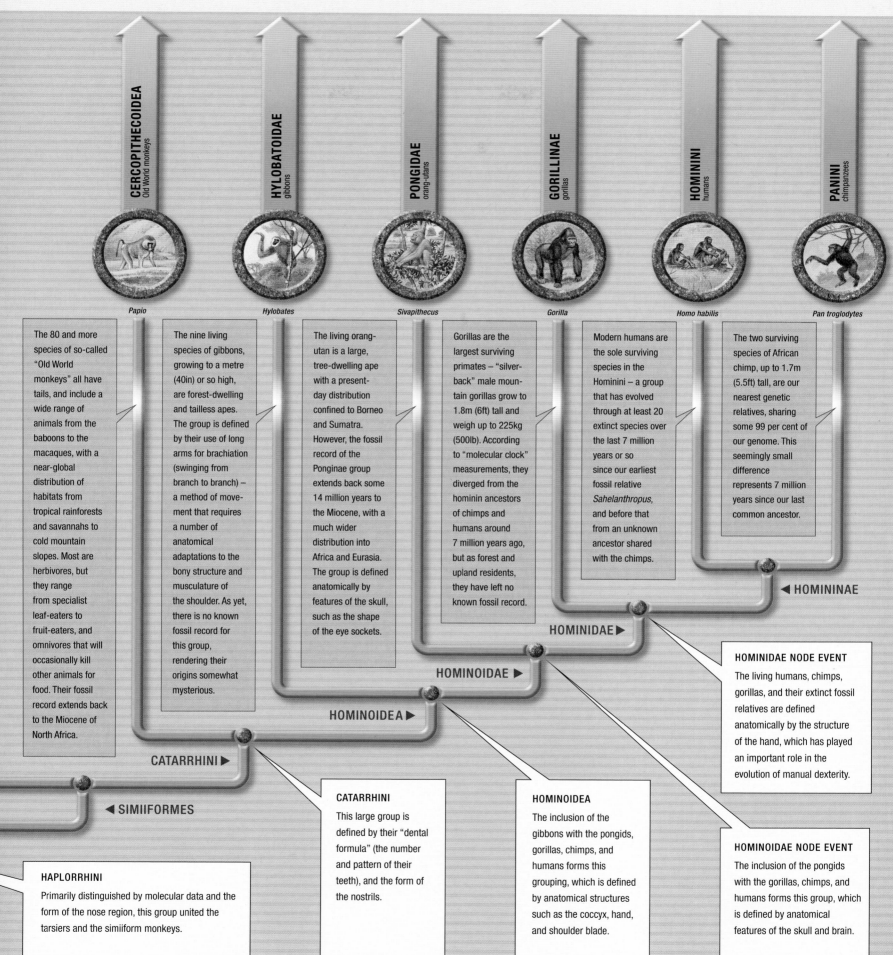

CERCOPITHECOIDEA Old World monkeys

HYLOBATOIDAE gibbons

PONGIDAE orang-utans

GORILLINAE gorillas

HOMININI humans

PANINI chimpanzees

Papio

Hylobates

Sivapithecus

Gorilla

Homo habilis

Pan troglodytes

The 80 and more species of so-called "Old World monkeys" all have tails, and include a wide range of animals from the baboons to the macaques, with a near-global distribution of habitats from tropical rainforests and savannahs to cold mountain slopes. Most are herbivores, but they range from specialist leaf-eaters to fruit-eaters, and omnivores that will occasionally kill other animals for food. Their fossil record extends back to the Miocene of North Africa.

The nine living species of gibbons, growing to a metre (40in) or so high, are forest-dwelling and tailless apes. The group is defined by their use of long arms for brachiation (swinging from branch to branch) – a method of movement that requires a number of anatomical adaptations to the bony structure and musculature of the shoulder. As yet, there is no known fossil record for this group, rendering their origins somewhat mysterious.

The living orang-utan is a large, tree-dwelling ape with a present-day distribution confined to Borneo and Sumatra. However, the fossil record of the Ponginae group extends back some 14 million years to the Miocene, with a much wider distribution into Africa and Eurasia. The group is defined anatomically by features of the skull, such as the shape of the eye sockets.

Gorillas are the largest surviving primates – "silverback" male mountain gorillas grow to 1.8m (6ft) tall and weigh up to 225kg (500lb). According to "molecular clock" measurements, they diverged from the hominin ancestors of chimps and humans around 7 million years ago, but as forest and upland residents, they have left no known fossil record.

Modern humans are the sole surviving species in the Hominini – a group that has evolved through at least 20 extinct species over the last 7 million years or so since our earliest fossil relative *Sahelanthropus*, and before that from an unknown ancestor shared with the chimps.

The two surviving species of African chimp, up to 1.7m (5.5ft) tall, are our nearest genetic relatives, sharing some 99 per cent of our genome. This seemingly small difference represents 7 million years since our last common ancestor.

◀ **HOMININAE**

HOMINIDAE ▶

HOMINIDAE NODE EVENT

The living humans, chimps, gorillas, and their extinct fossil relatives are defined anatomically by the structure of the hand, which has played an important role in the evolution of manual dexterity.

HOMINOIDEA ▶

HOMINOIDEA ▶

CATARRHINI ▶

◀ **SIMIIFORMES**

CATARRHINI

This large group is defined by their "dental formula" (the number and pattern of their teeth), and the form of the nostrils.

HOMINOIDEA

The inclusion of the gibbons with the pongids, gorillas, chimps, and humans forms this grouping, which is defined by anatomical structures such as the coccyx, hand, and shoulder blade.

HOMINOIDAE NODE EVENT

The inclusion of the pongids with the gorillas, chimps, and humans forms this group, which is defined by anatomical features of the skull and brain.

HAPLORRHINI

Primarily distinguished by molecular data and the form of the nose region, this group united the tarsiers and the simiiform monkeys.

LAURASIATHERIA

IT IS HARD TO BELIEVE THAT HEDGEHOGS AND
WHALES SHOULD BE INCLUDED IN THE SAME
GROUPING OF PLACENTAL MAMMALS, BUT THIS
IS WHAT THE MOLECULAR EVIDENCE SUGGESTS.
IN FACT, ABOUT HALF OF THE LIVING SPECIES OF
PLACENTAL MAMMALS ARE GROUPED IN THE
LAURASIATHERIA BY GENETIC STUDIES.

The mammals brought together in this group
range from shrews and bats to camels and blue
whales – an extraordinary diversity of body
form, size, and adaptation that includes some
of the smallest mammals and the largest. They
have radiated globally and adapted to habitats
that range from ocean depths to mountain tops,
underground and into the air. The fossil record
of the living groups extends back to Paleocene
times, but there are also a number of extinct
groups that take the Laurasiatheria back to the
the middle Cretaceous and a supposed origin,
as the name suggests, on the northern
supercontinent of Laurasia.

But there is still considerable debate over
the origin of many mammal groups, including
the various diverse laurasiatherians. The key
question is whether they really all evolved
across just a few million years of the early
Paleogene (in the aftermath of the Cretaceous–
Tertiary extinction), or whether they actually
have deeper and older origins. Unfortunately,
the fragmented nature of the fossil record (see
box opposite) offers little help in resolving
the problem.

Blue wildebeest (*Connochaetes
taurinus*) are highly successful African
members of the Ruminantia. Until
recently, they were classed as
Artiodactyla – "even-toed ungulates" –
since they effectively walk on two toes.

EULIPOTYPHLES
'insectivores'

Pholidocercus

The traditional
Linnaean Order
Insectivora was
never entirely
satisfactory, since it
could not be
grounded in
any substantial
anatomical
similarities. Recent
genetic data have
shown that it is in
fact polyphyletic –
the shrews, hedge-
hogs, and moles of
northern continents
have a separate
origin to the African
Insectiphilia. As a
result, these 300 or
so living species are
now united into the
Eulipotyphles group.
They have a fossil
record extending
back to Paleocene
times and, while
their anatomies are
as varied as ever,
molecular studies
of living animals
do at least support
this grouping.

CHIROPTERA
bats

Archaeonycteris

There are nearly
1000 species of
living bats, forming
one of the most
successful and
widely distributed
placental groups
with a fossil record
extending back to
Eocene times. Most
are small and
nocturnal, with a
variety of feeding
habits. Despite their
markedly different
appearances, small
"microchiropterans"
and larger fruit bats
or "megachiropter-
ans" seem to
have a shared
evolutionary history.
All are well adapted
to active flight, with
a skin membrane
(the patagium)
stretched between
the elongated digits
of the hand and
across the body,
hind limbs, and tail.

PERISSODACTYLA
odd-toed ungulates

Hyracotherium

This group of plant-
eaters, commonly
known as the "odd-
toed ungulates",
contains just
18 living species
ranging from zebras
and horses to
rhinocerses.
However, they
were much more
abundant, diverse,
and widely
distributed in the
past, following an
origin in Paleocene
times and
later Cenozoic
radiations. They are
characterized by the
structure of the hind
limb and molecular
data. Modern
species grow to 4m
(13ft) long, but the
extinct indricotheres
included the largest
land mammals
ever known.

CARNIVORA
carnivores

Amphicynodon

The 280 or so living
species in this
group have diverse
forms and habits
and range from the
walrus and bears to
hyenas, weasels,
and cats. They are
united by the
development of
carnassial teeth for
slicing the flesh of
their prey. Some
have diets adapted
for scavenging,
omnivorous, and
rarely plant-eating
lifestyles. When
their extinct "cimo-
lestan" relatives are
taken into account,
their fossil record
may extend back to
the middle
Cretaceous, but this
is controversial.

◀ FEREUNGULATA

SCROTIFERA ▶

LAURASIATHERIA ▶

FEREUNGULATA

Some 600 species ranging from horses to whales are united in this
grouping, which is supported by molecular data and the anatomical
features such as the structure of the placenta.

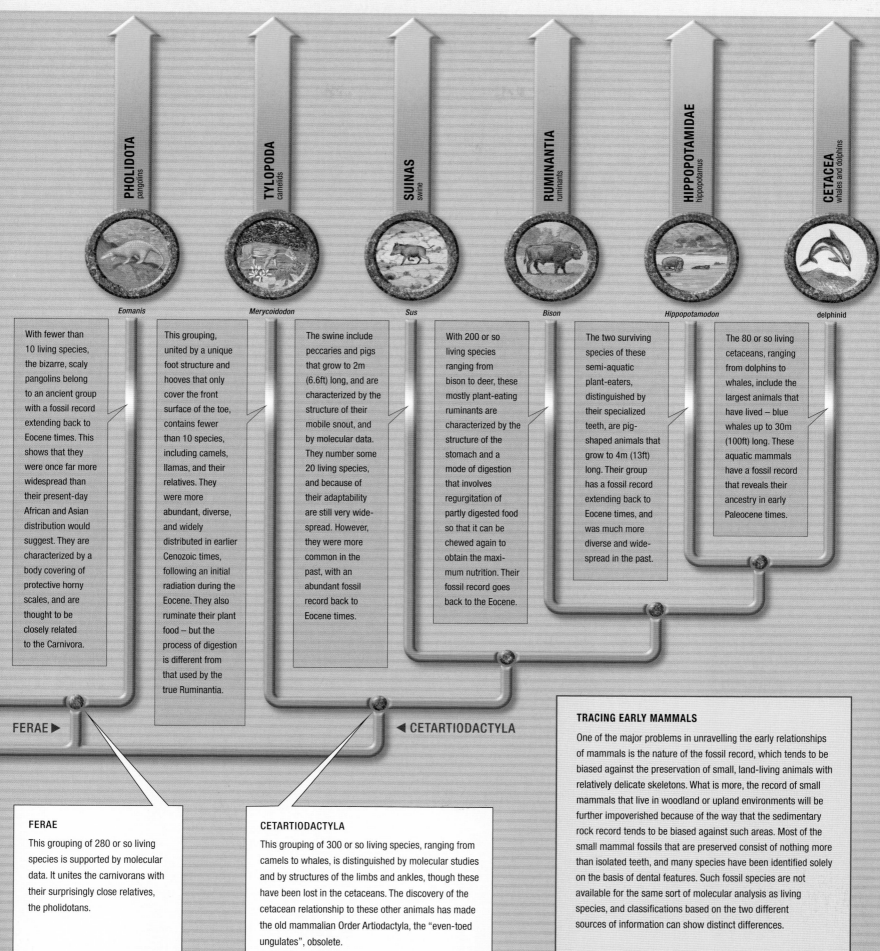

PHOLIDOTA pangolins

Eomanis

TYLOPODA camelids

Merycoidodon

SUINAS swine

Sus

RUMINANTIA ruminants

Bison

HIPPOPOTAMIDAE hippopotamus

Hippopotamodon

CETACEA whales and dolphins

delphinid

With fewer than 10 living species, the bizarre, scaly pangolins belong to an ancient group with a fossil record extending back to Eocene times. This shows that they were once far more widespread than their present-day African and Asian distribution would suggest. They are characterized by a body covering of protective horny scales, and are thought to be closely related to the Carnivora.

This grouping, united by a unique foot structure and hooves that only cover the front surface of the toe, contains fewer than 10 species, including camels, llamas, and their relatives. They were more abundant, diverse, and widely distributed in earlier Cenozoic times, following an initial radiation during the Eocene. They also ruminate their plant food – but the process of digestion is different from that used by the true Ruminantia.

The swine include peccaries and pigs that grow to 2m (6.6ft) long, and are characterized by the structure of their mobile snout, and by molecular data. They number some 20 living species, and because of their adaptability are still very widespread. However, they were more common in the past, with an abundant fossil record back to Eocene times.

With 200 or so living species ranging from bison to deer, these mostly plant-eating ruminants are characterized by the structure of the stomach and a mode of digestion that involves regurgitation of partly digested food so that it can be chewed again to obtain the maximum nutrition. Their fossil record goes back to the Eocene.

The two surviving species of these semi-aquatic plant-eaters, distinguished by their specialized teeth, are pig-shaped animals that grow to 4m (13ft) long. Their group has a fossil record extending back to Eocene times, and was much more diverse and widespread in the past.

The 80 or so living cetaceans, ranging from dolphins to whales, include the largest animals that have lived – blue whales up to 30m (100ft) long. These aquatic mammals have a fossil record that reveals their ancestry in early Paleocene times.

FERAE ▶

◀ CETARTIODACTYLA

FERAE

This grouping of 280 or so living species is supported by molecular data. It unites the carnivorans with their surprisingly close relatives, the pholidotans.

CETARTIODACTYLA

This grouping of 300 or so living species, ranging from camels to whales, is distinguished by molecular studies and by structures of the limbs and ankles, though these have been lost in the cetaceans. The discovery of the cetacean relationship to these other animals has made the old mammalian Order Artiodactyla, the "even-toed ungulates", obsolete.

TRACING EARLY MAMMALS

One of the major problems in unravelling the early relationships of mammals is the nature of the fossil record, which tends to be biased against the preservation of small, land-living animals with relatively delicate skeletons. What is more, the record of small mammals that live in woodland or upland environments will be further impoverished because of the way that the sedimentary rock record tends to be biased against such areas. Most of the small mammal fossils that are preserved consist of nothing more than isolated teeth, and many species have been identified solely on the basis of dental features. Such fossil species are not available for the same sort of molecular analysis as living species, and classifications based on the two different sources of information can show distinct differences.

ARCHOSAUROMORPHA

THE ARCHOSAUROMORPHA ARE DIAPSID REPTILES MARKED OUT BY THE DEVELOPMENT OF A THIRD OPENING IN THE SIDE OF THE SKULL, POSITIONED BETWEEN THE EYE SOCKET AND THE NOSTRIL OPENING AS AN "ANTORBITAL FENESTRA". THEY FIRST APPEARED IN THE LATE PERMIAN, AND ROSE TO BECOME THE DOMINANT LAND ANIMALS OF THE MESOZOIC ERA – HENCE A NAME THAT MEANS "RULING REPTILES".

Many of the features that helped archosaurs rise to prominence are associated with a predatory lifestyle – although they rapidly diversified. The additional skull opening in front of the eye allowed the snout to increase in length without adding greatly to the weight. Smaller openings develop in the lower jaw, and the teeth are anchored in sockets that make them far less likely to come loose while feeding. Another significant feature is the addition of an extra ridge for attaching muscles on the femur (upper leg bone). This may have been key to the development of bipedalism in archosaur groups such as the dinosaurs. Diversifying archosaur groups can often be distinguished by developments in the ankle structure.

Close relatives of the archosaurs began to appear towards the end of Permian times, alongside various large and successful synapsids and other diapsid reptiles, but the first true archosaurs arose in the Early Triassic and diversified rapidly into increasingly aquatic crocodilians, airborne pterosaurs, and the first dinosaurs. Various reasons have been put forward to explain their success – one is that they were well suited to the hostile conditions of continental interiors that would have become extremely hot and arid during the existence of the supercontinent "Pangea". Another is related to the possible archosaur metabolism (see box opposite).

Crocodilians such as the Nile crocodile (*Crocodylus niloticus*) are, along with the avialean dinosaurs, one of two archosaur groups that are still successful in the modern world.

PROLACERTIFORMES

Prolacerta

EUPARKERIA

Euparkeria

'SPHENOSUCHIA'

Hesperosuchus

Also known as the protorosaurs, these primtiive archosauromorphs are distinguished by shared features including their elongated neck vertebrae, slender ribs, and certain skull and jaw features. They include land-dwellers such as *Prolacerta*, fliers such as *Sharovipt-teryx*, and semi-aquatic long-necked forms such as *Tanystropheus*. Some 15 genera are known, from Permian and Triassic fossil deposits around the world, but it is now thought unlikely that the group is truly monophyletic – instead, various prolacertiformes are believed to have had different ancestries.

Known from just a single fossil locality in the Early Triassic of southern Africa, this intriguing reptile grew up to 60cm (24in) in length, with a light body, long tail, needle-like teeth, and unusually long hind legs. These have given rise to the idea that *Euparkeria* was at least partly bipedal and capable of running on its hind legs. However, its anatomy raises difficult questions about its cladistic relationships. Some scientists place it in the ancestry of dinosaurs, while others believe it is the sole known representative of a sister group to the Archosauria, as shown here.

This group of relatively small, basal crocodylomorphs, growing up to 1.5m (5ft) long, were mostly slender, gracile reptiles, with relatively long limbs and an erect posture. They evolved during the Triassic and are found as fossils through to the Middle Jurassic, with at least 15 genera known (and some other arguable members). Despite the fact that all its members share certain characteristics, there has been considerable argu-ment over whether the Sphenosuchia is a truly monophyletic group.

◄ ARCHOSAUROMORPHA

ARCHOSAUROMORPHA

The archosauromorpha unites the true archosaurs with their less specialized relatives that first appeared in the late Permian.

CROCODYLOMORPHA

First appearing in the Triassic, these long-limbed land predators developed over time into semi-aquatic hunters.

CROCODYLA crocodiles

PTEROSAURIA pterosaurs

ORNITHISCHIA bird-hipped dinosaurs
See page 290

SAUROPODOMORPHA sauropod dinosaurs

THEROPODA theropod dinosaurs
See page 292

Asiatosuchus

Pterodactylus

Muttaburrasaurus

Apatosaurus

Aucasaurus

Along with the birds, crocodiles and their relatiives are the only living archosaurs, with 23 living species growing up to several metres long. These semi-aquatic marine and fresh-water predators first evolved in the Early Cretaceous, diversifying into a wide range of forms during the Late Cretaceous and the Paleogene. Today they have a global distribution in the tropics as crocodiles, alligators, and gavials.

These well-known flying reptiles flourished in the skies of the Mesozoic Era, with wingspans from a few tens of centimetres to more than 10m (33ft). About 100 genera are known, traditionally divided into primitiive, long-tailed rhamphorynchoids and more advanced, short-tailed pterodactyloids (though the pterodactyloids undoubtedly had a rhamphorynchoid ancestor). As well as hollow bones, some developed hair, and may have been endothermic.

One of the two great branches of the dinosaurs, ornithischians have a name that means "bird-hipped" on account of the way their pubis, one of three major hip bones, points backward (a condition that evolved independently in birds). Arising in the Late Triassic, they flourished until the end of the Cretaceous, and included dinosaurs such as the armoured stegosaurs, and the duck-billed hadrosaurs.

This well-known group of saurischian herbivore dinosaurs had a large body, long neck, and a tiny head with weak, peg-like teeth. First appearing in the Late Triassic, the relatively small prosauropods are now thought to have been a sister group to the giant sauropods that flourished in the Jurassic and lived on in some parts of the world until the Late Cretaceous.

The second main branch of the Saurischia are the Theropoda – bipedal, mostly carnivorous dinosaurs that arose in the Late Triassic and flourished in various forms through to the end-Cretaceous extinction. After this cataclysm only the birds (Avialae), which emerged at some point in the Jurassic, lived on.

SAURISCHIA

The Saurischia are distinguished from the Ornithischia by their "lizard hips" – a pelvic structure in which the pubis points forwards as in a modern lizard. This bone later became rotated backwards and fused with the ischium in avialian dinosaurs, independently of similar developments in the ornithischians.

SAURISCHIA ▶

DINOSAURIA ▶

◀ CROCODYLOMORPHA

ARCHOSAURIA ▶

ARCHOSAUR METABOLISMS

Fossil evidence for the soft internal organs of extinct species rarely survives, but these are often the key to an animal's metabolism and lifestyle. Fortunately, we have two surviving groups of archosaurs – the birds and crocodiles – to study, and these reveal intriguing features. While birds are endothermic (warm-blooded) and crocodiles are ectothermic (cold-blooded), both have a four-chambered heart – an efficient organ normally associated with the fast metabolisms of mammals. Some paleontologists use this as evidence to argue that all archosaurs (including the dinosaurs and the slender, land-living ancestral crocodiles) were warm-blooded, and that this could have helped them to compete with the Triassic synapsids.

ARCHOSAURIA

The true archosaurs probably evolved in the late Permian Period, before splitting into the major divisions that flourished in the Mesozoic and, in some cases, through to the present day. Aside from their characteristic second hole in the skull, other archosaur features include a hole in the lower jawbone, teeth set in sockets, and a modified ankle joint.

DINOSAURIA

The dinosaurs are united as a group by various anatomical features including a more upright and vertical posture to the legs compared to other archosaurs. Cladistically, they are defined as the group containing the most recent common ancestor of the modern birds and the ornithischian dinosaur *Triceratops*, and all of its descendants.

ORNITHISCHIA

THE DINOSAURS WERE A MAJOR GROUP OF TERRESTRIAL ARCHOSAURS THAT ORIGINATED IN MIDDLE TO LATE TRIASSIC TIMES, RADIATED THROUGH THE MESOZOIC, AND MOSTLY BECAME EXTINCT IN THE END-CRETACEOUS EXTINCTION, EXCEPT FOR THE BIRDS – DESCENDANTS OF ONE GROUP OF FEATHERED THEROPODS. DINOSAURS ARE SUBDIVIDED INTO TWO MONOPHYLETIC GROUPS, THE ORNITHISCHIA AND SAURISCHIA.

The Ornithischia are chiefly marked by the superficially bird-like arrangement of their pelvic bones. However, they are also marked by several other features, including wider and more stable hips, and a somewhat smaller opening in the skull in front of the eye. As herbivores, however, perhaps the most significant addition is an additional "predentary" bone on the front of the lower jaw, which works in conjunction with the "premaxilla" of the upper jaw to form a beak for snipping vegetation.

The three most distinctive ornithischian groups are the thyreophorans, ornithopods, and ceratopsians. The heavy, quadrupedal thyreophorans bore an array of bony armour that often culminated in a club-shaped tail, and existed from the Early Jurassic to the Late Cretaceous. The bipedal ornithopods included the duck-billed hadrosaurs, often with ornate and puzzling skulls crests, which roamed the world in the Late Cretaceous. The ceratopsians, meanwhile, were largely quadrupedal and marked out by a range of increasingly elaborate head adornments including a rostral bone on the upper beak, facial horns of various sizes and configurations, and large bony frills that may have been used to protect the vulnerable neck, aid thermoregulation, or intimidate rivals.

HERRERASAURUS

PISANOSAURUS

HETERODONTOSAURIDAE

Herrerasaurus

Pisanosaurus

Heterodontosaurus

Discovered in the Late Triassic rocks of Argentina, this medium-sized (up to 4.5m or 15ft) bipedal carnivore is a controversial dinosaur thought by some to lie close to the origin of the saurischian group, and by others to pre-date the split between saurischians and ornithischians. Although considered more saurischian in its features, it shows few of the specialized characteristics they later developed. It was certainly a meat-eater, with a skull form that resembles non-dinosaurian archosaurs. Studies of its primitive remains have been used to support the idea that dinosaurs are a truly monophyletic group.

Generally believed to be the oldest ornithischian dinosaur yet discovered, this metre-long (40in), lightly built, bipedal plant-eater has proved controversial, since there is just a single fragmentary and incomplete specimen from Argentina, first described in 1967. Found in the same Late Triassic rock formation as the early carnivore *Herrerasaurus*, the far smaller *Pisanosaurus* may have been its prey. However, even recent studies have not been able to resolve whether it is a basal ornithischian or a slightly more advanced heterodontosaurid.

The position of this group of small, long-tailed plant-eaters within the ornithischians is highly controversial, with no current scientific consensus. However, a recent analysis of *Heterodontosaurus* and three other genera within the group gives some support to their position as the most primitive or basal ornithischians, based on features of their jaws, teeth, and hands. Growing up to 2m (6.6ft) long, they had a global distribution during the Early Jurassic, and a few survived into the Early Cretaceous.

ORNITHISCHIA ▶

DINOSAURIA ▶

The large ornithischian *Triceratops*, from the Late Cretaceous of North America, is one of the latest and most advanced representatives of the marginocephalian group.

DINOSAURIA

Because this group includes birds, it comprises a significant grouping of terrestrial vertebrates even today. They are placed within the larger group of archosaurs that also includes the crocodilians and pterosaurs.

THYREOPHORA

ORNITHOPODA

CERATOPSIA

PACHYCEPHALOSAURIA

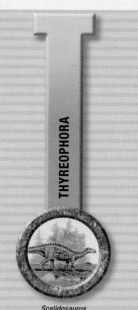

Scelidosaurus

Iguanodon

Triceratops

Stegoceras

This group of armoured plant-eaters, some of which grew up to 9m (30ft) long, forms a well-defined monophyletic group marked by features of the skull and skin armour. The most familiar thyreophorans are the armour-plated stegosaurians and ankylosaurs. However, the exact phylogenetic position of *Scelidosaurus*, first described in 1861, has proved controversial and is still not resolved.

This highly successful group originated among small and bipedal Jurassic plant-eaters that diversified and grew larger to become the dominant herbivores of the Cretaceous. Among the best known were the beaked and duck-billed hadrosaurs that grew to 15m (50ft) long. Recent analysis has cast doubt on the grouping, but this may be due to gaps in our knowledge of some of its members.

This group includes all marginocephalans closer to *Triceratops* than to *Pachycephalosaurus*. Ranging in age from the Late Jurassic to the Late Cretaceous, these herbivorous, beaked animals show a variety of forms, from small bipeds such as *Psittacosaurus* to large quadrupeds with elaborate horns and bony neck flanges, such as the 9m (30ft) *Triceratops*.

This well-established group, whose best known member is *Pachycephalosaurus*, was restricted to the Late Cretaceous. Medium-sized, bipedal, and largely plant-eating, they grew to 8m (26ft) long, and had a thick, flat, or dome-shaped skull roof that was probably used to butt predators or sexual rivals.

MARGINOCEPHALIA ▶

CERAPODA ▶

GENASAURIA ▶

CERAPODA

Ranging in age from Jurassic to Late Cretaceous, this group has a thick layer of enamel on the inside of the lower teeth, to aid chewing.

GENASAURIA

This group includes all the more advanced groups of ornithischians – both the cerapods (ornithopods, ceratopsians, and pachycephalosaurs) and the armoured thyreophorans.

ORNITHISCHIA

This long-established group of dinosaurs share a number of defining features of the skull, jaw, teeth, pelvis, and backbone (although some of the distinctive features of the teeth have also evolved several times within the archosaurs).

MARGINOCEPHALIA

This group includes *Triceratops*, *Pachycephalosaurus*, their most recent common ancestor, and all its descendants. They range in age from the Late Jurassic to the Late Cretaceous.

DISCOVERING THE DINOSAURS

The Dinosauria were first defined as a group by British anatomist Richard Owen in 1842, based on a few fragmentary fossil remains recovered from Mesozoic strata in southern England. The ambitious Owen used the name dinosaur, derived from Greek and meaning "terrible lizard", to define what he thought of as a "distinct tribe or sub-order of Saurian Reptiles" – and in doing so rather stole the limelight from William Buckland and Gideon Mantell, the pioneering paleontologists who had first described the extinct reptiles *Megalosaurus* and *Iguanodon*. But even Owen struggled to reconstruct his dinosaurs because of their incompleteness, and saw them as massive, lumbering quadrupeds.

It was only when more complete specimens were recovered, especially from North America, that a better understanding was achieved. In 1858, the discovery of a nearly complete 10m (33ft) skeleton from New Jersey, named as *Hadrosaurus* by Joseph Leidy, suggested for the first time that even big dinosaurs could move around on their hind limbs in a way that is not seen in living reptiles, and this helped revolutionize the whole study of dinosaurs.

Now, 150 years later, the field continues to be rewritten by further discoveries and new techniques of analysis. Some 500 genera are known (excluding the birds), but these are thought to account for just a quarter of the total number that will ultimately be discovered. Cladistics reveal them as a monophyletic group of diapsid archosaurs that diverged from the archosaurs in Middle to Late Triassic times, with dinosaurs such as *Eoraptor* lying close to their common ancestor, which was probably a small bipedal predator.

THEROPODA

THE SAURISCHIAN DINOSAURS WERE FIRST DEFINED IN THE LATE 19TH CENTURY BY HARRY GOVIER SEELEY, A STUDENT OF RICHARD OWEN, THE BRITISH ANATOMIST WHO HAD FIRST DEFINED THE DINOSAURS AS A SEPARATE GROUP IN 1842. WITHIN THE SAURISCHIANS, TODAY'S SCIENTISTS RECOGNIZE TWO DISTINCT GROUPS – THE FOUR-FOOTED SAUROPODS AND THE BIPEDAL THEROPODS.

The Sauropoda were first recognized in 1878 by the great US paleontologist Othniel C. Marsh. He classified them in Linnaean terms as a suborder of the Order Dinosauria, before promoting them in 1882 to an order in their own right, with the Dinosauria in turn promoted to subclass status. Since Marsh's time, a range of early sauropod relatives have been found, and along with the sauropods themselves, these are grouped in the Sauropodomorpha.

The theropods, meanwhile, were defined by Marsh in 1881, though since his time the extent of the group has expanded considerably. At first limited to a relatively small group of allosaurid dinosaurs and their relatives, it now encompasses animals ranging from the ceratosaurs to the modern birds. They appeared in the Late Triassic and were at first primarily bipedal carnivores, though some members, such as the oviraptorosaurs, evolved herbivorous feeding habits in Cretaceous times, and many of the Avialae (birds) have done the same. The group includes all species more closely related to the Avialae than to sauropods, and is defined by unique features of the jawbones and vertebrae. Their size range is extreme, with extinct species ranging from the largest ever land-living predators, such as the 14m (46ft) *Tyrannosaurus*, to the tiny *Epidexipteryx*, 25cm (10in) long. Living examples are often smaller, down to just 5cm (2in) for the tiny bee hummingbird.

The 14m (46ft) South American carnivore *Giganotosaurus* is an example of the Carcharodontosauridae group of theropod dinosaurs, which became the largest southern-hemisphere predators during the Cretaceous Period.

SAUROPODA

Dicraeosaurus

CERATOSAURIA

Elaphrosaurus

MEGALOSAURIDAE

Megalosaurus

This large group of plant-eating saurischians is renowned for the gigantic sizes attained by some of its members. They include the largest animals ever to have lived on land, such as the 28m (92ft) *Argentinosaurus*. Typically, they had long necks, small heads, massive bodies supported on pillar-like legs, and long, muscular, stiffened tails. The group is defined by unique features of their vertebrae and limbs – it excludes the prosauropods, a group of early sauropodomorphs such as *Plateosaurus* that were considerably smaller and sometimes bipedal.

The skeletal anatomy of this small group suggests that they had evolved less from the basal common ancestor of all theropods than from the more diverse Tetanurae, but there are no well-defined diagnostic features. Nevertheless, the ceratosaurs are taken to include dinosaurs such as *Elaphrosaurus*, *Ceratosaurus*, and *Abelisaurus*, which range from Jurassic to Late Cretaceous in age. They were medium-sized, bipedal carnivores growing up to 7.5m (25ft) long, and some species developed distinctive horns on top of the skull, probably for use in display.

This group of spinosauroids includes a range of bipedal carnivores from the Middle and Late Jurassic. First named by T.H. Huxley in 1869, it includes one of the first dinosaurs described, William Buckland's *Megalosaurus* of 1824. However, the fossil material was very fragmentary, consisting mainly of a jawbone and some teeth, and so neither the animal nor its group was well defined. Unfortunately, little further material has been discovered since to make matters clearer.

THEROPODA ▶

SAURISCHIA ▶

SAURISCHIA

This group is marked by the retention of the ancestral archosaur pelvic structure, with its forward-pointing pubis and backward-pointing ischium.

SPINOSAURIDAE

ALLOSAURIDAE

CARCHARODONTOSAURIDAE

COELUROSAURIA

See page 294

Irritator

Allosaurus

Giganotosaurus

Buitreraptor

DEFINING THE DINOSAURS

From his survey of the known dinosaur remains (just a few dozen species in his time), Harry Govier Seeley realized that there is a fundamental difference in the structure of the pelvis, with two consistent types – one like those of most other reptiles, which he named as "saurischian", and the other more like that of the birds, which he named "ornithischian". In the Saurischia the pubis was directed mostly forwards and the ischium backwards. By contrast in the Ornithischia the shaft of the pubis is turned around so that it points backwards alongside the ischium, and in most forms there is a new prong-like extension of the pubis that is directed forward and outward.

Seeley went further, claiming that instead of being divergences within a common group of dinosaurs, the two hip types derived from two different ancestors – in other words, dinosaurs were not a natural group. This debate rumbled on until the 1980s, when Jacques Gauthier used cladistic techniques to show that, while both the Ornithischia and Saurischia are monophyletic groups, so is the Dinosauria as a whole, defined by unique shared features of the skull, girdles, hind limbs, and hands.

This group of specialized Cretaceous theropods included large, bipedal carnivores up to 14m (46ft) long. They have long, crocodile-like skulls and blade-like teeth with fine serrations (if any), adapted for catching and eating fish. The type genus, *Spinosaurus*, carries a skin-covered "sail" along its back, 1.5m (5ft) tall and supported by extensions of the vertebrae. This may have acted as a thermoregulator or been for display.

These were originally described in 1877 by US paleontologst Othniel Marsh. The exact number of genera included in the broad group is not clear. They are a range of Late Jurassic, medium to large bipedal and carnivorous theropods, up to 14m (46ft) long. They have a similar body form to the carcharodontosaurids, with which they share a number of features of the skull, jaw, and pelvis.

This relatively short-lived Cretaceous group included some of the largest land predators such as the 14m (46ft) *Giganotosaurus*. They are defined by features of the skull, teeth, and vertebrae, and had heads up to 2m (6.6ft) long, with powerful jaws. The small forelimbs had three-clawed fingers, and a massive body was counterbalanced by a large muscular and stiffened tail.

This group of largely bipedal predators appeared in the Middle Jurassic and became largely extinct at the end of the Cretaceous. Over this time they diversified until they ranged in size from the tiny *Microraptor* to the giant *Tyrannosaurus*. Coelurosaur sub-groups include the tyrannosaurs, ornithomimosaurs and maniraptorans, including the surviving coelurosaurs, the birds.

◀ SPINOSAUROIDEA

ALLOSAUROIDEA ▶

AVETHEROPODA ▶

TETANURAE ▶

ALLOSAUROIDEA

This group includes the sinraptorids, carcharodontosaurids, and allosaurids, ranging in age from the Jurassic to the Late Cretaceous, and encompassing some of the largest land predators ever to have lived. The group is distinguished by features of the long, narrow skull.

TETANURAE

This large group includes most theropods and their descendants. The name means "stiff tails" and was first used in 1986, in one of the first dinosaur phylogenies using the cladistic method of analysis. The group is defined as including all those theropods more closely related to modern birds than to *Ceratosaurus*.

AVETHEROPODA

First recognized in 1988, this group comprises *Allosaurus* and all its descendants – the carnosaurs and coelurosaurs including modern birds. It is based on common features of the jaw and cervical vertebrae.

SPINOSAUROIDEA

First named in 1915 by German paleontologist Ernst Stromer, the Spinosauroidea (sometimes known as Megalosauroidea) was redefined as a clade in 1998. The group takes its name from its most famous member, the 16m (52ft) *Spinosaurus*, which may have been the largest carnivorous dinosaur of all.

COELUROSAURIA

THE COELUROSAURIA ENCOMPASS NOT ONLY SOME OF THE BEST KNOWN DINOSAUR SPECIES, BUT ALSO THOSE THAT ARE STILL WITH US TODAY AS BIRDS. THANKS TO A NUMBER OF REMARKABLE DISCOVERIES SINCE THE 1990S, THIS GROUP HAS RECENTLY BEEN THE SUBJECT OF INTENSE ATTENTION, AND OFFERS SOME OF THE BEST EXAMPLES OF EVOLUTION IN ACTION.

The group of dinosaurs that includes the largest land-living carnivore has attracted more attention than others ever since *Tyrannosaurus* was described by US paleontologist Henry Fairfield Osborn in 1905. Since then, numerous near-complete skeletons have been found.

In recent years, however, it is the smaller coelurosaurs that have attracted renewed interest to the group. The discovery of remarkably preserved predatory theropods with bird-like feathers at Liaoning in China (see pp.150–5) amazed paleontologists, and further discoveries have revealed a range of feather-like structures associated with some 20 genera.

As a result, a great deal of research has been carried out on the development of feathers. Most recent of all has been the discovery of filament-like structures associated with ornithischian dinosaurs such as *Psittacosaurus* and *Tianyulong* – though it is not yet known whether these structures have a common origin with coelurosaur feathers, or are simply an example of parallel evolution.

COMPSOGNATHIDAE

Compsognathus

This group of small, bipedal, and slender theropods, typically one or two metres (40–80in) from head to tail, had long legs, a thin flexible tail, a long neck and skull, and jaws armed with sharp teeth. Their arms were relatively short, with three long, clawed fingers. They are defined by features of the skull, vertebrae, and hands, and there is evidence that some of the compsognathids may have had a covering of hair-like down – though others seem to have had scales on the tail and legs.

SINOSAUROPTERYX

Sinosauropteryx

Closely related to *Compsognathus*, this bipedal, agile, and long-legged carnivore grew to 1.25m (50in) long. It had a very long, stiff tail (with 64 vertebrae), short arms, flexible wrists, and three-clawed hands. One of the most primitive dinosaurs of the coelurosaur group, *Sinosauropteryx* was the first non-avian theropod to be found with a covering of furry down and primitive feather-like structures. It is distinguished from *Compsognathus* by its skull and limb proportions.

TYRANNOSAUROIDEA

Gorgosaurus

This distinctive and well-known group of bipedal and carnivorous coelurosaurs, with sizes ranging from 1.6m to 12m (64in to 39ft), is characterized by large skulls, specialized "heterodont" teeth, reduced forelimbs with two or three digits, and elongated, powerfully built hindlimbs. Early forms were relatively small with longer forelimbs – for example the 4m (13ft) *Eotyrannus* from the Isle of Wight, England. The group is defined by features of the skull and pelvis. Primitive feather-like structures have been found on some forms.

ORNITHOMIMOSAURIA

Struthiomimus

This group of small theropods (growing to 4m/13ft long) is defined by features of the skull. However, their name translates as "ostrich mimics", on account of their obvious but superficial resemblance to modern ostriches. They were fast-moving herbivores with long and powerful hindlimbs that were well adapted for running, with a long foot and short-clawed toes. Their small skulls, meanwhile, were set on a long and slender neck, with large eyes and a generally toothless beak.

COELUROSAURIA ▶

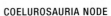

Velociraptor is one of the best known coelurosaurs, an agile dromaeosaurid hunter from Cretaceous Mongolia that is now known to have had feathered wing-like arms.

COELUROSAURIA NODE
This group includes all theropods closer to the avialae than to the carnosaurs, such as tyrannosaurs and maniraptorans.

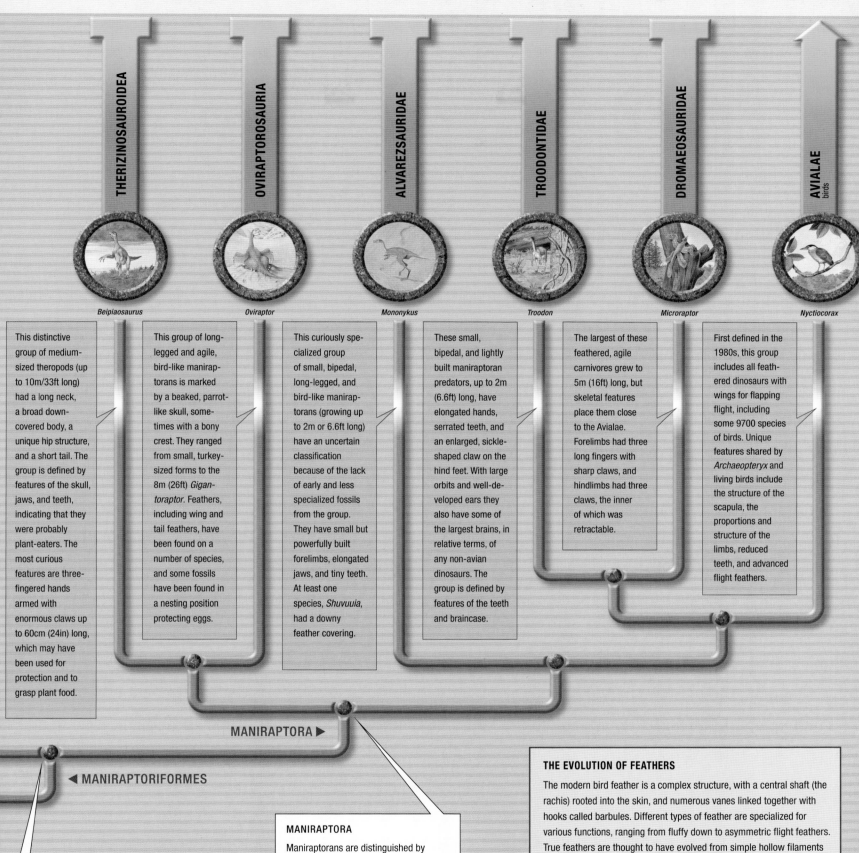

THERIZINOSAUROIDEA

Beipiaosaurus

This distinctive group of medium-sized theropods (up to 10m/33ft long) had a long neck, a broad down-covered body, a unique hip structure, and a short tail. The group is defined by features of the skull, jaws, and teeth, indicating that they were probably plant-eaters. The most curious features are three-fingered hands armed with enormous claws up to 60cm (24in) long, which may have been used for protection and to grasp plant food.

OVIRAPTOROSAURIA

Oviraptor

This group of long-legged and agile, bird-like maniraptorans is marked by a beaked, parrot-like skull, sometimes with a bony crest. They ranged from small, turkey-sized forms to the 8m (26ft) *Gigantoraptor*. Feathers, including wing and tail feathers, have been found on a number of species, and some fossils have been found in a nesting position protecting eggs.

ALVAREZSAURIDAE

Mononykus

This curiously specialized group of small, bipedal, long-legged, and bird-like maniraptorans (growing up to 2m or 6.6ft long) have an uncertain classification because of the lack of early and less specialized fossils from the group. They have small but powerfully built forelimbs, elongated jaws, and tiny teeth. At least one species, *Shuvuuia*, had a downy feather covering.

TROODONTIDAE

Troodon

These small, bipedal, and lightly built maniraptoran predators, up to 2m (6.6ft) long, have elongated hands, serrated teeth, and an enlarged, sickle-shaped claw on the hind feet. With large orbits and well-developed ears they also have some of the largest brains, in relative terms, of any non-avian dinosaurs. The group is defined by features of the teeth and braincase.

DROMAEOSAURIDAE

Microraptor

The largest of these feathered, agile carnivores grew to 5m (16ft) long, but skeletal features place them close to the Avialae. Forelimbs had three long fingers with sharp claws, and hindlimbs had three claws, the inner of which was retractable.

AVIALAE
birds

Nyctiocorax

First defined in the 1980s, this group includes all feathered dinosaurs with wings for flapping flight, including some 9700 species of birds. Unique features shared by *Archaeopteryx* and living birds include the structure of the scapula, the proportions and structure of the limbs, reduced teeth, and advanced flight feathers.

MANIRAPTORA ▶

◀ MANIRAPTORIFORMES

MANIRAPTORIFORMES

This group unites the maniraptorans with the ornithomimosaurs. They are thought to have evolved from a common Jurassic ancestor, but had separated by the Late Jurassic.

MANIRAPTORA

Maniraptorans are distinguished by elongated, three-clawed hands that increasingly tend towards the wing-like form of true birds. There is also a half-moon-shaped "semi-lunate" bone that adds articulation to the wrist, amongst other features.

THE EVOLUTION OF FEATHERS

The modern bird feather is a complex structure, with a central shaft (the rachis) rooted into the skin, and numerous vanes linked together with hooks called barbules. Different types of feather are specialized for various functions, ranging from fluffy down to asymmetric flight feathers. True feathers are thought to have evolved from simple hollow filaments that went through a series of adaptations to produce this complexity. Possible intermediary stages may have included multi-branched downy feathers or simple bristle-like structures developed for insulation, followed by plume feathers (possibly evolving for display), body contour feathers, and finally asymmetric flight feathers, as seen in *Archaeopteryx*, the earliest known bird.

THE
SITE
GAZETTEER

The world's great fossil sites,
illustrated throughout the
panoramic artwork pages,
range from publicly accessible
quarries to dangerous caves,
and from monumental, stark
geology to easily missed, buried
environments. The pages that
follow offer more information
on the history and present
state of these often fascinating
locations, so crucial to our
understanding of life's history
and evolution.

GAZETTEER OF SITES

THE DISTRIBUTION OF THE
WORLD'S IMPORTANT KNOWN
FOSSIL SITES IS DETERMINED IN
PART BY GEOLOGY, AND IN PART
BY ACCIDENT OF HISTORY. FOSSILS
WILL ONLY FORM IN LOCATIONS WHERE
SEDIMENTARY ROCK ACCUMULATES AND IS
PRESERVED OVER LONG PERIODS, BUT IT IS NO COINCIDENCE
THAT MANY OF THE BEST-STUDIED FOSSIL SITES ARE IN WESTERN
EUROPE AND AMERICA, WHERE THEY HAVE BEEN EASILY ACCESSIBLE TO
ACADEMIC AND AMATEUR PALEONTOLOGISTS. WHILE A FEW SITES IN OTHER
REGIONS HAVER BEEN KNOWN SINCE THE 19TH CENTURY, MANY MORE
HAVE BEEN DISCOVERED IN RECENT DECADES, AND MANY NEW TREASURES
UNDOUBTEDLY STILL AWAIT DISCOVERY IN REMOTE PARTS OF THE WORLD.

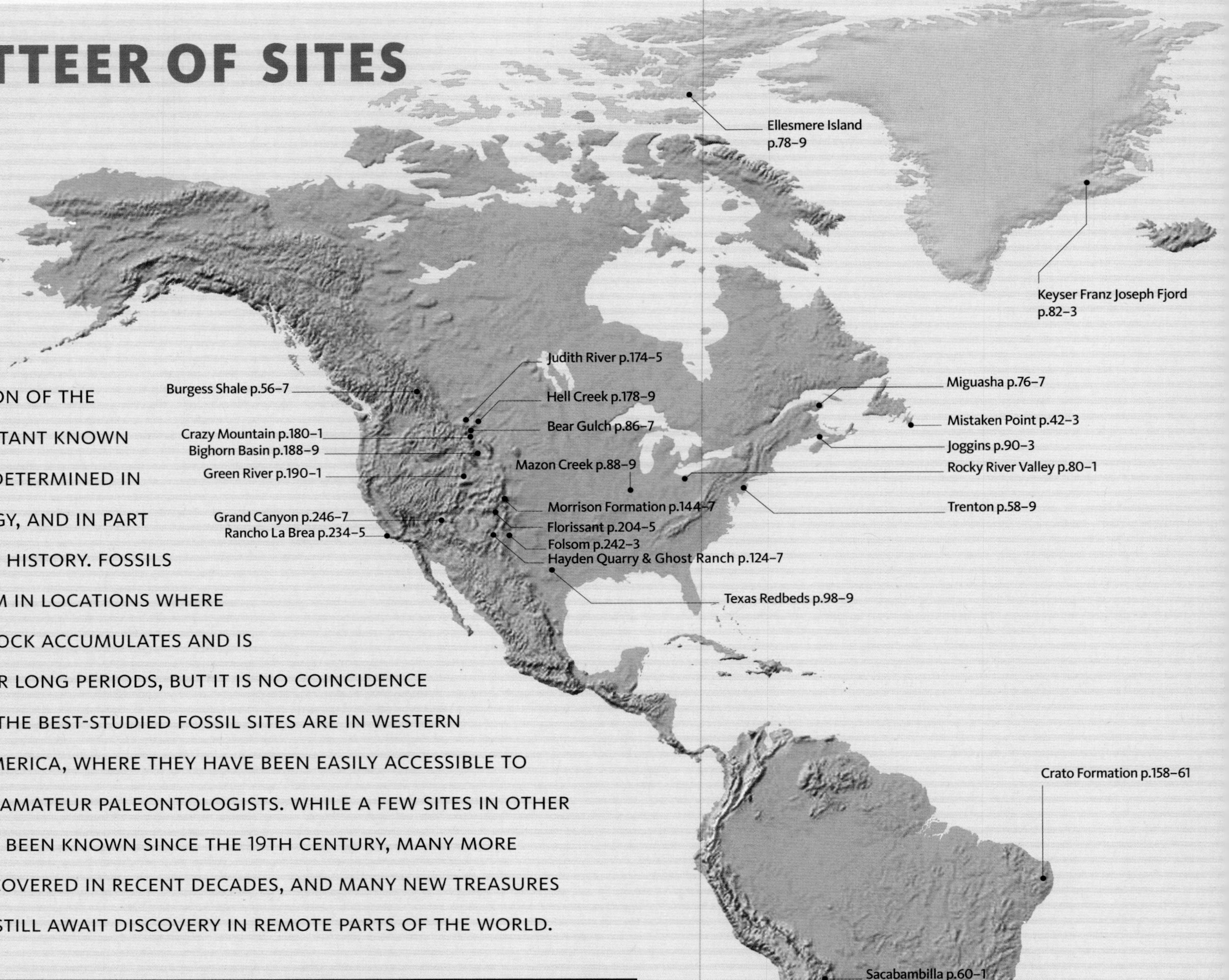

Ellesmere Island p.78–9

Keyser Franz Joseph Fjord p.82–3

Judith River p.174–5
Burgess Shale p.56–7
Hell Creek p.178–9
Miguasha p.76–7
Bear Gulch p.86–7
Mistaken Point p.42–3
Crazy Mountain p.180–1
Joggins p.90–3
Bighorn Basin p.188–9
Mazon Creek p.88–9
Rocky River Valley p.80–1
Green River p.190–1
Morrison Formation p.144–7
Trenton p.58–9
Grand Canyon p.246–7
Florissant p.204–5
Rancho La Brea p.234–5
Folsom p.242–3
Hayden Quarry & Ghost Ranch p.124–7
Texas Redbeds p.98–9

Crato Formation p.158–61

Sacabambilla p.60–1

Monte Verde p.240–1

Neuquén p.166–9

Valley of the Moon p.120–1

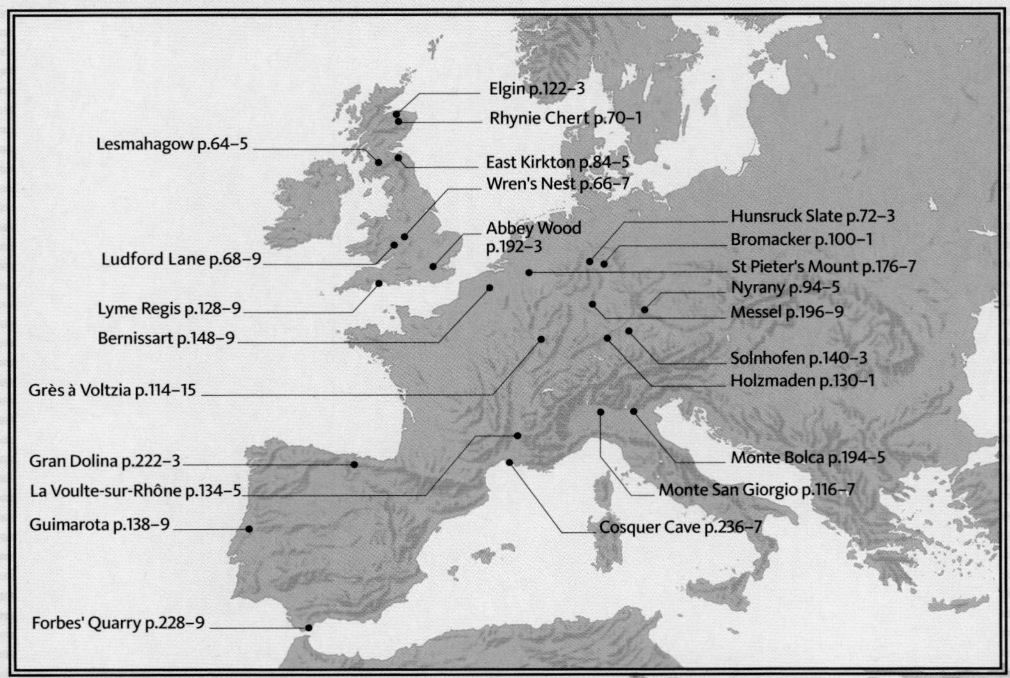

Elgin p.122–3
Rhynie Chert p.70–1
Lesmahagow p.64–5
East Kirkton p.84–5
Wren's Nest p.66–7
Abbey Wood p.192–3
Hunsruck Slate p.72–3
Bromacker p.100–1
Ludford Lane p.68–9
St Pieter's Mount p.176–7
Nyrany p.94–5
Lyme Regis p.128–9
Messel p.196–9
Bernissart p.148–9
Solnhofen p.140–3
Holzmaden p.130–1
Grès à Voltzia p.114–15
Gran Dolina p.222–3
Monte Bolca p.194–5
La Voulte-sur-Rhône p.134–5
Monte San Giorgio p.116–7
Guimarota p.138–9
Cosquer Cave p.236–7
Forbes' Quarry p.228–9

PALEONTOLOGISTS AT WORK

Crato Formation, Araripe Basin, Brazil

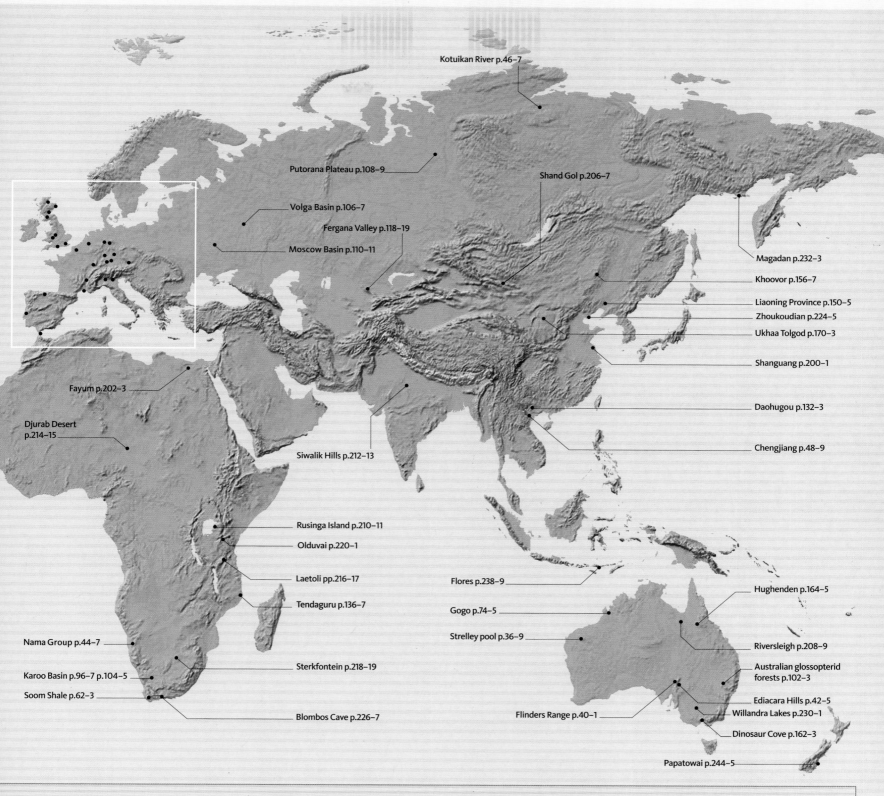

Kotuikan River p.46–7

Putorana Plateau p.108–9

Shand Gol p.206–7

Volga Basin p.106–7

Fergana Valley p.118–19

Moscow Basin p.110–11

Magadan p.232–3

Khoovor p.156–7

Liaoning Province p.150–5

Zhoukoudian p.224–5

Ukhaa Tolgod p.170–3

Shanguang p.200–1

Fayum p.202–3

Daohugou p.132–3

Djurab Desert
p.214–15

Chengjiang p.48–9

Siwalik Hills p.212–13

Rusinga Island p.210–11

Olduvai p.220–1

Laetoli pp.216–17

Flores p.238–9

Hughenden p.164–5

Tendaguru p.136–7

Gogo p.74–5

Strelley pool p.36–9

Riversleigh p.208–9

Nama Group p.44–7

Australian glossopterid
forests p.102–3

Sterkfontein p.218–19

Karoo Basin p.96–7 p.104–5

Ediacara Hills p.42–5

Soom Shale p.62–3

Willandra Lakes p.230–1

Flinders Range p.40–1

Blombos Cave p.226–7

Dinosaur Cove p.162–3

Papatowai p.244–5

Karoo Basin, South Africa

Messel, near Darmstadt, Germany

Gran Dolina, Atapuerca, Spain

Chengjiang, Yunnan Province, China

Graphite Peak p.112–13

ABBEY WOOD　192–3

London, England

54 MILLION YEARS AGO

The mammal remains of Abbey Wood are only part of a coastal deposit that also includes fish remains such as numerous shark's teeth. The pit is re-covered at the end of each season, with new trenches excavated each year. Most fossils are found through sieving with fine meshes, and the site welcomes amateur fossil hunters in organized parties.

SPECIES LIST

① *Coryphodon* Boreoeutheria
② *Ficus* Angiospermae
③ *Paramys* Euarchontoglires
④ *Palaeosinopa* Boreoeutheria
⑤ *Apatemys* Boreoeutheria
⑥ lauracean Angiospermae
⑦ marinavis Neognathae
⑧ *Ceriops* Angiospermae
⑨ *Palaeonictis* Laurasiatheria
⑩ *Cantius* Primates
⑪ *Oxyaena* Laurasiatheria
⑫ *Pliolophus* Laurasiatheria

FURTHER READING

BENTON, M.J. ET AL. 2005. *Mesozoic and Tertiary Fossil Mammals and Birds of Great Britain.* Joint Nature Conservation Committee, Peterborough
COCKBURN, H. and PALMER, D. 2008. *Fossil Detectives: Discovering Prehistoric Britain.* BBC Books, London

WEBSITES

www.abbeywood.ukfossils.co.uk/
www.jncc.gov.uk/pdf/grdb/GCRsiteaccount2903.pdf

AUSTRALIAN GLOSSOPTERID FORESTS　102–3

Queensland and NSW, Australia

267–260 MILLION YEARS AGO

Australia is the world's fourth largest producer of coal, much of which originated in the late Permian and occurs in vast deposits across Queensland and New South Wales. Many of the plants that made up the coal can also be found as fossils within the associated sandstones and shales throughout this region. Some of the finest specimens have found their way into the major museums of Australia, such as the Australian Museum in Sydney and Museum Victoria in Melbourne.

SPECIES LIST

① *Glossopteris* Gymnospermopsida
② *Phyllotheca* Sphenopsida
③ *Glossopteris linearis* Gymnospermopsida
④ *Plumsteadia* Gymnospermopsida
⑤ *Ebanaqua* Actinopterygii
⑥ *Sphenophyllum* Sphenopsida
⑦ *Austraglossa* Gymnospermopsida
⑧ *Dictyopterium* Gymnospermopsida
⑨ *Sphenopteris* Sphenopsida

FURTHER READING

WHITE, M. 1986. *The Greening of Gondwana: the 400 million year Story of Australia's Plants.* Kangaroo Press, Roseville, NSW

WEBSITES

www.ucmp.berkeley.edu/seedplants/
　pteridosperms/glossopterids.html
www.adonline.id.au/plantevol/tour/
　glossopterids.htm

BEAR GULCH　86–7

Montana, USA

320 MILLION YEARS AGO

The Bear Gulch limestones, famous for their well-preserved fossil fish, form an outcrop of about 70 square kilometres (217 square miles) in central Montana's Big Snowy Mountains. Many of the best fossils are on display at the University of Montana Paleontology Center in Missoula.

SPECIES LIST

① *Allenypterus* Sarcopterygii
② *Caridosuctor* Sarcopterygii
③ *Belantsea* Chondrichthyes
④ *Falcatus* Chondrichthyes
⑤ *Echinochimaera* Chondrichthyes
⑥ *Stethacanthus* Chondrichthyes
⑦ *Paratarrasius* Actinopterygii
⑧ *Crangopsis* Crustacea
⑨ *Harpagofututor* Chondrichthyes

FURTHER READING

GROGAN, E. and LUND, R. 2002. *The geological and biological environment of the Bear Gulch Limestone and a model for its deposition.* Geodiversitas 24/2, 295
BOTTJER, D.J. ET AL. (eds) 2001 *Exceptional Fossil Preservation.* Columbia University Press, New York

WEBSITES

www.sju.edu/research/bear_gulch/
www.cas.umt.edu/paleontology/rc_beargulchlime.htm

Bighorn Basin The "layer-cake" stratigraphy of Wyoming's sedimentary basins preserves successions of early Cenozoic strata and the remains of rapidly diversifying early mammals.

BERNISSART　148–9

Belgium

128–125 MILLION YEARS AGO

This underground mine is now flooded and inaccessible, but ten reconstructed *Iguanodon* dinosaur skeletons from the site are exhibited in the Royal Belgian Institute of Natural Sciences in Brussels, along with a display detailing the geology and paleontology of the site. Casts of *Iguanodon* from Bernissart are also on display in other museums, such as the Sedgwick Museum at the University of Cambridge, England, and London's Natural History Museum.

SPECIES LIST

① *Iguanodon atherfieldensis* Ornithischia
② *Ornithocheirus* Pterosauria
③ *Iguanodon bernissartensis* Ornithischia

FURTHER READING

NORMAN, D.B. 1980. *On the Ornithischian Dinosaur Iguanodon bernissartensis from the Lower Cretaceous of Bernissart (Belgium).* Memoire No. 178. L'Institut Royal des Sciences Naturelles de Belgique, Brussels

WEBSITE

www.naturalsciences.be/museum/dinosaurs

BIGHORN BASIN　188–9

Wyoming, USA

54 MILLION YEARS AGO

A succession of strata in Western North America record the evolution of Eocene mammal faunas from the Wasatchian faunas of the early Eocene (*c*.54MA), through the Bridger faunas

of the middle Eocene, to the more recent Uinta and Duchesne faunas. The Wasatchian biota is preserved in the Willwood Formation strata, in both Clarks Fork and the Central Bighorn basins of northwestern Wyoming. There are numerous localities scattered across these semi-arid landscapes with largely fragmentary skeletal material, especially teeth and bits of jawbone. Material is preserved in the Yale Peabody Museum in New Haven.

SPECIES LIST

1. *Miacis* Laurasiatheria
2. *Arfia* Laurasiatheria
3. *Hyopsodus* Laurasiatheria
4. *Platanus* Angiospermae
5. *Didymictis* Laurasiatheria
6. *Cantius* Primates
7. *Hyracotherium* Laurasiatheria
8. *Diacodexis* Laurasiatheria
9. *Celtis* Angiospermae
10. *Phenacodus* Laurasiatheria

FURTHER READING

HEINRICH, R.E. ET AL. 2008. *Earliest Eocene Miacidae (Mammalia: Carnivora) from Northwestern Wyoming.* Journal of Paleontology 82, 154–62.
WING, S.L. ET AL. 1995. *Plant and mammal diversity in the Paleocene to Early Eocene of the Bighorn Basin.* Palaeogeography, Palaeoclimatology, Palaeoecology 115, pp. 117–55

WEBSITE

www.geo-sciences.com/dinofossils.htm

BLOMBOS CAVE 226–7

South Africa

75,000 YEARS AGO

The Blombos cave, today just 100m (330ft) from the sea on South Africa's stormy southern coast, was discovered by Christopher Henshilwood of New York State University in 1991, and has been excavated in almost every season since the late 1990s. The cave deposits cover about 55 sq metres (600 sq feet), and are several metres deep, with the earliest deposits safely capped by a layer of wind-blown sterile sand that built up during a period when the cave was unoccupied and partly blocked by a large dune, from 70,000 years ago until around 2000 years ago. While the cave is not open to the public, some of the finds, including the famous engraved ochre fragment, are on display at the South African Museum in Cape Town.

SPECIES AND ARTEFACTS LIST

1. ochre crayon
2. bone tools
3. stone points
4. *Nassarius kraussianus* Mollusca
5. *Homo sapiens* Hominidae
6. *Procavia* Afrotheria
7. *Syncerus* Laurasiatheria
8. delphinid Laurasiatheria
9. *Raphicercus* Laurasiatheria
10. *Lepus* Euarchontoglires
11. *Antidorcas* Laurasiatheria
12. *Taurotragus* Laurasiatheria
13. *Arctocephalus pusillus* Laurasiatheria

WEBSITE

www.svf.uib.no/sfu/blombos/index.htm

BROMACKER 100–1

Gotha, Germany

290 MILLION YEARS AGO

Excavations at Bromacker have been under way in an abandoned sandstone quarry since 1993, providing new insights into the Permian ecosystem. Aside from the various unique specimens, many of the fossils were previously unknown outside the USA. For this reason, excavations are carried out as a joint project with the Carnegie Museum of Natural History and the University of California. Finds have been displayed at the Carnegie in Pittsburgh, and the Museum der Natur in Gotha.

SPECIES LIST

1. *Diadectes* 'Reptilia'
2. *Eudibamus* 'Reptilia'
3. *Dimetrodon* Synapsida
4. *Seymouria* 'Reptilia'
5. *Syscioblatta* Hexapoda

WEBSITES

www.carnegiemnh.org/research/eudibamus/index.html
www.epilog.de/dokumente/show/ausstellung/prehistoric/Gotha_ursaurier.htm (in German)

BURGESS SHALE 56–7

British Columbia, Canada

510 MILLION YEARS AGO

This World Heritage Site, high in the Yoho National Park and protected since 1981, is accessible only by arrangement through the Burgess Shale Geoscience Foundation. However, there is a display about the site in a nearby visitor centre. Major fossil collections are in the Royal Ontario Museum in Toronto, the Field Visitor Center, at Field, British Columbia, and the Smithsonian Institution, in Washington DC, USA.

SPECIES LIST

1. *Pikaia* Chordata
2. *Opabinia* Arthropoda
3. *Thaumaptilon* Cnidaria
4. *Yohoia* Arthropoda
5. *Odontogriphus* Mollusca
6. *Dinomischus* Unknown
7. *Marrella* Trilobita
8. *Aysheaia* Arthropoda
9. *Vauxia* Porifera
10. *Sidneyia* Arthropoda
11. *Olenoides* Trilobita
12. *Canadaspis* Arthropoda
13. *Anomalocaris* Arthropoda
14. *Wiwaxia* Mollusca
15. *Canadia* Polychaeta
16. *Pirania* Porifera
17. *Hallucigenia* Arthropoda
18. *Ottoia* Priapulida
19. *Eldonia* Unknown
20. *Ctenorhabdotus* Ctenophora

FURTHER READING

GOULD, S.J. 1989. *Wonderful Life.* Hutchinson Radius, London
CONWAY MORRIS, S. 1998. *Crucible of Creation.* Oxford University Press, Oxford

WEBSITES

www.burgess-shale.bc.ca
www.ucmp.berkeley.edu/cambrian/burgess.html

CHENGJIANG 48–9

Yunnan Province, China

520 MILLION YEARS AGO

Thousands of fossils, including many new species, have been obtained from numerous sites scattered over the Yangtze region of southwest China around Chengjiang. Many of the specimens are housed in the research centre of the Chengjiang Biota, Yunnan University, and the Nanjing Institute of Geology and Palaeontology, Academica Sinica, People's Republic of China.

SPECIES LIST

1. *Retifacies* Trilobita
2. *Kuamaia* Trilobita
3. *Quadrolaminiella* Porifera
4. *Eldonia* Unknown
5. *Anomalocaris* Arthropoda
6. *Longtancunella* Brachiopoda
7. *Archisaccophyllia* Cnidaria
8. *Canadaspis* Arthropoda
9. *Haikoucaris* Chelicerata

10. *Paraleptomitella* Porifera
11. *Myllokunmingia* Chordata
12. *Paucipodia* Arthropoda
13. *Paraselkirkia* Priapulida
14. *Hallucigenia* Arthropoda

FURTHER READING

JUNYAN CHEN ET AL. 1997. *The Cambrian Explosion and the Fossil Record.* National Museum of Natural Science, Taichung, Taiwan
HOU XIAN-GUANG ET AL. 1999. *The Chengjiang Fauna.* Yunnan Science and Technology Press, China.
HOU XIAN-GUANG ET AL. 2004. *The Cambrian Fossils of Chengjiang, China.* Blackwell Publishing, Oxford

WEBSITES

www.peripatus.gen.nz/paleontology/lagChengjiang.html
www.fossilmuseum.net/Fossil_Sites/Chengjiang.htm

COSQUER CAVE   236–7

Marseilles, France

18,500 YEARS AGO

The site of these beautiful cave paintings is today accessible only by experienced divers who travel along a 175m (575ft) tunnel from an entrance some 37m (120ft) below water. Fortunately, the site is well documented in website and book form.

SPECIES AND ARTEFACTS LIST

1. cave adornments
2. *Ibex* Laurasiatheria
3. *Equus* Laurasiatheria
4. *Megaloceros* Laurasiatheria
5. *Bison* Laurasiatheria
6. *Sterna paradisaea* Neognathae
7. *Homo sapiens* Hominidae
8. *Pinguinis* Neognathae
9. *Monachus* Laurasiatheria

FURTHER READING

CLOTTES, J. and COURTIN, J. 1996. *The Cave Beneath the Sea: Paleolithic Images at Cosquer.* Harry N. Abrams, Inc., New York

WEBSITE

www.bradshawfoundation.com/cosquer/

CRATO FORMATION   158–61

Araripe Basin, Northeastern Brazil

112 MILLION YEARS AGO

Named after the town of Crato, these rock strata outcrop on the flanks of the 800m high (2500ft) Chapada do Aripe tableland in northeastern Brazil. The limestones are mined commercially for cement manufacture and paving stones, and this quarrying activity focuses on slabby laminated limestones from the underlying unit, known as the Nova Olinda Member. Numerous small excavations have revealed an astonishing number of exquisitely preserved and valuable fossils that are exported all over the world. Their diversity records the mid-Cretaceous radiation of the flowering plants and the coevolution of their insect pollinators – it may turn out to be the most diverse Cretaceous terrestrial ecosystem in the world. In just over 20 years some 200 new species have been discovered and described from here. Fossils are found in major museums around the world and especially at the Museo de Paleontologia, Santana do Cariri, Ceara, Brazil.

SPECIES LIST 158–9

1. *Welwitschiostrobus* Gymnospermopsida
2. *Irritator* Theropoda
3. *Santanmantis* Hexapoda
4. *Ruffordia* Filicopsida
5. *Tapejara* Pterosauria
6. *Baeocossus* Hexapoda
7. nymphaealean Angiospermae
8. *Cretofedtschenkia* Hexapoda
9. belostomatid Hexapoda
10. *Lindleycladus* Coniferopsida
11. *Cretaraneus* Chelicerata
12. *Tettagalma* Hexapoda
13. *Ludodactylus* Pterosauria
14. *Dastilbe* Actinopterygii

SPECIES LIST 160–1

1. myrmeliontid Hexapoda
2. ephedroid (female) Gymnospermopsida
3. *Baisopardus* Hexapoda
4. *Cladocyclus* Actinopterygii
5. *Protoischnurus* Chelicerata
6. *Tapejara* Pterosauria
7. *Cratoraricrus* Uniramia
8. *Britopygus* Chelicerata
9. *Araripeliupanshania* Hexapoda
10. *Susisuchus* Crocodyliformes
11. ephedroid (male) Gymnospermopsida
12. *Arariphrynus* Lissamphibia
13. *Ruffordia* Filicopsida

FURTHER READING

MARTILL, D.M. ET AL. (eds) 2007. *The Crato Fossil Beds of Brazil: Window into an Ancient World.* Cambridge University Press, Cambridge
SELDEN, P. and NUDDS, J. 2004. *Evolution of Fossil Ecosystems.* Manson Publishing, London (see chapter 11)

WEBSITE

www.unb.br/ig/sigep/sitio005/sitio005english.htm

CRAZY MOUNTAIN   180–1

Montana, USA

63 MILLION YEARS AGO

Numerous localities, recording the evolution of terrestrial vertebrates across the transition from Cretaceous into Paleogene times, are exposed over large areas of northwestern North America, from Montana into Alberta. The Torrejonian mammal fauna, from Fort Union Beds in Montana's Crazy Mountain Basin, was first discovered in 1901, with large collections made at a number of localities. The basin's strata span much of the Paleocene's 10 million years, including the Torrejonian, Tiffanian, and Clarkforkian stages. The ages of the fossils are well constrained in datable rock units that span as little as 600,000–700,000 years, and in some places fossil mammals are very abundant – though preserved mainly as small scattered bones and teeth.

Fossils from the region are conserved in numerous museums, including the Carnegie Museum of Natural History in Pittsburgh, USA, the Museum of Natural History at the University of Kansas in Lawrence, USA, and the National Museum of Natural Sciences in Ottawa, Canada.

SPECIES LIST

1. *Conoryctes* Boreoeutheria
2. *Platanus* Angiospermae
3. *Ptilodus* Mammalia
4. *Stilpnodon* Boreoeutheria
5. *Rhamnus* Angiospermae
6. *Plesiadapis* Primates
7. *Didymictis* Laurasiatheria
8. *Vitis* Angiospermae
9. *Chriacus* Laurasiatheria
10. *Taxodium* Coniferopsida
11. *Pantolambda* Boreoeutheria
12. *Prodiacodon* Boreoeutheria

FURTHER READING

SIMPSON, G.G. 1937. *The Fort Union of the Crazy Mountain Field Montana and its Mammal Faunas.* Smithsonian Institution Bulletin 169, Washington.

DAOHUGOU 132–3

Inner Mongolia, China

171–164 MILLION YEARS AGO

An extraordinary series of fossil mammals, amphibians, and insects, with their soft tissues wonderfully preserved, has recently been uncovered from the freshwater lake and stream deposits of Daohugou in Inner Mongolia. Most scientists studying the site believe that they are of mid-Jurassic age (*c.*164 million years old), but dating is controversial, and some argue that the fossils may be as recent as the Early Cretaceous (*c.*130 million years old). Specimens are housed at the Jinzhou Museum of Paleontology in Jinzhou City, Liaoning Province, China.

SPECIES LIST

1. *Leptolingia* Hexapoda
2. *Pseudotribos* Mammalia
3. *Volaticotherium* Mammalia
4. *Mongolbittacus* Hexapoda
5. cryptobranchoid Lissamphibia
6. *Quadraticossus* Hexapoda
7. *Chunerpeton* Lissamphibia
8. *Castorocauda* Mammalia
9. *Grammolingia* Hexapoda
10. *Pedopenna* Theropoda

FURTHER READING

ZHE-XI LUO ET AL. 2007. Nature (1 November) pp.93–7
MENG, J. ET AL. 2006. *A Mesozoic Gliding Mammal from Northeastern China.* Nature (14 December) pp.889–93
QIANG JI ET AL. 2006. *A Swimming Mammaliaform from the Middle Jurassic.* (24 February) Science pp.1123–7

WEBSITE

en.wikipedia.org/wiki/daohugou_beds

DINOSAUR COVE 162–3

Victoria, Australia

110 MILLION YEARS AGO

The fascinating fossil deposits of Dinosaur Cove are inaccessible to all but professional paleontologists, on account of their location in high sea-facing cliffs. During the 1980s and 1990s, the excavations were supervised by Thomas and Patricia Rich of Museum Victoria. Mining and blasting equipment was used to open out the deposits, and they are now considered to be exhausted. However, many of the fossils of these unusual Australian dinosaurs are displayed at Museum Victoria in Melbourne.

SPECIES LIST

1. sphenopterid Filicopsida
2. *Bishops* Monotremata
3. *Gingkoites australis* Ginkgoales
4. *Leaellynasaura* Ornithischia
5. sphenopterid Filicopsida
6. *Koolasuchus* Temnospondyli
7. *Taeniopteris* Cycadophyta

FURTHER READING

RICH, T. H. and VICKERS-RICH, P.. 2000. *The Dinosaurs of Darkness.* University of Indiana Press, Bloomington

WEBSITES

www.museumvictoria.com.au/history/dinosaur.htm
www.museumvictoria.com.au
(follow links to the vertebrate paleontology collection)

Crato Formation Brazil's Crato and associated Sanatana strata contain one of the greatest fossil treasure troves in the world, with many organisms preserved whole, and some soft tissues intact.

DJURAB DESERT 214–15

Chad

7 MILLION YEARS AGO

The fragmentary fossils of *Sahelanthropus tchadensis* were discovered at three sites in the remote Djurab Desert of western Chad between July 2001 and March 2002, during a long series of scientific expeditions to the desert organized by French paleontologist Alain Beauvilain. Thanks to desert processes of constant erosion and re-burial, fossils from an earlier, more verdant age can be found lying exposed on the surface, though it takes an expert eye to spot them and even more expertise to read their stratigraphic context – just one reason why the *Sahelanthropus* finds have remained controversial.

SPECIES LIST

1. *Ictitherium* Laurasiatheria
2. *Sahelanthropus* Hominidae
3. *Machairodus* Laurasiatheria
4. *Anancus* Afrotheria
5. "*Macrotermes*" Hexapoda
6. *Hipparion* Laurasiatheria
7. *Orycteropus* Afrotheria
8. *Hexaprotodon* Laurasiatheria
9. *Nyanzochoerus* Laurasiatheria
10. *Kobus* Laurasiatheria
11. cercopithecoid Anthropoidea
12. *Sivatherium* Laurasiatheria

FURTHER READING

BEAUVILAIN, A. 2003. *Toumaï, l'aventure humaine.* La Table Ronde, Paris (in French)
GIBBONS, A. 2006. *The First Human: the Race to Discover Our Earliest Ancestors.* Doubleday, New York

WEBSITES

www.toumai.site.voila.fr/index.html
www.sahelanthropus.com/

EDIACARA HILLS 42–5

Flinders Range, South Australia, Australia

575–545 MILLION YEARS AGO

This classic locality, 400km (250 miles) north of Adelaide, is now the Ediacara Fossil Reserve Palaeontological Site, in Flinders Chase National Park. It was from here many of the ancient and fascinating fossils of the organisms now known as "Ediacarans" were first described. Many of these fossils can be seen in the South Australian Museum in Adelaide and the Western Australia Museum in Perth.

FURTHER READING

MCMENAMIN, A.S. 1998. *The Garden of Ediacara*. Columbia University Press, New York

WEBSITES

www.ucmp.berkeley.edu/vendian/mistaken.html
www.environment.gov.au/heritage/places/national/ediacara/
www.ucmp.berkeley.edu/vendian/ediacara.html

ELGIN 122–3

Moray, Scotland

220 MILLION YEARS AGO

Most of the Elgin site is now preserved within a legally protected Site of Special Scientific Interest, but fossil finds, and in particular tetrapod tracks, continue to be found in the area. Commercial quarrying activity occasionally unearths whole new trackways, but unfortunately some of the unprotected ones have been vandalized by the attempts of "collectors" to remove individual prints. Finds from Elgin, and some preserved trackways, can be viewed at the Royal Edinburgh Museum, the Hunterian Museum in Glasgow, and the Elgin Museum itself.

SPECIES LIST

❶ *Ornithosuchus* Archosauria
❷ *Stagonolepis* Archosauria
❸ *Hyperodapedon* Archosauria
❹ *Scleromochlus* Archosauria

FURTHER READING

FRASER, N. 2006. *Dawn of the Dinosaurs: Life in the Triassic*. Indiana University Press, Bloomington (see chapter 8)

WEBSITES

www.gla.ac.uk/~gxha14/elgin.html
www.elginmuseum.org.uk/

ELLESMERE ISLAND 78–9

Arctic Canada

380 MILLION YEARS AGO

The recently discovered early tetrapod *Tiktaalik* was found at a remote and inaccessible location in Ellesmere Island in the Canadian Arctic Peninsula. The discovery was the result of an intensive and deliberate search of ancient Devonian rocks exposed here, with a team of paleontologists from several US institutions, led by Neil Shubin of the University of Chicago, visiting the site multiple times over several years.

SPECIES LIST

❶ *Tiktaalik* Sarcopterygii
❷ *Laccognathus* Sarcopterygii
❸ *Asterolepis* Gnathostomes

FURTHER READING

DAESCHLER, E.B., SHUBIN, N.H. and JENKINS, F.A. JR. 2006. *A Devonian tetrapod-like fish and the evolution of the Tetrapod body plan.* Nature (6 April), pp.757–63
HOLMES, B. 2006. *Meet your ancestor.* New Scientist (9 September) pp.35–9
SHUBIN, N. 2008. *Your Inner Fish.* Allen Lane, London

WEBSITES

www.devoniantimes.org/Order/re-tiktaalik.html
www.geolsoc.org.uk/template.cfm?name=Daeschler

Flinders Range Despite their great antiquity, the Precambrian strata of the Ediacara Hills in the Flinders Range preserve sediment moulds of many strange soft-bodied creatures.

FAYUM 202–3

Egypt

34 MILLION YEARS AGO

The fossils of the Fayum depression have been known since the mid-19th century, and land mammals, including a number of important early primates, have been unearthed here since 1901. The Wadi al-Hitan site ("Valley of the Whale"), where remains of the primitive cetacean *Basilosaurus* were discovered, is a designated UNESCO World Heritage Site, but other areas of the basin have recently been earmarked for tourist development. Fossils from the region are displayed at the Cairo Geological Museum, and are also in the collections of the Yale Peabody Museum in New Haven, Connecticut, USA, and Duke University Primate Center, North Carolina, USA.

SPECIES LIST

❶ *Ardea* Neognathae
❷ *Sarothrura* Neognathae
❸ *Apidium* Anthropoidea
❹ *Phiomia* Afrotheria
❺ *Aegyptopithecus* Primates
❻ *Epipremnum* Angiospermae
❼ *Arsinotherium* Afrotheria
❽ *Pandion* Neognathae
❾ *Actophilornis* Neognathae
❿ *Moeritherium* Afrotheria
⓫ *Nyctiocorax* Neognathae
⓬ *Balaeniceps* Neognathae
⓭ *Haliaeetus* Neognathae

FURTHER READING

BEARD, C. 2004. *The Hunt for the Dawn Monkey: Unearthing the Origins of Monkeys, Apes, and Humans*. California University Press, Berkeley (see chapter 4)

WEBSITE

www.fossils.duke.edu/research/egypt.html

FERGANA VALLEY 118–19

Kyrgyzstan

228 MILLION YEARS AGO

One of the world's richest deposits of Triassic-age fossil insects lies in the Fergana region of Kyrgyzstan in Central Asia. Since the 1960s, Russian paleontologists have made numerous expeditions to this remote and largely inaccessible desert region. They recovered thousands of fossil plants and insects but only a handful of tetrapods, now mostly housed at the Palaeontological Institute of the Russian Academy of Sciences in Moscow.

SPECIES LIST

❶ *Notocupoides* Hexapoda
❷ *Hadeocoleus* Hexapoda
❸ *Saurichthys* Actinopterygii
❹ *Longisquama* Diapsida
❺ *Podozamites* Coniferopsida
❻ *Sharovipteryx* Archosauria
❼ *Axioxyela* Hexapoda
❽ *Madygenia* Cynodontia
❾ *Gigatitan* Hexapoda

FURTHER READING

FRASER, N. 2006. *Dawn of the Dinosaurs*. Indiana University Press, Bloomington (see Chapter 6)

FLINDERS RANGE 40–1

South Australia, Australia

647–635 MILLION YEARS AGO

Some of the first widely accepted evidence for the "Snowball Earth" glaciation of Precambrian (Marinoan) times comes from sites in South Australia. Unusual geological features preserved in Flinders Chase National Park show that, some 640 million years

ago, a climate crisis saw glacial ice sheets expand to cover the entire planet down to low latitudes.

PRESERVED GEOLOGICAL EVIDENCE

❶ dropstone

FURTHER READING

WALKER, G. 2003. *Snowball Earth*. Bloomsbury, London

WEBSITES

www.environment.sa.gov.au/parks/sanpr/
flindersranges/
www.ga.gov.au/servlet/BigObjFileManager?
bigobjid=GA7517

FLORES   238–9

Indonesia

18,000–15,000 YEARS AGO

The discovery of "Flores man" by an Australian/Indonesian team working on the island of Flores in 2003 has proved to be one of the most controversial fossil finds of recent years. Twelve-metre deep (40ft) strata in the limestone floor of Liang Bua cave revealed the remains of nine individuals and an array of other bones and stone tools scattered through a 75,000-year history of occupation (up to about 18,000 years ago). The bones were not fossilized, but instead were steeped in groundwater, making them extremely fragile and difficult to excavate. Following formal announcement of the discovery, the scientific community was effectively split between those who sided with the theory that "Flores man" was effectively a new species, and those who thought that *Homo floresiensis* was merely a stunted form of *Homo sapiens*. Matters were not helped when the fossils were effectively "kidnapped" by Indonesian paleoanthropologist Teuku Jacob, who

allegedly damaged them during attempts to make casts. Political sensitivities have meant access to the fossils is closely fought over, and the argument over their status continues. The same issues led to the closure of the Liang Bua cave site itself, with excavators only allowed back on site in 2007.

SPECIES LIST

❶ *Varanus komodoensis* Lepidosauria
❷ *Hipposideros* Laurasiatheria
❸ *Stegodon* Afrotheria
❹ *Homo floresiensis* Hominidae
❺ *Papagomys armandvillei* Euarchontoglires

FURTHER READING

MORWOOD, M. and VAN OOSTERZEE, P. 2007. *A New Human: The Startling Discovery and Strange Story of the "Hobbits" of Flores, Indonesia*. Collins, New York

WEBSITES

erl.wustl.edu/research/imseg/hobbit.html
biology.plosjournals.org/perlserv/
?request=getdocument&doi=10.1371%2Fjournal.pbio.
0040440 &ct=1

FLORISSANT   204–5

Colorado, USA

34 MILLION YEARS AGO

Excavated since the 19th century, this site in Colorado is now designated the Florissant Fossil Beds National Monument (not to be confused with the city in Missouri). Many of the petrified redwood stumps are exposed for viewing, and the site has a visitor centre offering guided tours and orientation information. Excavation is ongoing but carefully controlled, with daily demonstrations for visitors throughout the tourist season. Many of the beautifully preserved fossil insects are on display in the site museum.

SPECIES LIST

❶ charadriid Neognathae
❷ *Mesohippus* Laurasiatheria
❸ *Herpetotherium* Marsupialia
❹ *Mahonia* Angiospermae
❺ *Labiduromma* Hexapoda
❻ *"Bledius"* Hexapoda
❼ *Ephemera* Hexapoda
❽ *Oligodonta* Hexapoda
❾ *Rosa* Angiospermae
❿ *Marquettia* Hexapoda
⓫ *Prodryas* Hexapoda
⓬ *Heriades* Hexapoda
⓭ *Florissantia* Angiospermae
⓮ *Aphodius* Hexapoda
⓯ *Microstylum* Hexapoda
⓰ *Koelreuteria* Angiospermae
⓱ *Nephila* Chelicerata
⓲ *Merycoidodon* Laurasiatheria
⓳ *Miopodagrion* Hexapoda
⓴ *Amelanchier* Angiospermae
㉑ *Vanessa* Hexapoda
㉒ *Palaeovespa* Hexapoda
㉓ *Holcorpa* Hexapoda
㉔ *Megacerops* Laurasiatheria
㉕ *Syrphus* Hexapoda

FURTHER READING

MEYER, H. 2003. *The Fossils of Florissant*. Smithsonian Books, Washington

WEBSITE

www.nps.gov/flfo/

FOLSOM   242–3

New Mexico, USA

10,500 YEARS AGO

The "bone pit" at Folsom was discovered in 1908 by a black cowboy and self-educated fossil enthusiast. George McJunkin noticed the dismembered skeletons after a flash flood carved a large gulley through land on Wild Horse Arroyo. He

attempted to gain the interest of other fossil collectors, but the remoteness of the site meant that McJunkin had it to himself until his death in 1922, when motor transport suddenly brought the site within reach. Two of McJunkin's associates now continued occasional excavations at the site, but it was not until 1926 that they delivered a bag of bones and other finds to the Colorado Museum of Natural History in Denver, and the site's true importance became clear. The Folsom fossils became the focus of a heated debate about exactly when humans had settled the Americas, and it was only when arrowheads were found embedded in a bison skeleton that a full scientific expedition was undertaken by the Colorado Museum and the American Museum of Natural History. Further excavations took place in the 1970s and 1990s, and "Clovis" arrowheads continue to be found across the surrounding area. Fossils and arrowheads from the site are displayed at the local Folsom Museum, the AMNH in New York, and the Denver Museum of Nature and Science.

SPECIES AND ARTEFACTS LIST

❶ *Homo sapiens* Hominidae
❷ *Bison antiquus* Laurasiatheria
❸ Folsom point/spear

FURTHER READING

MELTZER, D.J. 2006. *Folsom: New Archaeological Investigations of a Classic Paleoindian Bison Kill*. University of California Press, Berkeley
PRESTON, D. 1997. *Fossils and the Folsom Cowboy*. Natural History Magazine, February

WEBSITE

www.dmns.org/main/en/General/Exhibitions/
content/folsompoint.htm

FORBES' QUARRY   228–9

Gibraltar

50,000 YEARS AGO

The small rocky peninsular of Gibraltar on the southern Spanish coast is hemmed by steep limestone cliffs. Here, among the numerouus caves, bones and stone tools have been recovered from Ice Age deposits since the 19th century. Specimens are housed in the Gibraltar Museum and in the Natural History Museum, London, UK.

SPECIES AND ARTEFACTS LIST

1. *Homo neanderthalensis* Hominidae
2. stone tools
3. *"Oryctolagus"* Euarchontoglires
4. *Testudo graeca* "Reptilia"
5. *Mytilus* Mollusca
6. *Capra ibex* Laurasiatheria
7. *"Haliaeetus"* Neognathae
8. *Rhinoceros* Laurasiatheria
9. *Elephas* Afrotheria

FURTHER READING

STRINGER, C. and ANDREWS, P. 2005. *The Complete World of Human Evolution.* Thames & Hudson, London
STRINGER, C. ET AL. (eds) 2000. *Neanderthals on the Edge: 150th Anniversary Conference of the Forbes' Quarry discovery, Gibraltar.* Oxbow Books, Oxford
WHYBROW, P. (ed.) 2000. *Travels with the Fossil Hunters.* Cambridge University Press, Cambridge

WEBSITES

www.gib.gi/museum/gib%20neanderthals.htm
www.gibraltar.costasur.com/en/neanderthal.html

GHOST RANCH   126–7

New Mexico, USA

213 MILLION YEARS AGO

Hundreds of *Coelophysis* skeletons have been excavated from the Whitaker Quarry at Ghost Ranch, New Mexico, and are to be seen in museums across the USA, such as the American Museum of Natural History in New York, the Carnegie Institute in Pittsburgh, Pennsylvania, the Museum of North Arizona in Flagstaff, Arizona, and the Yale Peabody Museum at New Haven, Connecticut. The site itself now hosts a conference centre and the Ruth Hall Museum with displays about the fossils and the local geology.

SPECIES LIST

1. *Coelophysis* Theropoda
2. *Araucarioxylon* Coniferopsida
3. *Semionotus* Actinopterygii
4. *Chinlea* Sarcopterygii
5. *Hesperosuchus* Crocodyliformes
6. *Rutiodon* Archosauria

FURTHER READING

COLBERT, E.H. 1995. *The Little Dinosaurs of Ghost Ranch.* Columbia University Press, New York
FRASER, N. 2006. *Dawn of the Dinosaurs.* Indiana University Press, Bloomington

WEBSITE

geoinfo.nmt.edu/tour/landmarks/ghost_ranch/ home.html

GOGO   74–5

Canning Basin, Western Australia, Australia

384 MILLION YEARS AGO

The landscapes of the Canning Basin in the remote far north of Western Australia reveal a resurrected seascape dotted with coral reefs from the late Devonian. The beautifully preserved three-dimensional fossil fish can only be revealed through lengthy chemical preparation, and some of the best specimens can be seen at the West Australian Museum in Perth and Museum Victoria in Melbourne.

SPECIES LIST

1. *Bothriolepis* Gnathostomes
2. *Gogonasus* Sarcopterygii
3. *Moythomasia* Actinopterygii
4. *Eastmanosteus* Gnathostomes
5. *Campbellodus* Gnathostomes
6. *Holodipterus* Sarcopterygii
7. *Griphognathus* Sarcopterygii
8. *Onychodus* Sarcopterygii
9. stromatoporoid Porifera

FURTHER READING

LONG, J.A., 1995. *The Rise of Fishes.* The Johns Hopkins University Press, Baltimore
LONG, J.A., 2006. *Swimming in Stone – the Amazing Gogo Fossils of the Kimberley.* Fremantle Arts Centre Press, Fremantle, Western Australia

WEBSITE

www.ahc.gov.au/publications/geofossil/gogo.html
museumvictoria.com.au/Collections-Research/ Our-Collections/Science-Collections/ Palaeontology/

GRAN DOLINA   222–3

Atapuerca, Spain

780,000 YEARS AGO

The 780,000-year-old cave deposits at Gran Dolina are just one phase in the history of the cave system that was occupied across several hundred thousand years. The sites were discovered during the excavation of railway cuttings, and scientific investigation began in 1964. In 2000, Atapuerca was designated as a World Heritage Site, and an Archeological Park is now established on the site, offering organized tours and a specially built reception centre.

SPECIES AND ARTEFACTS LIST

1. *Homo antecessor* Hominidae
2. *"Equus altidens"* Laurasiatheria
3. *Bison voigtstedtensis* Laurasiatheria
4. tools
5. *Stephanorhinus etruscus* Laurasiatheria
6. *Ursus* Laurasiatheria
7. *Marmota* Euarchontoglires
8. *Lynx* Laurasiatheria
9. *Eucladoceras giulii* Laurasiatheria
10. *Sus scrofa* Laurasiatheria
11. *Vulpes praeglacialis* Laurasiatheria
12. *Dama* Laurasiatheria

FURTHER READING

CERVERA, J. 2001. *Atapuerca.* Plot Ediciones, Madrid (in Spanish)

WEBSITES

www.atapuerca.com
www.amnh.org/exhibitions/atapuerca/index.php

GRAND CANYON 246–7

Arizona, USA

1.8 BILLION YEARS AGO

The Grand Canyon is one of the world's great natural wonders, and is protected within the Grand Canyon National Park established in 1919. Immense in scale, the canyon is 448 km (278 miles) long, an average of 16km (10 miles) wide and up to 1.6km (1 mile) deep. Its oldest rocks are the Vishnu Schist, exposed at the base of the inner gorge. As a major tourist attraction, the canyon has numerous visitor centres and other facilities, generally more numerous along the more accessible South Rim. Many firms offer guided tours, some with a specialist geological perspective.

PRESERVED GEOLOGICAL EVIDENCE

1. Quaternary surface with sloth fossils
2. Kaibab Limestone (Mid-Permian)
3. Coconino Limestone (Mid-Permian)
4. Supai Group (Pennsylvanian–Early Permian)
5. Redwall Limestone (Mississippian)
6. Temple Butte Limestone (Devonian)
7. Tonto Group (Cambrian)
8. Precambrian metamorphics and strata with microfossils

FURTHER READING

BEUS, S. and MORALES, M. 2002. *Grand Canyon Geology (2nd edition)*. Oxford University Press, New York

WEBSITES

www.nps.gov/grca/
3dparks.wr.usgs.gov/grandcanyon/

GRAPHITE PEAK 112–13

Antarctica

250 MILLION YEARS AGO

At 85°S, Graphite Peak in the central Transantarctic Mountains of Antarctica is one of the most inhospitable and inaccessible fossil sites on Earth. It was also the site of the first discovery of earliest Triassic vertebrates in Antarctica.

SPECIES LIST

1. *Thrinaxodon* Cynodontia
2. *Lystrosaurus* Therapsida
3. brachyopoid Temnospondyli
4. *Voltziopsis* Coniferopsida
5. *Prolacerta* Archosauria
6. *Procolophon* "Reptilia"

FURTHER READING

RETALLACK, G.J. and KRULL, E.S. 1999. *Landscape ecological shift at the Permian–Triassic Boundary in Antarctica*. Australian Journal of Earth Sciences 46, 785–812

WEBSITE

www.nsf.gov/funding/pgm_summ.jsp?pims_id=8173&org=ANT&from=home (and follow links to Antarctic Mountains and related URLs)

GREEN RIVER 190–1

Wyoming, USA

54 MILLION YEARS AGO

The Green River Formation is one of the largest known accumulations of Paleogene lacustrine sediments in the world, covering some 65,000 sq km (25,000 sq miles) of the American Midwest. Fossil Butte National Monument, its Visitor Center at Kemmerer, Wyoming, and some nearby commercial galleries are the best places to see the extraordinary abundance and diversity of Green River fossils, ranging from turtles, snakes, and birds to plants and insects. However, many fossils have been collected and are to be seen in collections all over North America and indeed elsewhere in the world.

The deposits are still commercially quarried for their fossils, but collecting licences can be bought locally.

SPECIES LIST

1. *Rhus* Angiospermae
2. *Limnofregata* Neognathae
3. *Sabalites* Angiospermae
4. *Heliobatis* Chondrichthyes
5. *Knightia* Actinopterygii
6. *Trionyx* "Reptilia"
7. *Borealosuchus* Crocodyliformes
8. *Amia* Actinopterygii
9. *Gallinuloides* Neognathae
10. *Presbyornis* Neognathae
11. *Typha* Angiospermae
12. *Icaronycteris* Laurasiatheria
13. *Onychonycteris* Laurasiatheria
14. *Boavus* Lepidosauria
15. *Uintatherium* Laurasiatheria
16. *Ailanthus* Angiospermae

FURTHER READING

GRANDE, L. 1980. *Paleontology of the Green River Formation with a review of the Fish Fauna*. Geological Survey of Wyoming Bulletin 63

NUDDS, J.R. and SELDEN, P.A. 2008. *Fossil Ecosystems of North America: A guide to the sites and their extraordinary biotas*. Manson Publishing, London (see chapter 11)

WEBSITES

www.nps.gov/fobu/expanded/index.htm
www.ucmp.berkeley.edu/tertiary/eoc/greenriver.html
www.fossilmuseum.net/Fossil_Sites/GreenRiverSite.htm

GRÈS À VOLTZIA 114–15

Vosges, France

242 MILLION YEARS AGO

Thousands of fossil insects, plants, and other freshwater organisms have been collected from 242-million-year-old Middle Triassic strata exposed at numerous localities along the western flank of the Vosges Mountains in northeastern France. There are several quarries still working in the region, and permission must be sought before entering any of them.

Some of the best fossil specimens are conserved in the the Grauvogel-Gall Collection in the University of Louis Pasteur, Strasbourg, France, but are sadly not on public display.

SPECIES LIST

1. *Rosamygale* Chelicerata
2. blattodean Hexapoda
3. *Voltzia* Coniferopsida
4. scorpionid Chelicerata
5. myriapod Uniramia
6. *Antrimpos* Crustacea
7. *Progonionemus* Cnidaria
8. *Anomopteris* Filicopsida
9. *Limulitella* Chelicerata
10. *Dipteronotus* Actinopterygii
11. *Voltziaephemera* Hexapoda

FURTHER READING

SELDEN, P. and NUDDS, J. 2004. *Evolution of Fossil Ecosystems*. Manson Publishing, London (see chapter 7)

Grand Canyon The majestic panoramic sweep of the Colorado plateau appears to be rent apart by the chasm of the Grand Canyon, which descends through more than a kilometre of sedimentary strata and back through a billion years of Earth's history.

GUIMAROTA 138–9

Leiria, Portugal

152 MILLION YEARS AGO

Tens of thousands of fossils, including some 10,000 isolated teeth, have been recovered by German paleontologists from strata containing Late Jurassic coals, formerly mined near Leiria in central Portugal.

Most of the fossils – which include abundant plants, shellfish, fish, amphibians, lizards, and primitive mammals – were found in the 1970s, when the mine was reopened and worked for a decade purely for its fossils. Much of the fossil material is still being studied by experts at the Institute of Paleontology in the Freie Universitat Berlin.

SPECIES LIST

1. *Saurillodon* Lepidosauria
2. *Haldanodon* Mammalia
3. *Compsognathus* Theropoda
4. *Celtedens* Lissamphibia
5. *Phlebopteris* Filicopsida
6. *Archaeopteryx* Avialae
7. *Henkelotherium* Mammalia
8. *Pagiophyllum* Coniferopsida
9. *Baiera* Ginkgoales
10. *Klukia* Filicopsida
11. *Lycopodium* Lycopsida
12. paulchoffatiid Mammalia
13. *Machimosaurus* Crocodyliformes
14. *Rhamphorhynchus* Pterosauria

FURTHER READING

MARTIN, T. and KREBS, B. (eds) 2000. *Guimarota: A Jurassic Ecosystem*. Pfeil Verlag, Munich.

WEBSITE

www.geocities.com/trevor_dykes/ guimarotaexcavations.htm

HAYDEN QUARRY 124–5

New Mexico, USA

216 MILLION YEARS AGO

Hayden Quarry forms one part of the larger Ghost Ranch locality, and its sand- and siltstones have been excavated since the days of the infamous "Bone Wars" in the late 19th century, when the great fossil collector Edward Drinker Cope dug here. Recent excavations have been carried out by paleontologists from the University of California at Berkeley, and finds are displayed at the UC Museum of Paleontology, California, USA, as well as at the site's own paleontology museum.

SPECIES LIST

1. *Dromomeron* Dinosauria
2. pseudopalatine Archosauria
3. *Chindesaurus* Saurischia
4. *Eucoelophysis* Dinosauria
5. *Postosuchus* Archosauria
6. *Typothorax* Archosauria

FURTHER READING

IRMIS, R.B. ET AL. 2007. *A Late Triassic Dinosauromorph Assemblage from New Mexico and the Rise of Dinosaurs*. Science (20 July), pp.358–61

WEBSITES

geoinfo.nmt.edu/tour/landmarks/ghost_ranch home.html

www.ucmp.berkeley.edu/science/profiles/ irmis_0705.php

Joggins These tilted coastal strata expose a magnificent section through Carboniferous "coal measure" deposits, some of which retain sediment-filled casts of the original trees whose organic remains now constitute the coal seams.

HELL CREEK 178–9

Montana, Dakotas, and Wyoming, USA

66 MILLION YEARS AGO

The Hell Creek Formation lies in the heart of the Rocky Mountains, spreading from eastern Montana into North and South Dakota and Wyoming. It preserves one of the finest collections of late Mesozoic fossils in North America. The formation outcrops across the region, and fossil-rich sites have been excavated here since the early 1900s. John R. Horner of Montana State University has made many of the most impressive recent discoveries, including some of the best fossils of *Tyrannosaurus*, and new genera such as *Maiasaura*. Heavy winter rains reveal new fossil-bearing sediments each year, and the digging season takes place in the heat of high summer. Fossils from Hell Creek are displayed in many major US museums, but the Museum of the Rockies in Bozeman, Montana, probably has the best collection. The Marmath Research Foundation, meanwhile, offers participation in organized digs for paying volunteers.

SPECIES LIST

1. *Magnolia* Angiospermae
2. *Edmontonia* Ornithischia
3. *Cimolestes* Eutheria
4. *Edmontosaurus* Ornithischia
5. *Triceratops* Ornithischia
6. *Tyrannosaurus* Theropoda

FURTHER READING

DINGUS, L. 2004. *Hell Creek, Montana: America's Key to the Prehistoric Past*. St. Martins Press, New York

HORNER, J.R. 2001. *Dinosaurs Under the Big Sky*. Mountain Press Publishing Company, Missoula

WEBSITES

www.mrfdigs.com/

www.museumoftherockies.org/

HOLZMADEN 130–1

Baden-Württemberg, Germany

182 MILLION YEARS AGO

Fossils from the famous Posidonia shales of Holzmaden have been exported around the world, but now only two quarries operate in the locality. Of these, one is open for public excavation (for a fee). Fossil finds are monitored by staff from the nearby Hauff Museum, which houses a spectacular display of Holzmaden fossils. Finds of scientific importance must be handed over for a "finder's fee", but many smaller fossils can be taken away.

SPECIES LIST

1. *Harpoceras* Mollusca
2. *Passaloteuthis* Mollusca
3. *Rhomaleosaurus* Plesiosauria
4. *Steneosaurus* Crocodyliformes

⑤ *Lepidotes* Actinopterygii
⑥ *Stenopterygius* Ichthyosauria
⑦ *Palaeospinax* Chondrichthyes
⑧ *Dorygnathus* Pterosauria
⑨ *Pentacrinus* Echinodermata

FURTHER READING

SELDEN, P. and NUDDS, J. 2004. *Evolution of Fossil Ecosystems.* Manson Publishing, London (see chapter 8)

WEBSITE

www.urweltmuseum.de/

HUGHENDEN — 164–5

Queensland, Australia

105 MILLION YEARS AGO

Central Queensland marks the site of a former inland sea that was at its maximum extent some 117 million years ago, and had retreated from the area by around 97 million years ago. Fossils from this shallow sea and the shores around it are plentiful, and the area has been known for its preserved record of ancient life since the 1920s, when the first giant skull of the marine pliosaur *Kronosaurus* was discovered and described. Fossils continue to be found, especially around the towns of Hughenden and Richmond – which has a visitor centre and museum, and various nearby sites are open to the public. The original *Kronosaurus* skull is on display at the Queensland Museum in Brisbane, along with many other local fossils.

SPECIES LIST

① *Minmi* Ornithischia
② *Muttaburrasaurus* Ornithischia
③ *Kronosaurus* Plesiosauria
④ *Platypterigius* Ichthyosauria

FURTHER READING

VICKERS-RICH, P. and RICH, T.H. 1999. *Wildlife of Gondwana.* Indiana University Press, Bloomington

WEBSITES

www.kronosauruskorner.com.au/
www.qm.qld.gov.au/features/dinosaurs/

HUNSRÜCK SLATE — 72–3

Koblenz, Germany

407 MILLION YEARS AGO

The slate mines from which the Early Devonian marine fossils of Hunsrück have been recovered are no longer working, but some of the spoil tips are accessible to collectors with permission. Specimens can be seen at the Hunsrück-Fossilienmuseum in Bundenbach, Germany.

SPECIES LIST

① *Furcaster* Echinodermata
② *Drepanaspis* "Agnatha"
③ *Palaeosolaster* Echinodermata
④ *Imitatocrinus* Echinodermata
⑤ *Bundenbachochaeta* Polychaeta
⑥ *Nahecaris* Crustacea
⑦ acanthodian Chondrichthyes
⑧ *Palaeoisopus* Chelicerata
⑨ *Gemuendina* Gnathostomes
⑩ *Mimetaster* Arthropoda

FURTHER READING

BARTELS, C. BRIGGS, D.E.G. and BRASSEL, G. 1998. *The Fossils of the Hunsrück Slate.* Cambridge University Press, Cambridge
SELDEN, P. AND NUDDS, J. 2004. *Evolution of Fossil Ecosystems.* Manson Publishing, London

WEBSITE

www.fossilmuseum.net/Fossil_Sites/Bundenbach/Bundenbach.htm

JOGGINS — 90-3

Nova Scotia, Canada,

314–313 MILLION YEARS AGO

These accessible coastal strata have recently been designated as a protected World Heritage Site. Thousands of fossils have been found here, and many are housed in the Redpath Museum at McGill University in Montreal, Quebec. A local visitors' centre opened in 2007.

SPECIES LIST 90–1

① *Baphetes* Tetrapoda
② scorpionid Chelicerata
③ *Arthropleura* Uniramia
④ *Cordaites* Gymnospermopsida
⑤ *Neuropteris* Pteridospermales
⑥ *Protodiscus* Mollusca
⑦ *Xyloiulus* Uniramia
⑧ *Hylonomus* "Reptilia"
⑨ *Sigillaria* Lycopsida

SPECIES LIST 92–93

① *Hylerpeton* Lepospondyli
② *Coryphomartus* Chelicerata
③ *Amynilyspes* Uniramia
④ *Sphenophyllum* Sphenopsida

⑤ *Dendrerpeton* Temnospondyli
⑥ *Sigillaria* Lycopsida
⑦ megasecopteran Hexapoda
⑧ *Graeophonus* Chelicerata

FURTHER READING

FALCON-LANG, H.J. 2006. *A history of research at the Joggins Fossil Cliffs of Nova Scotia, Canada, the world's finest Pennsylvanian section.* Proceedings of the Geologists' Association 117, 377–392

WEBSITES

museum.gov.ns.ca/fossils/sites/joggins/index.htm
www.ucmp.berkeley.edu/carboniferous/joggins.html

JUDITH RIVER — 174–5

Montana, USA, and Alberta, Canada

79–74 MILLION YEARS AGO

The Judith River Formation (known in Canada as the Belly River Formation) stretches from Montana into southern Alberta and Saskatchewan. Like the nearby and slightly later Hell Creek Formation, it contains a spectacular selection of fossilized Late Cretaceous flora and fauna. The richest deposits are in the Dinosaur Park Formation, a subdivision that preserves a lush subtropical swamp forest. Today, the deposits outcrop as typically harsh "badlands" territory and, as at Hell Creek, paying volunteers can take part in organized scientific excavations. Fossils have been found here since 1884, when Joseph B. Tyrrell discovered the fossil skull of *Albertosaurus* emerging from a hill overlooking the Red Deer River, and many of the finest discoveries are preserved in Alberta's Royal Tyrrell Museum.

SPECIES LIST
1. *Struthiomimus* Theropoda
2. *Quetzalcoatlus* Pterosauria
3. *Chasmosaurus* Ornithischia
4. *Deltatheridium* Mammalia
5. *Stegoceras* Ornithischia
6. *Gorgosaurus* Theropoda
7. *Lambeosaurus* Ornithischia
8. *Troodon* Theropoda

FURTHER READING
CURRIE, P.J. and LOPPELUS, E.B. 2005. *Dinosaur Provincial Park: A Spectacular Ancient Ecosystem Revealed.* Indiana University Press, Bloomington

WEBSITES
www.tyrrellmuseum.com/peek/
 index2.php?strSection=1
www.montanadinosaurdigs.com/
www.tpr.alberta.ca/parks/dinosaur/flashindex.asp

KAROO BASIN   96–7 and 104–5

South Africa

299 AND 260 MILLION YEARS AGO

The 300-million-year-old Dwyka strata outcrop across the northern Karoo Basin of South Africa and preserve evidence of the late Carboniferous Ice Age. In places, dropstones from floating ice are common within these ancient sedimentary rocks, but ice-scratched rock surfaces are much rarer – one classic example can be seen near Douglas.

   Fossils of Permian age have been collected from across the Karoo Basin since the 19th century, and have revealed much about the early evolution of the cynodonts and other mammal-like reptiles. The National Museum at Bloemfontein houses the most important collection of Karoo fossils, and various tour operators offer fossil-hunting expeditions in the area.

SPECIES LIST AND PRESERVED GEOLOGICAL EVIDENCE 96–7
1. ophiacodontid Synapsida
2. embedded dropstone
3. glacial pavement

SPECIES LIST 104–5
1. *Owenetta* "Reptilia"
2. *Diictodon* Therapsida
3. *Moschorhinus* Therapsida
4. *Lystrosaurus* Therapsida
5. *Youngina* Diapsida
6. *Procynosuchus* Cynodontia
7. *Cyanosaurus* Cynodontia

FURTHER READING
JOHNSON, M.R. ET AL. 2005. *The Geology of South Africa.* Council for Geoscience, Pretoria

WEBSITE
www.nasmus.co.za/palaeo/palaeo1.htm

KEYSER FRANZ JOSEPH FJORD   82–3

Greenland

366 MILLION YEARS AGO

This site is on a remote and inaccessible mountainside in East Greenland. Most of the tetrapod fossils recovered from here are in the collections of the Geological Museum at the University of Copenhagen, Denmark. Replicas and an interpretive display can also be seen in the Zoological Museum of the University of Cambridge, England.

SPECIES LIST
1. *Drepanophycus* Lycopsida
2. *Ichthyostega* Tetrapoda
3. *Groenlandaspis* Gnathostomes
4. *Acanthostega* Tetrapoda
5. *Serracaulis* Lycopsida

FURTHER READING
CLACK, J.A. 2002. *Gaining Ground.* Indiana University Press, Bloomington

WEBSITES
www.mnhn.fr/museum/front/medias/
 publication/9801_go7n1a4.pdf

KHOOVOR   156–7

Ovorhangay, Mongolia

112 MILLION YEARS AGO

This remote locality in the Mongolian desert, with its rich fauna of small EarlyCretaceous mammals, was discovered by joint Soviet Mongolian expeditions that recovered thousands of teeth and bones from its waterlain sediments. Specimens from the site are now housed in the Paleontological Institute of the Russian Academy of Sciences, Moscow; the Institute of Geology at the Academy of Science, Ulaanbaatar, Mongolia, and the Institute of Paleobiology, Polish Academy of Sciences, Warsaw.

SPECIES LIST
1. *Nilssoniopteris* Bennettitales
2. euriconodontid Mammalia
3. *Baiera* Ginkgoales
4. *Kielantherium* Mammalia
5. *Araucaria* Coniferopsida
6. *Shamosaurus* Ornithischia
7. *Iguanodon* Ornithischia
8. *Psittacosaurus* Ornithischia
9. *Sparganium* Angiospermae
10. *Prokennalestes* Eutheria
11. *Pterophyllum* Bennettitales
12. *Lycoptera* Actinopterygii

Karoo Basin The eroded hills of this vast South African basin expose an immense thickness of late Carboniferous and Permian strata, and preserve the fossil skeletons of the strange land animals of the time.

FURTHER READING
BENTON, M.J. ET AL (eds) 2000. *The Age of Dinosaurs in Russia and Mongolia.* Cambridge University Press, Cambridge
KIELAN-JAWOROWSKA, Z. ET AL. 2004. *Mammals from the Age of Dinosaurs: Origins, Evolution, and Structure.* Columbia University Press, New York

KIRKTON, EAST   84–5

West Lothian, Scotland

328 MILLION YEARS AGO

This disused limestone quarry lies near Bathgate, 27km (17 miles) west of Edinburgh, West Lothian. It is now a legally protected Site of Special Scientific Interest, and collecting is restricted and requires permission. Many of the site's finest fossils are held in the collections of the Royal Scottish Museum in Edinburgh, the Hunterian Museum in Glasgow, and the Zoology Museum of the University of Cambridge.

SPECIES LIST
1. *Sphenopteridium* Pteridospermales
2. *Westlothiana* "Reptilia"
3. *Ophiderpeton* Lepospondyli
4. *Archaeocalamites* Sphenopsida
5. opilionid Chelicerata
6. *Eldeceeon* "Reptilia"
7. *Balanerpeton* Temnospondyli
8. *Acanthodes* Chondrichthyes
9. *Pulmonoscorpius* Chelicerata
10. millipede Uniramia

FURTHER READING
CLACK, J.A. 2002. *Gaining Ground.* Indiana University Press, Bloomington

WEBSITE
www.jncc.gov.uk ‹http://www.jncc.gov.uk›
 (follow links to East Kirkton)

KOTUIKAN RIVER 46–7

Arctic Siberia, Russia

548–535 MILLION YEARS AGO

The Early Cambrian strata of the Kotuikan River outcrop in northern Siberia across a largely inaccessible forested wilderness well inside the Arctic Circle (70°N, 105°E). The Kotuikan River is more than 100km (60 miles) from the nearest habitation, but Russian paleontologists have nevertheless recovered hundreds of millimetre-sized marine shelly fossils – the earliest known remains of shelled organisms – from its limestones.

SPECIES LIST
4. chancellorid Unknown
5. *Cambrocyathellus* Archaeocyathida
6. radiocyathid Archaeocyathida
7. coralomorph Cnidaria
8. *Okulitchicyathus* Archaeocyathida

FURTHER READING
KNOLL, A. 2003. *Life on a Young Planet.* Princeton University Press, Princeton
FEDONKIN, M.A. ET AL. 2007. *The Rise of Animals: Evolution and Diversification of the Kingdom Animalia.* The Johns Hopkins University Press, Baltimore

WEBSITE
www.mit.edu/~watters/grotz_etal_paleobio_2000.pdf

LAETOLI 216–17

Tanzania

3.8 MILLION YEARS AGO

The Laetoli site lies just 50km (30 miles) south of Olduvai Gorge, within the Ngorongo Crater Conservation Area. It preserves some of the most evocative fossils of all – the footprints of ancient bipedal hominids that may have been our own ancestors.

Although the site was visited and surveyed by the famous archeologists Louis and Mary Leakey from the 1930s onward, it was not until 1976 that Mary Leakey's team discovered the flattened platform of ash preserving a wide variety of animal footprints from about 3.6 million years ago. As trace fossils, the Laetoli trackways present a unique preservational problem, in that so much of their value lies in their context. So, rather than attempt to remove them from the site, Leakey's team re-buried the layer at the end of each dig season. Despite this, acacia trees began to take root in the disturbed soil above the tracks, and so the site was re-excavated in the mid-1990s, and buried once again in more sterile conditions. While the footprints themselves will remain buried for the foreseeable future, replicas of the Laetoli trackways are displayed at the nearby Olduvai Museum, the National Museum in Dar es Salaam, Tanzania, and in some other major museums worldwide.

SPECIES AND ARTEFACTS LIST
1. *Madoqua* Laurasiatheria
2. *Numida* Neognathae
3. *Loxodonta* Afrotheria
4. *Diceros* Laurasiatheria
5. *Giraffa* Laurasiatheria
6. *Australopithecus afarensis* Hominidae
7. animal tracks
8. hominid footprints

FURTHER READING
MORRELL, V. 1995. *Ancestral Passions: The Leakey Family and the Quest for Humankind's Beginnings.* Touchstone, New York

WEBSITE
www.ngorongorocrater.org/olduvai.html

LA VOULTE-SUR-RHÔNE 134–5

Ardeche, France

163 MILLION YEARS AGO

First recognized in the 19th century, the remarkably preserved shallow marine ecosystem of La Voulte-sur-Rhône is exposed for excavation in a series of quarries to the west of the town itself. Excavation continues to the present, and many of the finest fossils were in the collection of the Musée de Paléontologie in La Voulte itself, which closed down in 2006. However, efforts are being made to display them elsewhere in the near future.

SPECIES LIST
1. *Dollocaris* Crustacea
2. *Rhomboteuthis* Mollusca
3. *Aeger* Crustacea
4. *Ophiopinna* Echinodermata
5. *Pholidophorus* Actinopterygii
6. *Proteroctopus* Mollusca
7. *Bositra* Mollusca
8. *Eryma* Crustacea

FURTHER READING
BOTTJER, D.J. ET AL. (eds) 2002. *Exceptional Fossil Preservation.* Columbia University Press, New York (see chapter 16)

WEBSITE
www.musee-fossiles.com/

LESMAHAGOW 64–5

Lanarkshire, Scotland

430 MILLION YEARS AGO

A number of localities across the Midland Valley of Scotland have produced exceptional fossils of jawless fish and arthropods from the Early Silurian. Many of these sites were discovered in the last century and are now legally protected Sites of Special Scientific Interest. Their fossils are scattered throughout many museums, such as Glasgow's Hunterian Museum.

SPECIES LIST
1. *Ainiktozoon* Crustacea
2. *Loganellia* "Agnatha"
3. *Ceratiocaris* Crustacea
4. *Pterygotus* Chelicerata
5. *Jamoytius* "Agnatha"
6. *Birkenia* "Agnatha"

FURTHER READING
ALDRIDGE, R.J. ET AL. 2000. *British Silurian Stratigraphy.* Joint Nature Conservation Committee, Peterborough

WEBSITE
www.hmag.gla.ac.uk/neil/Lesmahagow/Lesmahagow.html

LIAONING PROVINCE 150–5

China

128 MILLION YEARS AGO

The astonishing feathered dinosaurs found around the city of Chaoyang in China's Liaoning province are part of the Early Cretaceous Jiofotang and Yixian formations, which outcrop across the region and particularly in the Songling mountains. Japanese scientists working in occupied Manchuria during the early 20th century were the first to report that the area (then known as Jehol) was rich in fossils, but the defeat of Japan in World War II, followed by the turbulent conditions of the Chinese Revolution itself, and the later "Cultural Revolution" under Mao Zedong, meant that Chinese scientists and their international collaborators did not return to the region in earnest until the 1990s. Since then, the stunning discoveries made in the area have caused a sensation far beyond the paleontological community, and the precious feathered fossils have frequently toured major museums. Highlights of the Liaoning finds are displayed at the recently refurbished Geological Museum of China in Beijing, while the American Museum of Natural History in New York also has an extensive display about the site and its impact.

SPECIES LIST 150–1
1. iguanodontid Ornithischia
2. *Callobatrachus* Lissamphibia
3. *Ischnidium* Hexapoda
4. *Sinosauropteryx* Theropoda
5. *Confuciusornis* Avialae
6. *Protarchaeopteryx* Theropoda

SPECIES LIST 152–3
1. cicadomorph Hexapoda
2. *Manchurochelys* "Reptilia"
3. *Jinzhousaurus* Ornithischia
4. *Microraptor* Theropoda
5. *Ephemeropsis* Hexapoda
6. *Repenomamus* Mammalia
7. *Psittacosaurus* Ornithischia
8. *Jeholopterus* Pterosauria
9. *Psittacosaurus* hatchlings Ornithischia
10. *Beipiaosaurus* Theropoda
11. *Dilong* Theropoda

Lyme Regis The unstable cliffs of southern England's "Jurassic Coast" are famous for marine fossils from the time of the dinosaurs and are now a World Heritage Site as well as a popular holiday resort.

SPECIES LIST 154–5
1. *Eomaia* Eutheria
2. *Sapeornis* Avialae
3. *Protarchaeopteryx* Theropoda
4. *Caudipteryx* Theropoda
5. *Archaefructus* Angiospermae
6. *Sinodelphys* Marsupialia
7. *Tetraphalerus* Hexapoda
8. *Czekanowskia* Gymnospermopsida
9. *Sinornithosaurus* Theropoda
10. *Sinosauropteryx* Theropoda
11. *Jeholodens* Mammalia

FURTHER READING
NORELL, M.A. and ELLISON, M. 2005. *Unearthing the Dragon: The Great Feathered Dinosaur Discovery*. Pi Press, New Jersey

WEBSITES
www.fossilmuseum.net/Fossil_Sites/
    LiaoningSite.htm
www.amnh.org/exhibitions/dinosaurs/diorama/

LUDFORD LANE 68–9
Shropshire, England
419 MILLION YEARS AGO

This legally protected Site of Special Scientific Interest is a historic locality from which a great variety of fossils have been recovered since the 19th century. They include many different kinds of shallow-sea-dwelling organism, especially brachiopods and gastropods, along with the remains of early land plants and arthropods.

SPECIES LIST
1. *Lingula* Brachiopoda
2. *Cooksonia* Rhyniophyta
3. *Steganotheca* Rhyniophyta
4. *Palaeotarbus* Chelicerata
5. *Eoarthropleura* Uniramia
6. *Strophochonetes* Brachiopoda

FURTHER READING
ALDRIDGE, R.J. ET AL. 2000. *British Silurian Stratigraphy*. Joint Nature Conservation Committee, Peterborough

WEBSITE
palaeo.gly.bris.ac.uk/Palaeofiles/Lagerstatten/
    ludford/florafauna.html
www.jncc.gov.uk/pdf/gcrdb/GCRsiteaccount1642.pdf

LYME REGIS 128–9
Dorset, England
195 MILLION YEARS AGO

The vast amounts of fossil material found in the shale cliffs of Lyme Bay on the southern coast of England were brought to the world's attention when a 12-year-old Mary Anning unearthed a fossil ichthyosaur in 1811, and visitors have been coming to hunt for fossils ever since. Storms regularly wash fresh supplies of fossils out of the cliffs – ammonites and belemnites are particularly ubiquitous, but larger animals are still regularly discovered by professionals and amateurs alike.

However, the instability that replenishes the fossils along the beach also makes the area prone to landslips, some of which can be dangerous to life

and property. Fossils from Lyme are proudly exhibited in local shops and at the Dinosaurland Museum, and many are displayed in London's Natural History Museum, the Bristol City Museum, and other major collections.

SPECIES LIST
1. *Caturus* Actinopterygii
2. *Ichthyosaurus* Ichthyosauria
3. *Microderoceras* Mollusca
4. *Scelidosaurus* Ornithischia
5. *Dapedium* Actinopterygii
6. *Hybodus* Chondrichthyes
7. *Pentacrinites* Echinodermata
8. *Passaloteuthis* Mollusca
9. *Plesiosaurus* Plesiosauria
10. *Dimorphodon* Pterosauria

FURTHER READING
DINELEY, D.L. and METCALF, S.J. 1999. *Fossil Fishes of Great Britain*. Joint Nature Conservation Committee, Peterborough

WEBSITES
www.jncc.gov.uk (follow links to Lyme Regis)
www.lymeregismuseum.co.uk
www.lulworth.com/education/heritage_centre.htm

MAGADAN 232–3
Siberia, Russia
40,000 YEARS AGO

The frozen ground (permafrost) of Siberia's remote and inhospitable Arctic Ocean coastlands and river valleys have exposed most of the world famous frozen "mummies" of Ice Age mammals. Ever since the early 1900s, expeditions have struggled to recover the remains of mammoths, woolly rhinos, and other animals, and efforts have been renewed thanks to the recent possibility of recovering ancient DNA from these frozen bodies and perhaps one day cloning a mammoth.

The baby nicknamed "Dima" was recovered from terrace deposits of the Kirgiliakh river, a tributary of the Kolyma, north of Magadan in the far northeast of the Russian Republic. Along with several other Siberian specimens, it is on display at the Zoological Museum of the State University of St Petersburg, Russia.

SPECIES LIST
1. *Coelodonta antiquitatis* Laurasiatheria
2. *Equus caballus* Laurasiatheria
3. *Canis lupus* Laurasiatheria
4. *Mammuthus primigenius* Afrotheria
5. *Bison priscus* Laurasiatheria

FURTHER READING
GUTHRIE, R.D. 1990. *Frozen Fauna of the Mammoth Steppe.* University of Chicago Press, Chicago
LISTER, A. and BAHN, P. 1994. *Mammoths.* Macmillan, New York

MAZON CREEK   88–9
Illinois, USA
314 MILLION YEARS AGO

Mazon Creek is just one locality in the wider Carboniferous Francis Creek Shale. Here fossils are preserved in concretions of siderite (iron carbonate) that are frequently unearthed during the strip-mining of coal. This unusual form of deposition is particularly good at preserving soft-bodied organisms, and amateurs have successfully discovered many new species, including the famous *Tullimonstrum* on display at Illinois State Museum.

SPECIES LIST
1. *Belotelson* Crustacea
2. *Tullimonstrum* Mollusca
3. *Aviculopecten* Mollusca
4. myxinoid "Agnatha"
5. *Essexella* Cnidaria
6. *Calamites* Sphenopsida
7. *Geralinura* Chelicerata
8. *Sphenophyllum* Sphenopsida
9. *Latzelia* Uniramia
10. *Eucenus* Hexapoda
11. blattodean Hexapoda
12. *Gerarus* Hexapoda
13. *Xyloiulus* Uniramia
14. *Neuropteris* Pteridospermales

FURTHER READING
NITECKI, M.H. 1979. *Mazon Creek Fossils Symposium.* Academic Press, New York
NUDDS, J.R. and SELDEN, P.A. 2008. *Fossil Ecosystems of North America.* Manson Publishing, London (see chapter 7)

WEBSITES
www.museum.state.il.us/exhibits/mazon_creek
www.ucmp.berkeley.edu/carboniferous/mazon.html

MESSEL   196–9
Near Darmstadt, Germany
48 MILLION YEARS AGO

The shale beds of Messel, near Frankfurt am Main, have been mined since 1859, but paleontologists only grew aware of the extraordinary and beautifully preserved 50-million-year-old fossils emerging from the quarry in the early 20th century. Serious scientific excavations did not begin until the 1970s, and these revealed shale deposits some 130m (400ft) deep, above a sandstone bed. The lake that had once occupied this site had deep, anoxic bottom waters that periodically turned over, releasing noxious gases that suffocated life in its upper layers, and even in the surrounding shores and forest. As organisms settled to the bottom, they were preserved with very little decay. By 1991, the quarry was commercially exhausted, and it was then purchased by the Hessian government to secure continued access. It has been designated as a UNESCO World Heritage Site since 1995. The finest collections of fossils from the site are displayed at the Darmstadt LandesMuseum and the Senckenberg NaturMuseum.

SPECIES LIST 196–7
1. *Palaeopython* Lepidosauria
2. *Propalaeotherium* Laurasiatheria
3. *Paroodectes* Laurasiatheria
4. *Messelobunodon* Laurasiatheria
5. *Miacis* Laurasiatheria
6. *Formicium* Hexapoda
7. *Palaeoglaux* Neognathae
8. cicada Hexapoda
9. *Archaeonycteris* Laurasiatheria
10. *Primozygodactylus* Neognathae
11. *Darwinius masillae* Primates
12. *Eomanis* Laurasiatheria
13. *Hyrachyus* Laurasiatheria

SPECIES LIST 198–9
1. *Palaeochiropteryx* Laurasiatheria
2. buprestid Hexapoda
3. *Rhynchaeites* Neognathae
4. *Eurotamandua* Xenarthra
5. *Zantedeschia* Angiospermae
6. *Eopelobates* Lissamphibia
7. *Kopidodon* Boreoeutheria
8. *Leptictidium* Boreoeutheria
9. *Trionyx* "Reptilia"
10. *Gastornis* Neognathae
11. *Eocoracias* Neognathae
12. *Cephalotaxus* Coniferopsida
13. *Pholidocercus* Laurasiatheria
14. *Buxolestes* Boreoeutheria
15. *Asiatosuchus* Crocodyliformes
16. *Eurohippus* Laurasiatheria
17. *Aegialornis* Neognathae
18. *Typha* Angiospermae

FURTHER READING
SCHAAL, S. ET AL. 1992. *Messel: An Insight into the History of Life and of the Earth.* Clarendon Press, Oxford

WEBSITES
www.unep-wcmc.org/sites/wh/pdf/Messel%20Pit.pdf
www.hlmd.de (Flash site, in German)

MIGUASHA   76–7
Escuminac Bay, Quebec, Canada
384–374 MILLION YEARS AGO

The famous Devonian strata of Escuminac Bay, source of so many beautifully preserved fossil fish, lie within the protected area of the Miguasha National Park in Quebec's Gaspé Peninsula. The park was nominated as a World Heritage Site in 1993. Some of the fossils can be seen in the local Miguasha Museum, and others at the Swedish Museum of Natural History in Stockholm.

SPECIES LIST
1. *Cheirolepis* Actinopterygii
2. *Endeiolepis* "Agnatha"
3. *Escuminaspis* "Agnatha"
4. *Eusthenopteron* Sarcopterygii
5. *Bothriolepis* Gnathostomes
6. *Scaumenacia* Sarcopterygii
7. *Archaeopteris* Gymnospermopsida

FURTHER READING
SCHULTZE, H-P. and CLOUTIER, R. (eds) 1996. *Devonian fishes and plants of Miguasha, Quebec, Canada.* Pfeil Verlag, Munchen
LONG, J.A. 1995. *The Rise of Fishes.* The Johns Hopkins University Press, Baltimore
CLACK, J.A. 2002. *Gaining Ground.* Indiana University Press, New York

WEBSITES
www.wcmc.org.uk/protected_areas/data/wh/miguasha.html
www.canadianencyclopedia.ca/index.cfm?PgNm=TCE&Params=A1ARTA0010404

MISTAKEN POINT   42–3

Avalon Peninsula, Newfoundland, Canada

575–565 MILLION YEARS AGO

This protected coastal site in south-eastern Newfoundland, Canada, is remote but accessible, and one of the few places where Precambrian fossils of the ancient and mysterious "Ediacarans" can be seen (with permission) emerging from the rocks.

SPECIES LIST

❶ *Charniodiscus* "Ediacara"
❷ *Thectardis* "Ediacara"
❸ "spindles" "Ediacara"
❹ *Bradgatia* "Ediacara"
❺ *Charnia wardi* 'Ediacara'

WEBSITE

www.ucmp.berkeley.edu/vendian/ediacara.html

MONTE BOLCA   194–5

Verona, Italy

49 MILLION YEARS AGO

More than 500 species of terrestrial and marine organisms have been recovered from the mid-Eocene strata of Monte Bolca. Several localities

Morrison Formation The arid "badlands" of Utah and Wyoming may be agriculturally unproductive, but the rock strata of the Morrison Formation are one of the world's most productive sources of Jurassic dinosaur fossils.

around the village of Bolca, near Verona in northern Italy – especially Pesciara, Monte Postale, Monte Vegroni, Praticini, Loschi, Le Pozzette, and Zovo e Valleco – have been incorporated into a fossil park where the quarries can be visited.

One of the best fossil collections from the locality, the Gazola Collection, is now housed in the Natural History Museum of Verona. However, during the Napoleonic Wars of the late 18th and early 19th centuries, Count Gazola sold much of his collection to the French, and it is now in the Natural History Museum in Paris. There is also a collection at Padua University Museum.

SPECIES LIST

❶ *Exellia* Actinopterygii
❷ *Lophius* Actinopterygii
❸ *Ceratoicthys* Actinopterygii
❹ *Eomyrophis* Actinopterygii
❺ *Eoholocentrum* Actinopterygii
❻ *Psettopsis* Actinopterygii
❼ *Trygon* Chondrichthyes
❽ *Eobothus* Actinopterygii
❾ *Mene* Actinopterygii

FURTHER READING

BOTTJER, D.J. ET AL. 2002. *Exceptional Fossil Preservation*. Columbia University Press, New York (see chapter 20)
STANGHELLINI, E. 1979. *Bolca and its Fossils*. Espro, Verona

WEBSITES

en.wikipedia.org/wiki/bolca
www.palaeo.gly.bris.ac.uk/palaeofiles/lagerstatten/monte_bolca/index.html

MONTE SAN GIORGIO   116–17

Switzerland

233 MILLION YEARS AGO

Since the late 19th century, the world's most spectacular and best preserved marine vertebrate fauna from Triassic times has been extracted from quarries near the southern Alpine villages of Besano and Meride, on both sides of the Italian–Swiss border. The area has been submitted as a proposed World Heritage Site in 2006. Hundreds of entire reptile skeletons can now be seen in the Natural History Museum of Milan and the Paleontological Institute and Museum of the University of Zurich.

SPECIES LIST

❶ *Shastasaurus* Ichthyosauria
❷ *Mixosaurus* Ichthyosauria
❸ *Tanystropheus* Archosauria
❹ *Saurichthys* Actinopterygii
❺ *Paraplacodus* Placodontia
❻ *Eoprotrachyceras* Mollusca
❼ *Askeptosaurus* Diapsida
❽ *Ceresiosaurus* Diapsida
❾ *Birgeria* Actinopterygii

FURTHER READING

BURGIN, T. ET AL. 1989. *The Fossils of Monte San Giorgio*. Scientific American (June), pp.74–81

WEBSITES

whc.unesco.org/en/list/1090
www.worldheritagesite.org/sites/montesangiorgio.html

MONTE VERDE   240–1

Southern Chile

14,500 YEARS AGO

The controversial site of Monte Verde in Chile was discovered in 1975, when a veterinary student recovered a mastodont bone from the banks of the Chinchihuapi Creek. Excavation got under way in earnest during 1977, and archeologists Mario Pino and Tom Dillehay soon found evidence for a human settlement here, which they controversially argued pre-dated the Clovis culture known from sites such as Folsom, New Mexico. It took 20 years for this dating to overcome the established view of prehistory in the Americas, and it was only in 2008 that radiocarbon dating of new discoveries conclusively settled the argument, placing the latest occupation of the Monte Verde site a full 1000 years before the earliest known Clovis layers. Arguments about just how far back the signs of human habitation at Monte Verde stretch currently remain unresolved.

SPECIES AND ARTEFACTS LIST

❶ cooking hearths
❷ *Homo sapiens* Hominidae
❸ skin shelters
❹ *Vultur* Neognathae
❺ tools/weapons
❻ *Lama* Laurasiatheria
❼ *Mammut americanum* Afrotheria

FURTHER READING

BONNICHSEN, R., LEPPER, B.T. and STANFORD, D. 2006. *Paleoamerican Origins: Beyond Clovis.* Texas A&M University Press, College Station

WEBSITE

whc.unesco.org/en/tentativelists/1873/

MORRISON FORMATION 144–7

Western USA

155–148 MILLION YEARS AGO

The Morrison Formation covers a vast area around what was once an inland sea, separating the present-day Rockies and western seaboard of the USA from the eastern half of the country. The fossil dinosaurs found here are some of the best known in the world, and can be found in many US and international museums, most famously at the Carnegie Museum in Pittsburgh, Pennsylvania, and at the American Museum of Natural History in New York. Sites of interest within the formation include Dinosaur National Monument in Utah, and Como Bluff and Fossil Butte National Monument in Wyoming.

SPECIES LIST 144–5

① *Nilssonia* Cycadophyta
② cycadalean Cycadophyta
③ *Czekanowskia* Gymnospermopsida
④ *Goniopholis* Crocodyliformes
⑤ *Diplodocus* Sauropodomorpha
⑥ osmundacean Filicopsida
⑦ *Fruitafossor* Mammalia
⑧ *Camarasaurus* Sauropodomorpha

SPECIES LIST 146–7

① *Camptosaurus* Ornithischia
② *Stegosaurus* Ornithischia
③ *Fruitafossor* Mammalia
④ *Glyptops* "Reptilia"
⑤ *Diplodocus* Sauropodomorpha
⑥ *Coniopteris* Filicopsida
⑦ *Apatosaurus* Sauropodomorpha
⑧ *Ceratodus* Sarcopterygii
⑨ *Allosaurus* Theropoda

FURTHER READING

FOSTER, J. 2007. *Jurassic West: The Dinosaurs of the Morrison Formation and Their World.* Indiana University Press, Bloomington

WEBSITES

www.nps.gov/dino/
www.nps.gov/fobu/

MOSCOW AND VOLGA BASINS 106–7 and 110–11

Russia

260 AND 251 MILLION YEARS AGO

The vast plains that stretch northeast from Moscow towards the Urals and Barents Sea are largely floored with Permian strata. Rivers such as the Dvina have cut sections through the low-lying strata and exposed fossil-bearing layers such as the Sokolki fauna of latest Tatarian (Guadalupian/Lopingian) age, near Kotlas on the banks of the Little Northern Dvina, Archangel'sk Province, some 700 km (440 miles) northeast of Moscow. Their fossil riches were first explored by Polish geologist Vladimir Amalitskii around the turn of the 20th century.

This exploration was followed up some decades later by Russian paleontologist Ivan Efremov, who found hundreds of skeletons that included a whole new array of amphibians and reptiles including the amphibious *Dvinosaurus* and *Kotlassia*, and the sabre-toothed gorgonopsian *Inostrancevia*. Specimens are to be seen in the Palaeontological Institute of the Russian Academy of Sciences in Moscow.

SPECIES LIST 106–7

① *Microphon* "Reptilia"
② *Arctotypus* Hexapoda
③ *Dicynodon* Therapsida
④ *Kotlassia* "Reptilia"
⑤ *Inostrancevia* Therapsida
⑥ *Scutosaurus* "Reptilia"
⑦ *Dvinia* Cynodontia

SPECIES LIST 110–11

① *Benthosuchus* Temnospondyli
② *Wetlugasaurus* Temnospondyli
③ *Lystrosaurus* Therapsida
④ *Contritosaurus* "Reptilia"
⑤ *Chasmatosuchus* Archosauria

FURTHER READING

BENTON, M.J. ET AL. (eds) 2000. *The Age of Dinosaurs in Russia and Mongolia.* Cambridge University Press, Cambridge

NAMA GROUP 44–7

Namibia

548–535 MILLION YEARS AGO

The youngest of the known "Vendian" localities has yielded many superbly preserved fossils of the mysterious "Ediacara" and the equally puzzling organisms that succceeded them. There are a number of different sites scattered across the region – typically in remote locations and mostly on private land.

SPECIES LIST 44–5

⑤ *Namalia* "Ediacara"
⑥ *Pteridinium* "Ediacara"
⑦ *Ernietta* "Ediacara"

SPECIES LIST 46–7

① thrombolite Stromatolites
② *Namacalathus* Unknown
③ *Cloudina* Unknown

FURTHER READING

MCMENAMIN, M.A.S. 1998. *The Garden of Ediacara.* Columbia University Press, New York

KNOLL, A.H. 2003. *Life on a Young Planet.* Princeton University Press, Princeton

WEBSITES

www.mit.edu/~watters/watters_etal_paleo
bio_2001.pdf
www.palaeos.com/Proterozoic/Neoproterozoic/
Ediacaran

NEUQUÉN 166–9

Northwestern Patagonia, Argentina

98 MILLION YEARS AGO

The Late Cretaceous marine and terrestrial deposits of Argentina's Neuquén Basin preserve the remains of abundant new tetrapods, ranging from exquisitely articulated skeletons of small to medium-sized reptiles, mammals, and dinosaurs, to their fragile nests and eggs.

This region of Patagonian Argentina now attracts considerable numbers of tourists. Some private fossil galleries have been opened up and there are commercial dinosaur "hunting" expeditions. Specimens may be seen in the Museo Municipal Ernesto Bachmann, Villa El Chocon, and the Museo "Carmen Funes", Paleontologia de Vertebrados, Plaza Huincul, Neuquén, Argentina.

SPECIES LIST 166–7

❶ *Buitreraptor* Theropoda
❷ *Andesaurus* Sauropodomorpha
❸ *Giganotosaurus* Theropoda
❹ *Anabisetia* Ornithischia

SPECIES LIST 168–9

❶ *Argentinosaurus* Sauropodomorpha
❷ *Aucasaurus* Theropoda

Putorana Plateau The seemingly endless flat-lying lavas of Siberia record a momentous event in Earth's history, when some 80 per cent of all life was wiped out at the end of Permian times.

FURTHER READING

CHIAPPE, L. and DINGUS, L. 2001. *Walking on Eggs: Discovering the astonishing secrets of the world of dinosaurs.* Little, Brown, London

GASPARINI, Z. ET AL. 2007. *Patagonian Mesozoic Reptiles.* Indiana University Press, Bloomington

MAKOVICKY, P.J. ET AL. 2005. *The Earliest Dromaeo-saurid Theropod from South America.* Nature (13 October), pp.1007–11

WEBSITE

en.wikipedia.org/wiki/neuquen_group

NYRANY 94–5

Czech Republic

310 MILLION YEARS AGO

Hundreds of fossil tetrapods were extracted from coal deposits around the small mining town of Nyrany in the Czech Republic in the 19th century. The fossils are now in major museums around the world such as the American Museum of Natural History in New York, the Field Museum of Natural History in Chicago, and the National Museum in Prague, Czech Republic.

SPECIES LIST

❶ palaeodictyopteridan Hexapoda
❷ *Gephyrostegus* "Reptilia"
❸ *Archaeothyris* Synapsida
❹ *Sauropleura* Lepospondyli
❺ *Aornerpeton* Lepospondyli
❻ *Namurotypus* Hexapoda
❼ *Microbrachis* Lepospondyli
❽ branchiosaur Temnospondyli
❾ *Stigmaria* Lycopsida
❿ blattodean Hexapoda
⓫ amblypygid Chelicerata

FURTHER READING

MILNER, A.H. 1980. *The Tetrapod Assemblage from Nyrany, Czechoslovakia.* In *Systematics Association Special Volume 15.* Academic Press, New York

PANCHEN, A.L. (ed.) 1980. *The terrestrial environment and the Origin of Land Vertebrates.* In *Systematics Association Special Volume 15.* Academic Press, New York

OLDUVAI 220–1

Tanzania

1.8 MILLION YEARS AGO

Olduvai is a dry river gorge, some 100km (60 miles) long, cut into the high Serengeti Plain to the northwest of the Eastern Rift Valley, with its large volcanoes and the Ngorongoro Crater. Over the last 2 million years, more than 100m (330ft) of horizontal sediments have been deposited here by rivers, lakes, and eruptions of volcanic ash.

The Olduvai Gorge Museum is situated on the rim of the gorge and has exhibits about the prehistory of the area. However, most of the fossils are housed in the Kenya National Museum in Nairobi.

SPECIES AND ARTEFACTS LIST

❶ *Paranthropus boisei* Hominidae
❷ *Hipparion* Laurasiatheria
❸ *Panthera leo* Laurasiatheria
❹ *Deinotherium* Afrotheria
❺ *Homo habilis* Hominidae
❻ stone tools
❼ *Sivatherium* Laurasiatheria
❽ *Pelorovis* Laurasiatheria
❾ *Crocuta* Laurasiatheria

FURTHER READING

LEAKEY, M.D. 1971. *Olduvai Gorge: Excavation in Beds I and 2. 1960–63.* Cambridge University Press, Cambridge

WEBSITES

en.wikipedia.org/wiki/Olduvai_Gorge
www.sfu.ca/archaeology/museum/olduvai/index2.html

PAPATOWAI   244–5

Otago, South Island, New Zealand

700 YEARS AGO

Around 700 years ago, the last of New Zealand's Ice Age megafauna, the flightless moas, were quickly hunted to extinction by newly arrived voyagers from Oceania. Numerous kill sites have been discovered among dunes and caves, on the coasts of New Zealand and extending some way inland – signs of how this rich avian food resource was driven to extinction.

Specimens are to be seen in various museums, most notably the Otago Museum, Dunedin, and the Museum of New Zealand, Wellington.

SPECIES LIST

1. *Petroica* Neognathae
2. *Apteryx* Palaeognathae
3. *Dinornis giganteus* Palaeognathae
4. *Euryapteryx* Palaeognathae
5. *Arctocephalus forsteri* Laurasiatheria
6. *Homo sapiens* Hominidae
7. *Emeus* Palaeognathae

FURTHER READING

ANDERSON, A. 1989. *Prodigious Birds.* Cambridge University Press, Cambridge

DARBY, J. ET AL. 2003. *The Natural History of Southern New Zealand.* University of Otago Press, Dunedin

HOLDAWAY, R.N. and JACOMB, C. 2000. *Rapid extinction of the Moas (Aves: Dinornithiformes): Model, Test and Implications.* Science (24 March) pp.2250–4.

WEBSITE

www.nzbirds.com/birds/moa.html

PUTORANA PLATEAU   108–9

Siberia, Russia

252–1 MILLION YEARS AGO

Today, flood basalts cover about 675,000 sq km (265,000 sq miles) of Siberia – and originally they may have extended over as much as 7 million sq km (2.7 million sq miles). These lavas erupted from the Earth within less than a million years, from at least four distinct centres of volcanism. One of the biggest outpourings of lava in Earth's history, this coincides with the end-Permian extinction event – also the biggest known in Earth's history.

The most easily accessible and most studied area lies in the Noril'sk region, west of the Putorana Plateau in northwest Siberia, where volcanic material is up to 3700m (12,000ft) thick, with at least 11 discrete eruptive sequences and 45 separate flows. These individual flows may be tens to hundreds of metres deep, and alternate with layers of ash deposits. One such ash layer, some 15–25m (50–80ft) thick, coveres more than 30,000 sq km (11,700 sq miles). In the Maymecha Kotuy region of northeast Siberia, meanwhile, the volcanic layers are more than 6500m (21,300ft) deep.

SPECIES LIST

1. *Archosaurus* Archosauria

WEBSITE

www.le.ac.uk/gl/ads/SiberianTraps/Index.html

RANCHO LA BREA   234–5

Los Angeles, California, USA

20,000 YEARS AGO

The famous La Brea tar pits were first recorded by Spanish explorers in 1769, although local Native Americans had long been aware of them, using the asphalt as an adhesive. Early white settlers soon began to discover bones embedded within the hardened bitumen, and in 1877 paleontologist William Denton first recognized that some of these remains came from extinct animals. However, the site was forgotten about until John C. Merriam of the University of California began to excavate in earnest from 1912.

Today, the tar pits lie in Hancock Park, surrounded by the urban sprawl of downtown Los Angeles. Most of the hundred or so pits have been exhausted of large fossils, but research continues into microfossils such as pollen and insects, which can reveal

much about the environment of the time, and scientists continue to make new discoveries by reanalysis of the original finds. The nearby Page Museum displays many spectacular fossils, and during the excavation season visitors can watch paleontologists working on Pit 91. Other La Brea fossils are featured in various major US collections.

SPECIES LIST

❶ *Teratornis merriami* Neognathae
❷ *Neotoma* Euarchontoglires
❸ *Equus occidentalis* Laurasiatheria
❹ *Smilodon* Laurasiatheria
❺ *Bison antiquus* Laurasiatheria
❼ *Paramylodon* Xenarthra
❽ *Camelops* Laurasiatheria
❻ *Panthera leo atrox* Laurasiatheria
❾ *Arctodus* Laurasiatheria
❿ *Canis dirus* Laurasiatheria

FURTHER READING

HARRIS, J.M. and JEFFERSON, G.T. 1985. *Rancho La Brea: treasures of the tar pits.* Natural History Museum of Los Angeles County, Science Series 31
NUDDS, J.R. and SELDEN, P.A. 2008. *Fossil Ecosystems of North America.* Manson Publishing, London (see chapter 14)
STOCK, C. and HARRIS, J.M. 1992. *Rancho La Brea: a record of Pleistocene Life in California.* Natural History Museum of Los Angeles, Science Series 37

WEBSITES

www.tarpits.org/
www.ucmp.berkeley.edu/quaternary/labrea.html

Rhynie Chert The rolling fields of Rhynie in Aberdeenshire, Scotland, hide and protect Devonian strata and one of the world's most important fossil sites – preserving evidence for the early evolution of land plants.

RHYNIE CHERT   70–1

Aberdeenshire, Scotland

408 MILLION YEARS AGO

This legally protected Site of Special Scientific Interest is not even visible on the surface, since its rocks and fossils are below ground. Periodic excavations have been carried out for scientific purposes ever since the first fossils were found here in 1913. The early land community of plants and animals preserved here is internationally famous for the quality of its preservation, and the National Museum of Scotland in Edinburgh has a display on the Rhynie Chert.

SPECIES LIST

❶ *Asteroxylon* Lycopsida
❷ *Palaeocharinus* Chelicerata
❸ *Protacarus* Chelicerata
❹ *Leverhulmia* Crustacea
❺ *Aglaophyton* Rhyniophyta
❻ *Horneophyton* Rhyniophyta
❼ *Lepidocaris* Crustacea
❽ *Rhyniella* Hexapoda
❾ *Rhynia* Rhyniophyta
❿ *Nothia* ?Rhyniophyta

FURTHER READING

SELDEN, P. and NUDDS, J. 2004. *Evolution of Fossil Ecosystems.* Manson Publishing, London
CLEAL, C.J. and THOMAS, B.A. 1995. *Palaeozoic Palaeobotany of Great Britain.* Chapman & Hall, London

WEBSITES

www.abdn.ac.uk/rhynie/
www.uni-muenster.de/GeoPalaeontologie/Palaeo/ Palbot/erhynie.html
www.ucmp.berkeley.edu/devonian/rhynie.html

RIVERSLEIGH   208–9

Queensland, Australia

23 MILLION YEARS AGO

This remarkable Australian site preserves a wide range of Oligocene and Miocene animals, including mammals, birds, and reptiles, in hard limestone lake deposits and caves formed in what was once a rich tropical forest on the way to becoming semi-arid grassland. Scientific excavations at the site are ongoing, and new discoveries are frequent, but only one site at the locality itself is open for visitors to observe. The Riversleigh Fossil Centre, 250km (150 miles) from the site itself at Mount Isa, displays many of the finds.

SPECIES LIST

❶ *Litokoala* Marsupialia
❷ *Namilamadeta* Marsupialia
❸ *Priscileo* Marsupialia
❹ *Bullockornis* Neognathae
❺ *Montypythonoides* Lepidosauria
❻ *Distioechurus* Marsupialia
❼ *Hypsiprymnodon* Marsupialia
❽ *Yalkaparidon* Marsupialia
❾ *Burramys* Marsupialia
❿ *Strigocuscus* Marsupialia
⓫ *Litoria* Lissamphibia
⓬ *Nimbacinus* Marsupialia
⓭ meliphagid Neognathae
⓮ *Paljara* Marsupialia
⓯ paradisaeid Neognathae
⓰ *Pseudochirops* Marsupialia
⓱ *Ekaltadeta* Marsupialia
⓲ *Brachipposideros* Laurasiatheria
⓳ *Neohelos* Marsupialia
⓴ *Physignathus* Lepidosauria

FURTHER READING

ARCHER, M., HAND, S.J. and GODTHELP, H. 2001. *Australia's Lost World: Prehistoric Animals of Riversleigh.* Indiana University Press, Bloomington

WEBSITES

www.outbackatisa.com.au/
whc.unesco.org/pg.cfm?cid=31&id_site=698
www.rivsoc.org.au

ROCKY RIVER VALLEY   80–1

Ohio, USA

370 MILLION YEARS AGO

Incomplete fossil remains of the giant arthrodire fish *Dunkleosteus* have been found in the Late Devonian shales of Ohio's Rocky River Valley, and as far away as Poland and Morocco. The genus is named after Dr David Dunkle, former curator of the Cleveland Museum of Natural History, which has a permanent display devoted to *Dunkleosteus*. Casts of the head are to be seen in other museums, such as the Queensland Museum in Brisbane, Australia.

SPECIES LIST

❶ *Cladoselache* Chondrichthyes
❷ *Dunkleosteus* Gnathostomes
❸ *Ctenacanthus* Chondrichthyes

WEBSITES

cmnh.org/site/ResearchandCollections/ VertebratePaleontology/Collections.aspx
www.ucmp.berkeley.edu/vertebrates/basalfish/ placodermi.html
www.fossilmuseum.net/Fossil_Galleries/ Fish_Devonian/Dunkleosteous/Dunkleosteus.htm

RUSINGA ISLAND 210–11
Lake Victoria, Kenya

18 MILLION YEARS AGO

Rusinga Island is almost completely made up of Miocene fossil-bearing sediments formed between 20 and 17 million years ago. Numerous fossils of woodland-dwelling bovids, rodents, and apes are preserved close to the environments in which they lived. Specimens are conserved at the Natural History Museum, London, and in the National Museum of Kenya, Nairobi.

SPECIES LIST

1. *Chalicotherium* Laurasiatheria
2. *Gymnurechinus* Laurasiatheria
3. *Proconsul* Primates
4. *Paranomalurus* Euarchontoglires
5. *Masritherium* Laurasiatheria
6. *Rhynchocyon* Afrotheria
7. *Python* Lepidosauria
8. *Gomphotherium* Afrotheria
9. *Dendropithecus* Anthropoidea
10. *Hyainailourus* Laurasiatheria
11. *Dorcatherium* Laurasiatheria

FURTHER READING

STRINGER, C. and ANDREWS, P. 2005. *The Complete World of Human Evolution.* Thames & Hudson, London
WALKER, A. and SHIPMAN P. 2005. *The Ape in the Tree.* Belknap, Cambridge, MA
WALKER, A. and TEAFORD, M. 1989. *The hunt for Proconsul.* Scientific American (January) pp.76–82.

WEBSITE

en.wikipedia.org/wiki/Rusinga_island

SACABAMBILLA 60–1
Cochabamba, Bolivia

455 MILLION YEARS AGO

Some of the oldest known fossils of jawless fish ("agnathans") were found in 1986 by a team of French paleontologists working Late Ordovician strata in the eastern cordillera of the Bolivian Andes. The rocks are exposed around the 4000m-high (13,000ft) mountain village of Sacabambilla, near Cochamba.

SPECIES LIST

1. *Sacabambaspis* "Agnatha"
2. lingulid Brachiopoda

FURTHER READING

JANVIER, P. 1996. *Early Vertebrates.* Clarendon Press, Oxford

WEBSITE

www.tolweb.org/Arandaspida

SHAND GOL 206–7
Mongolia

30 MILLION YEARS AGO

The fossils from this Early Oligocene locality, which lies some 480km (300 miles) southwest of Ulaanbataar, Mongolia, were first found and described by paleontologists working with Roy Chapman Andrews' famous Central Asian expeditions of the 1920s, and further collected by expeditions from the Soviet Academy of Science in the late 1940s. While the most spectacular find was that of the giant *Indricotherium*, it was only one of many new mammals to have been found here – though most are small and only known from their teeth and a few other skeletal elements.

Specimens are housed at the American Museum of Natural History, New York, the Institute of Paleontology at the Russian Academy of Sciences in Moscow, and the Paleontology–Stratigraphy Section of the Mongolian Academy of Science in Ulaanbaatar.

SPECIES LIST

1. *Plesictis* Laurasiatheria
2. *Hyaenodon* Laurasiatheria
3. *Palaeoprionodon* Laurasiatheria
4. *Amphicynodon* Laurasiatheria
5. *Indricotherium* Laurasiatheria
6. *Cricetops* Euarchontoglires
7. *Ochonta* Euarchontoglires
8. *Nimravus* Laurasiatheria
9. *Tupaiodon* Laurasiatheria

FURTHER READING

MELLETT, J.S. 1968. *The Oligocene Hsanda Gol Formation, Mongolia: A Revised Faunal List (American Museum Novitiates no. 2318).* American Museum of Natural History, New York
WANG ET AL. 2005. *American Museum Novitiates no. 3483.* American Museum of Natural History, New York

SHANGHUANG 200–1
South Jiangsu Province, China

45 MILLION YEARS AGO

Tiny teeth and other anthropoid skeletal elements have been found in Middle Eocene fissure fillings within limestone strata near Shanghuang, South Jiangsu, China. Since these limestones are actively quarried, the fossils can only be recovered when paleontologists manage to collect the fissure deposits.

Some of the fossils of these intriguing early anthropods are held at the Carnegie Museum of Natural History in Pittsburg, Pennsylvania, USA.

SPECIES LIST

1. eosimiid Anthropoidea
2. *Eosimias* Anthropoidea
3. *Adapoides* Primates

FURTHER READING

BEARD, C. 2004. *The hunt for the Dawn Monkey: unearthing the origins of monkeys, apes, and humans.* University of California Press, Berkeley

WEBSITES

www.en.wikipedia.org/wiki/eosimias

SIWALIK HILLS   212–13

India and Pakistan

12 MILLION YEARS AGO

The Middle to Late Miocene strata of the Siwalik Hills, outcropping across Pakistan and northwestern India, contain some of the most fascinating Cenozoic fossil sites in Asia. In particular, the Chinji-Nagri Formations, ranging in age from 14 to 9 million years old, contain many rich fossil-bearing deposits. Since 1973, David Pilbeam of the Yale Peabody Museum and his colleagues have collected and catalogued some 40,000 specimens from more than 900 sites on the Potwar Plateau. Specimens are conserved in the Natural History Museum, London.

SPECIES LIST

1. *Hippopotamodon* Laurasiatheria
2. *Percrocuta* Laurasiatheria
3. *Platybelodon* Afrotheria
4. *Protragocerus* Laurasiatheria
5. *Hyainailourus* Laurasiatheria
6. *Sivapithecus* Hominidae
7. *Gomphotherium* Afrotheria
8. *Chalicotherium* Laurasiatheria
9. *Eutamius* Euarchontoglires
10. *Giraffokeryx* Laurasiatheria

FURTHER READING

STRINGER, C. AND ANDREWS, P. 2005. *The Complete World of Human Evolution*. Thames & Hudson, London
ANDREWS, P. AND CRONIN, J. 1982. *The relationships of Sivapithecus and Ramapithecus and the evolution of the orang utan.* Nature (17 June), pp.541–6
PILBEAM, D., IN BERNOR, D. ET AL. 1996. *Evolution of Western Eurasian Neogene Mammals.* Columbia University Press, p.96 et seq.
BARRY, J.C., IN VRBA, E.S. ET AL. 1995. *Paleoclimate and Evolution.* Yale University Press, New Haven, CT, pp.115–34

WEBSITE

www.portal.gsi.gov.in/pls/portal/url/page/gsi_static/ gsi_stat_geo_tourism_monuments

SOLNHOFEN   140–3

Bavaria, Germany

151 MILLION YEARS AGO

Over the centuries, several quarries have worked the smooth, fine-grained "Plattenkalk" limestones of southern Germany for lithographic stone and tiles, and some of these quarries can still be visited. Thousands of fossils have been collected and sold both to private individuals and to museums – including *Archaeopteryx*, one of the most famous fossils in the world. Fossils from Solnhofen can be seen at the Humboldt Museum für Naturkunde in Berlin, the Jura Museum in Eichstatt, the Burgermeister Muller Museum in Solnhofen itself, and the Munich Museum. Further afield, the Teyler Museum in Haarlem, the Netherlands, and the Natural History Museum in London, UK, also hold prized specimens.

SPECIES LIST 140–1

1. *Aegirosaurus* Ichthyosauria
2. *Compsognathus* Theropoda
3. *Bavarisaurus* Lepidosauria
4. *Archaeopteryx* Avialae
5. *Rhamphorhynchus* Pterosauria
6. *Ginkgoites* Ginkgoales
7. *Mesolimulus* Chelicerata
8. *Leptolepides* Actinopterygii

SPECIES LIST 142–3

1. *Tarsophlebia* Hexapoda
2. *Aeger* Crustacea
3. *Pterodactylus* Pterosauria
4. *Caturus* Actinopterygii
5. *Acanthoteuthis* Mollusca
6. *Cycleryon* Crustacea
7. *Aspidorhynchus* Actinopterygii

FURTHER READING

BARTHEL, K.W. SWINBURNE, N.H.M. and CONWAY MORRIS, S, 1990. *Solnhofen: A Study in Mesozoic Palaeontology.* Cambridge University Press, Cambridge
SELDEN, P. and NUDDS, J. 2004. *Evolution of Fossil Ecosystems.* Manson Publishing, London

WEBSITES

palaeo.gly.bris.ac.uk/Palaeofiles/Lagerstatten/ Solnhofen/
home.arcor.de/ktdykes/altmuehl.htm
www.fossilmuseum.net/Fossil_Sites/solnhofen/ Solnhofen_Lagerstatt.htm
www.yale.edu/ypmip/locations/solnhofen.htm

SOOM SHALE   62–3

South Africa

450 MILLION YEARS AGO

The Soom Shale localities are not accessible to the general public. Outcrops occur in the Cederberg mountains in the Western Cape Region of South Africa and fossils are rare.

SPECIES LIST

1. *Onychopterella* Chelicerata
2. nautiloid Mollusca
3. orbiculoid Brachiopoda
4. *Promissum* Conodonta
5. *Soomaspis* Trilobita

FURTHER READING

SELDEN, P. and NUDDS, J. *Evolution of Fossil Ecosystems.* Manson Publishing, London

WEBSITES

www.palaeo.gly.bris.ac.uk/palaeofiles/lagerstatten/ soom/fauna.html

ST PIETER'S MOUNT   176–7

Maastricht, The Netherlands

68 MILLION YEARS AGO

The famous skull of the Maastricht mosasaur found at St Pieter's Mount near the Meuse river is still displayed in the French Musée d'Histoire Naturelle in Paris, whence it was taken by Napoleon's soldiers in the late 18th century. However, Maastricht has

remained a rich source of fossils through to the present day, and the Maastricht Natural History Museum has an extensive collection ranging from the Carboniferous to the Quaternary, including a mounted mosasaur reconstruction.

SPECIES LIST

1. *Mosasaurus* Lepidosauria
2. *Belemnitella* Mollusca
3. *Scaphites* Mollusca
4. *Squalicorax* Chondrichthyes
5. *Marsupites* Echinodermata
6. *Temnocidaris* Echinodermata
7. *Hyotissa* Mollusca
8. *Plagiostoma* Mollusca

WEBSITE

http://www.nhmmaastricht.nl

STERKFONTEIN   218–19

South Africa

2.8 MILLION YEARS AGO

Fossils were first discovered in the limestone caves of Sterkfontein during commercial mining operations that began in the 1890s, but it was not until the 1930s that a team led by Raymond Dart and Robert Broom began excavations here. Dart had already discovered the first australopithecine fossils (the famous "Taung child"), but remains from the Sterkfontein area

included the first adult *Australopithecus africanus*, and the more robust genus *Paranthropus*. Finds have continued to the present day, and about a thousand hominid fossils have now been discovered on the site, revealing a fragmented story of several million years of human history.

Today Sterkfontein and its neighbouring caves are designated as the "Cradle of Humankind" World Heritage Site, and excavations are continuing at several sites within the area. Sterkfontein lies close to Johannesburg and the new Maropeng visitor complex opened in 2005, including a changing display of original fossils.

SPECIES LIST

1. *Protea* Angiospermae
2. *Megalotragus* Laurasiatheria
3. *Panthera pardus* Laurasiatheria
4. *Canis mesomelas* Laurasiatheria
5. *Australopithecus africanus* Hominidae
6. *Stephanoeatus* Neognathae
7. *Hippotragus* Laurasiatheria
8. *Antidorcas* Laurasiatheria
9. *Homotherium* Laurasiatheria
10. *Papio* Anthropoidea

WEBSITES

www.maropeng.co.za/
whc.unesco.org/en/list/915

STRELLEY POOL   36–9

**Pilbara Craton,
Western Australia, Australia**

3430 MILLION YEARS AGO

This remote site in Western Australia is home to one of the earliest and best preserved ancient stromatolite outcrops in the world.

PRESERVED GEOLOGICAL EVIDENCE 36–7

1. "egg carton" form Stromatolites
2. "cuspate swale" form Stromatolites
3. "encrusting/domical" form Stromatolites

SPECIES LIST 38–9

1. *Primaevifilum* Stromatolites
2. *Archaeoscillatoriopsis* Stromatolites

FURTHER READING

KNOLL, A.H. 2003. *Life on a Young Planet.* Princeton University Press, Princeton

WEBSITES

www.dmp.wa.gov.au/5243.aspx

TENDAGURU   136–7

Tanzania

152 MILLION YEARS AGO

Between 1907 and 1931, more than 200 tonnes of Late Jurassic fossils, including many dinosaurs, were excavated firstly by German and later by British scientists, in remote

excavations around Tendaguru in southeastern Tanzania. Shipped back to Europe, most of these, including *Brachiosaurus*, one of the most complete sauropods ever found, are housed at Berlin's Museum für Naturkunde. The British specimens are stored at the Natural History Museum in London.

SPECIES LIST

1. paramacellodid Lepidosauria
2. *Dicraeosaurus* Sauropodomorpha
3. gleicheniacean Filicopsida
4. *Brachiosaurus* Sauropodomorpha
5. *Rhamphorhynchus* Pterosauria
6. *Tendagurutherium* Mammalia
7. *Dryosaurus* Ornithischia
8. *Elaphrosaurus* Theropoda
9. *Barosaurus* Sauropodomorpha
10. *Kentrosaurus* Ornithischia
11. cheirolepidacean Coniferopsida

FURTHER READING

MAIER, G. 2003. *African Dinosaurs Unearthed: the Tendaguru Expeditions.* Indiana University Press, Bloomington

WEBSITE

www.museum.hu-berlin.de/ausstellungen/
ausstellungen_2201.asp?lang=1

Strelley Pool The stromatolitic rocks of this barren landscape in Western Australia contain some of the best preserved evidence for early microbial life, flourishing in tropical waters some 3.4 billion years ago.

TEXAS REDBEDS   98–9
Texas, USA
295 MILLION YEARS AGO

The abundant Permian fossils embedded in the rock outcrops around Seymour, Texas, were discovered in 1877 by the great dinosaur hunter Edward Drinker Cope, and have been excavated occasionally ever since. Fossils from this locality, and particularly the ubiquitous synapsid *Dimetrodon*, are displayed in museums around the world, but one of the most significant collections is at the Houston Museum of Natural Science, Texas.

SPECIES LIST
1. *Meganeuropsis* Hexapoda
2. *Araeoscelis* Diapsida
3. *Dunbaria* Hexapoda
4. *Captorhinus* "Reptilia"
5. dipnoan Sarcopterygii
6. *Dimetrodon* Synapsida
7. *Diplocaulus* Lepospondyli
8. *Eryops* Temnospondyli
9. *Edaphosaurus* Synapsida

WEBSITE
www.hmns.org/exhibits/permanent_exhibits/
  paleontology.asp

TRENTON   58–9
New Jersey, USA
461 MILLION YEARS AGO

Fossil-bearing reef limestones and shales of mid-to-late Ordovician age are exposed at many historic sites scattered over an extensive region of eastern North America from New York State to Illinois. Among the many invertebrate groups represented by these well-preserved fossils, the trilobites and crinoids are probably the best known. The discovery of the Late Ordovician trilobite *Triarthrus* in the Utica Shale at Rome, New York State, in the late 19th century provided the first good information about the soft-part appendages of these arthropods.

Fossils from these strata are to be found in many major American museums, but some of the best are held at New York State Museum in Albany, the Field Museum in Chicago, the American Museum of Natural History in New York, and the Yale Peabody Museum, in New Haven, Connecticut.

SPECIES LIST
1. *Sinuites* Mollusca
2. *Megalograptus* Chelicerata
3. *Balanacrinus* Echinodermata
4. *Homotelus* Trilobita
5. *Orthoceras* Mollusca
6. *Salteraster* Echinodermata
7. *Sowerbyella* Brachiopoda

FURTHER READING
LEVI-SETTI, R. 1993. *Trilobites*. University of Chicago Press, Chicago
WHITTINGTON, H.B. 1992. *Trilobites. Fossils Illustrated volume 2*. The Boydell Press, Woodbridge, Suffolk

WEBSITES
www.peabody.yale.edu/collections/ip/
www.nysm.nysed.gov/ (and follow links)

UKHAA TOLGOD   170–3
Mongolia
80 MILLION YEARS AGO

Since the 1920s and the American Museum of Natural History expeditions led by Roy Chapman Andrews, the fossil treasures of Mongolia's Gobi Basin have been investigated by Soviet and Polish paleontologists, and lately revisted by the American Museum of Natural History. Many thousands of fossils, including dinosaurs, lizards, and small mammals, have been found at numerous sites scattered across the arid landscape, from localities such as Ukhaa Tolgod and the Campanian age (*c*.80 million year old) Djadokhta Formation strata.

Specimens are on display at the American Museum of Natural History in New York, the Palaeontological Institute of the Polish Academy of Sciences in Warsaw, and the Palaeontological Institute of the Mongolian Academy of Science in Ulaanbataar.

SPECIES LIST 170–1
1. *Pinacosaurus* Ornithischia
2. *Zalambdalestes* Eutheria
3. *Kryptobataar* Mammalia
4. *Protoceratops* Ornithischia
5. *Estesia* Lepidosauria
6. *Mononykus* Theropoda
7. *Saurornithoides* Theropoda

SPECIES LIST 172–3
1. *Oviraptor* Theropoda
2. *Velociraptor* Theropoda
3. *Nemegtbataar* Mammalia

FURTHER READING
NOVACEK, M. 1996. *Dinosaurs of the Flaming Cliffs*. Anchor Books, New York
KIELAN-JAWOROWSKA, Z. 1969. *Hunting for Dinosaurs*. The Maple Press Company, York, PA

WEBSITE
digitallibrary.amnh.org/
  (search for articles by M. Norell)
www.geosciences.unl.edu/~dloope/pdf/
  NatHistory.pdf

VALLEY OF THE MOON   120–1
La Rioja Province, Argentina
227 MILLION YEARS AGO

Some of the earliest known dinosaurs have been found in the Late Triassic strata of the Ischigualasto/Talampaya Natural Parks of Argentina, listed World Heritage Sites since 2000. This remote but accessible region near the foothills of the Andes is a tourist attraction on account of the barren hilly landscape alluded to in its name. Fossils are rare, and most of the specimens are housed in the Natural History Museum in Buenos Aires.

SPECIES LIST
1. *Hyperodapedon* Archosauria
2. *Exaeretodon* Cynodontia
3. *Herrerasaurus* Saurischia
4. *Saurosuchus* Archosauria
5. *Protojuniperoxylon* Coniferopsida
6. *Eoraptor* Saurischia

WEBSITES
www.en.wikipedia.org/wiki/Ischigualasto
www.ischigualasto.org/nuevoischi/ischigualasto/
  flash/default.htm (in Spanish)

VOLGA BASIN   106–7
Russia
260 MILLION YEARS AGO

See Moscow and Volga Basins entry.

WILLANDRA LAKES 230–1
New South Wales, Australia

40,000 YEARS AGO

The so-called "Mungo Man" burial is just one of several fossil hominids found alongside marsupial remains in the Willandra Lakes World Heritage Site since the 1970s. Nearby were found the remains of the oldest ritual cremation in the world, dated to 40,000 years old and therefore of similar age to the ochre-sprinkled burial. The site is seen as very important to the Australian Aborigines and is therefore culturally sensitive, but many of the artefacts found here are displayed at the Mungo National Park Visitor Centre, and guided tours with Aboriginal rangers are also available.

SPECIES AND ARTEFACTS LIST
1. fish bones
2. *Velesunio* Mollusca
3. *Homo sapiens* Hominidae
4. *Macropus* Marsupialia

FURTHER READING
BOWLER, J.M. ET AL. 2003. *New ages for human occupation and climatic change at Lake Mungo, Australia.* Nature (20 February), pp.837–40
BURENHULT, G. (ed.) 2003. *People of the Past.* Fog City Press, Sydney, pp.147–70

WEBSITES
www.visitnsw.com/Mungo_National_Park_P629.aspx
www.whc.unesco.org/en/list/167

WREN'S NEST 66–7
Dudley, England

425 MILLION YEARS AGO

This legally protected Site of Special Scientific Interest and National Nature Reserve is a historic limestone quarry from which thousands of fossils of marine invertebrates, from more than 600 species, have been recovered. These are now to be seen in British museums including the nearby Dudley Museum, and the Sedgwick Museum at the University of Cambridge. They represent the shelled organisms that lived around a small coral reef, especially trilobites, brachiopods, corals, and bryozoans.

SPECIES LIST
1. *Protochonetes* Brachiopoda
2. *Halysites* Cnidaria
3. *Monograptus* Graptolithina
4. *Calymene* Trilobita
5. *Gissocrinus* Echinodermata
6. *Favosites* Cnidaria
7. *Heliolites* Cnidaria
8. *Atrypa* Brachiopoda
9. *Cyrtograptus* Graptolithina
10. *Ketophyllum* Cnidaria
11. *Dalmanites* Trilobita

FURTHER READING
ALDRIDGE, R.J. ET AL. 2000. *British Silurian Stratigraphy.* Joint Nature Conservation Committee, Peterborough

WEBSITES
www.english-nature.org.uk/special/nnr/nnr_details.asp?NNR_ID=170
www.dudley.gov.uk/leisure-and-culture/museums--galleries/dudley-museum--art-gallery/heritage/wrens-nest-national-nature-reserve

ZHOUKOUDIAN 224–5
Beijing, China

540,000 YEARS AGO

Dragon Bone Hill, the famous home of "Peking Man", lies near Zhoukoudian in the southwest suburbs of modern Beijing. The cave system is some 140m (460ft) long, and the sediments accumulated on the cave floor during its long history of occupation are, in places, up to 40m (130ft) deep. The history of the site's excavation is long and troubled, with many of the original finds lost during World War II, but intermittent excavations since then have recovered a few more fragments of ancient *Homo erectus*, and accurate casts of the lost material fortunately survive. A museum at the site preserves many of the relics, and there is a visitor trail to see the various caves. Zhoukoudian was classified as a World Heritage Site in 1987, and an international programme of research has been ongoing since the 1990s.

SPECIES AND ARTEFACTS LIST
1. tools
2. *Homo erectus/ergaster* Hominidae
3. *Pseudaxis* Laurasiatheria
4. *Megantereon* Laurasiatheria
5. *Megaloceros* Laurasiatheria
6. *Macaca* Anthropoidea
7. *Canis* Laurasiatheria
8. *Celtis* Laurasiatheria
9. *Pachycrocuta* Laurasiatheria

FURTHER READING
BOAZ, N.T. and CIOCHON, R.L. 2004. *Dragon Bone Hill.* Oxford University Press, Oxford
VAN OOSTERZEE, P. 1999. *Dragon Bones.* Allen & Unwin, Sydney

WEBSITES
whc.unesco.org/en/list/449
www.unesco.org/ext/field/beijing/whc/pkm-site.htm

Valley of the Moon Strange-looking wind-sculpted rocks have given this arid region of Argentina its popular name, but the early dinosaur fossils that lie hidden within these Late Triassic strata are even more fascinating.

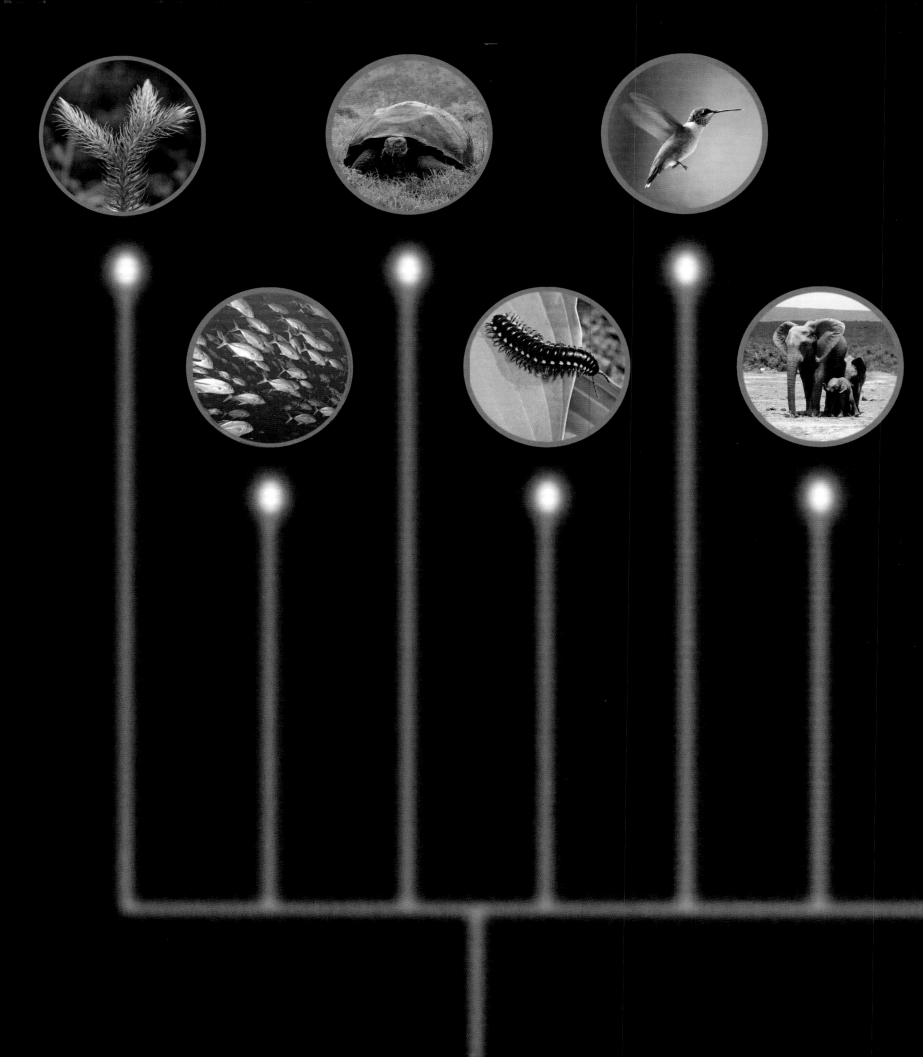

THE
SPECIES
INDEX

There are an estimated
10 million different species
alive on Earth today, of which
barely one fifth have been
described. Countless millions
of other species have lived and
died throughout the history of
the Earth, and yet the vast
majority of organisms, living
and extinct, fall into the same
few dozen distinctive groups.
The pages that follow list all
the species featured in the
panoramic artworks, and put
them in the context of the
major groups of life on the
planet both today, and in
the distant past.

SPECIES INDEX

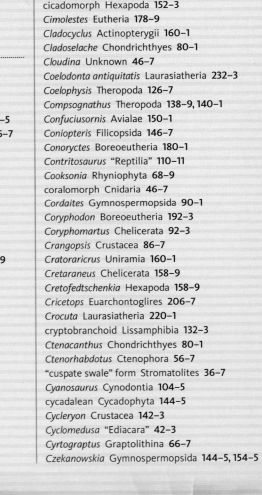

SPECIES LISTING

THROUGHOUT THE 542 MILLION YEARS SINCE THE BEGINNING OF THE PALEOZOIC ERA, LIFE ON EARTH HAS DIVERSIFIED TO FILL A HUGE RANGE OF ECOLOGICAL NICHES AND TAKE ON A GREAT VARIETY OF FORMS. NEVERTHELESS, THE VAST MAJORITY OF ORGANISMS FALL WITHIN A RELATIVELY SMALL NUMBER OF OVERARCHING GROUPS, DISTINGUISHED FROM OTHERS BY THEIR SHARED CHARACTERISTICS AND ANCESTRY.

The species listings in Evolution put the animals and plants shown in the panoramic artworks into context within these major living and extinct groups. Pages 326–9 feature a complete listing of the species found in the artworks, which acts as an index and also identifies the group within which they lie. Pages 330–59 include detailed accounts of the history and characteristics of these groups (listed in alphabetical order). Each entry is followed by a detailed breakdown that reveals the major subdivisions within the group and helps put the individual species into further context.

SAMPLE ENTRY

Latin and common name Entries are arranged by Latinate scientific name.

Time period The period during which the group survived.

Context These listings show the group's origins and its sibling groups.

Subdivisions This section lists the major subdivisions within the group.

Description The main text describes the group's distinguishing features.

Subgroups This level of heading indicates a major subdivision.

Lower level subdivisions These subheadings often correspond to Linnaean "families".

Species listing Individual listings refer back to the panoramic artworks.

ARCHOSAURIA (ARCHOSAURIAN REPTILES)

Triassic – Extant

WITHIN Diapsida
COMPARE Lepidosauria, Ichthyosauria, Plesiosauria, Placodontia
SEE ALSO Pterosauria, Crocodyliformes, Dinosauria (including Ornithischia and Saurischia)

Crocodiles and birds are the extant representatives of the Archosauria, a group that also includes extinct lineages such as the pterosaurs and the non-avian dinosaurs, and whose ancestry can be traced back to Early Triassic times. The archosaurs are diapsid reptiles characterized by two openings in the skull behind the eye.

● **ARCHOSAUROMORPHA**

RHYNCHOSAURIDAE
Hyperodapedon 120–1, 122–3

PROLACERTIDAE
Prolacerta 112–13
Tanystropheus 116–17

ACTINOPTERYGII (RAY-FINNED FISH)

Late Silurian – Extant

WITHIN Osteichthyes
COMPARE Sarcopterygii

Today, the actinopterygians form the majority of what we call bony "fish" (**Osteichthyes**), excluding the cartilaginous **Chondrichthyes** (sharks and rays) and the **Sarcopterygii** (lobe-finned fish).

With some 23,700 living species (most of which are teleosts), the actinopterygians comprise nearly a half of all vertebrates. From marine origins in late Silurian times, they diversified enormously over several tens of millions of years, and have now colonized all aquatic environments from ocean depths to mountain lakes, and from rivers to hot springs. Typically, they have scale-covered bodies, a gas-filled swim bladder for buoyancy, and "ray fins" for swimming.

Their evolution went through three phases, with the basal actinopterygians and "chondrosteans" of late Palaeozoic times followed by a Mesozoic neopterygian or "holostean" radiation, and a teleost radiation that has continued from the Jurassic to the present day. Over time, the structure and size of scales has reduced from thick and heavy armour to very thin, flexible and overlapping teleost scales, accompanied by reduced skeletons and advanced jaw structures. The basic actinopterygian body structure has proved remarkably adaptable to different habitats and lifestyles, evolving into shapes as diverse as flatfish, eels, and bizarre deep-sea fish.

● **BASAL ACTINOPTERYGII**

CHEIROLEPIDIDAE
Cheirolepis 76–7

STEGOTRACHELIDAE
Moythomasia 74–5

POLYPTERIDAE
Paratarrassius 86–7 (birchir)

● **CHONDROSTEI**

BIRGERIIDAE
Birgeria 116–17

BOBASATRINIDAE
Ebanaqua 102–3

Bigeye trevally (*Caranx sexfasciatus*). These widespread marine fish are perciforms, part of a group containing 40 per cent of modern bony fish.

SAURICHTHYIFORMES
Saurichthys 116–17, 118–19

PERLEIDIFORMES
Dipteronotus 114–15

● **NEOPTERYGII (HOLOSTEANS)**

SEMINIONOTIDAE
Semionotus 126–7
Lepidotes 130–1

DAPEDIIDAE
Dapedium 128–9

AMIIDAE
Caturus 128–9, 142–3
Amia 190–1

● **TELEOSTEI**

ASPIDORHYNCHIDAE
Aspidorhynchus 142–3

PHOLIDOPHORIDAE
Pholidophorus 134–5

LEPTOLEPIDIDAE
Leptolepides 140–1

ICHTHYODECTIDAE
Cladocyclus 160–1

OSTEOGLOSSIFORMES
Lycoptera 156–7

ANGUILLIFORMES
Eomyrophis 194–5

CLUPEIFORMES
Knightia 190–1

GONORHYNCHIFORMES
Dastilbe 158–9

● **EUTELEOSTEI**

LOPHIIFORMES
Ceratoicthys (angelfish) 194–5
Lophius (anglerfish) 194–5

BERYCIFORMES
Eoholocentrum (squirrelfish) 194–5

PERCIFORMES
Exellia (spadefish) 194–5
Mene 194–5
Psettopsis (perciform) 194–5

PLEURONECTIFORMES
Eobothus (flounder) 194–5

AFROTHERIA (AFROTHERIAN MAMMALS)

Late Cretaceous – Extant

WITHIN *Eutheria*
COMPARE *Xenarthra, Boreoeutheria*

The first main branch of eutherian placental mammals (ie those whose babies are connected to the mother's uterus with a placenta) that evolved, perhaps as much as 100 million years ago, was an African group – hence the name "Afrotherian". Their descendants include the living tenrecs, aardvarks, golden moles, elephant shrews. and, somewhat surprisingly, a related group of elephants, sea cows, and hyraxes.

Of all the Afrotheria, elephants (proboscideans) have the best fossil record, stretching back to early Eocene times with animals such as the metre-long (3ft), hippo-like *Moeritherium*. The proboscideans spread across Africa, Eurasia, and into the Americas, producing a range of spectacular elephant-like animals including the extinct gomphotheres, mastodonts, mammutids, mammoths, and the surviving elephants of Africa and Asia.

● TUBULIDENTATA

ORYCTEROPODIDAE
Orycteropus (anteater) 214–15

● MACROSCELIDEA

MACROSCELIDAE
Rhynchocyon (elephant shrew) 210–11

● HYRACOIDEA

PROCAVIIDAE
Procavia (hyrax) 226–7

● EMBRITHOPODA

ARSINOTHERIIDAE
Arsinotherium 202–3

MOERITHERIIDAE
Moeritherium 202–3

DEINOTHERIIDAE
Deinotherium 220–1

STEGODONTIDAE
Stegodon 238–9

● PROBOSCIDEA

MAMMUTIDAE
Mammut americanum (mastodon) 240–1

GOMPHOTHERIIDAE
Gomphotherium 210–11, 212–13
Phiomia 202–3
Platybelodon 212–13
Anancus 214–15

ELEPHANTIDAE
Mammuthus primigenius
  (woolly mammoth) 232–3
Elephas 228–9
Loxodonta 216-17

"AGNATHA" (JAWLESS FISH)

Early Cambrian – Extant

WITHIN *Chordata*
COMPARE *Gnathostomes*
SEE ALSO *Conodonta*

This ancient group of strange jawless fish includes the first dominant vertebrates. They were most diverse in Silurian and Devonian times, and first colonized freshwaters in the Devonian, but are today reduced to the specialized hagfish and lampreys. Initially agnathans were filter and deposit feeders that sucked water through simple terminal mouths. Gills removed food particles and oxygen before the depleted water was flushed out through gill openings in the throat region. In the absence of teeth, many agnathans were protected from predators with armoured bony scales and plates, but these limited their movement and made them look most ungainly.

● BASAL AGNATHANS

MYXINOIDEA
myxinoid (hagfish) 88–9

● PTERASPIDOMORPHI

ARANDASPIDA
Sacabambaspis 60–1

HETEROSTRACI
Drepanaspis 72–3

ANASPIDA
Endeiolepis 76–7
Jamoytius 64–5

● CEPHALASPIDOMORPHI

BIRKENIIFORMES
Birkenia 64–5

THELODONTIFORMES
Loganellia 64–5

OSTEOSTRACI
Escuminaspis 76–7

AMNIOTA (EGG-LAYING TETRAPODS)

Carboniferous – Extant

WITHIN *Tetrapoda*
SEE ALSO "Reptilia", Synapsida, Diapsida (including Lepidosauria, Archosauria, Ichthyosauria, Plesiosauria, Placodontia)

One of the most important advances in the evolution of the tetrapods was the internal fertilization of the egg while still inside the mother, and the subsequent enclosure of the fertilized and developing embryo in a protective membrane prior to birth. This ability is a fundamental feature of a group called amniotes, which unites the "reptiles", birds, and mammals. The further internal secretion of a shell around the embryo and its membrane is characteristic of the egg-laying birds, "reptiles", and a few primitive mammals. Unfortunately, eggs do not fossilize well unless the shell is mineralized with calcium carbonate, and even then are rarely preserved – the oldest fossil eggs are Triassic in age, but there is no doubt that egg-laying occurred in amniotes long before this.

African elephants (*Loxodonta africana*) are the largest and best known surviving descendants of the ancient Afrotherian mammal lineage.

So, identification of fossil amniotes has to rely on other skeletal features, especially those of the skull. Those of the earliest reptiles only have openings for the eyes (the "anapsid" condition), as seen in living chelonians such as the turtles and in several extinct fossil groups from Permian and Triassic times. Two other major groups can be defined on the basis of skull type – the synapsids and diapsids. Synapsids, such as the mammals and their extinct ancestral groups, have one extra pair of skull openings, while the diapsids have two pairs. This latter condition is characteristic of the majority of living "reptiles" such as the crocodiles, snakes, and lizards, and is also found in the birds. Important extinct diapsid groups include the pterosaurs and dinosaurs. Finally, a fourth skull type, called "euryapsid", is seen in a number of extinct marine reptiles such as the ichthyosaurs and plesiosaurs, and is a development from the diapsid condition in which the lower pair of openings has been suppressed.

ANGIOSPERMAE (FLOWERING PLANTS)

Jurassic – Extant

WITHIN *Gymnospermopsida*
COMPARE *Pteridospermales, Bennettitales, Cycadophyta, Ginkgoales, Coniferopsida*
SEE ALSO *Graminae*

Today's terrestrial vegetation is mostly dominated by 250,000 species of flowering plants that range from aquatics to grasses and trees. Classified into some 450 families, they are probably more diverse than any other group of plants has ever been, and have achieved their dominance fairly rapidly since eclipsing the cycads (**Cycadophyta**), conifers (**Coniferopsida**), and **Bennettitales** at the end of Cretaceous times. A number of features characterize the flowering plants, but none are unique and they are not all present in all flowering plants. The most important angiosperm feature that is preservable in the fossil record is the outer covering or carpel that encloses the unfertilized seed to form an ovule. This protects fertilization and to some extent shields the developing embryo from predation. Similar but partial enclosure of ovules also occurred in some Mesozoic non-angiosperms such as the cycads. Similarly, while flower

Flowering magnolia (*Magnolia* sp.). These modern flowers are part of the Order Magnoliales, long considered the most primitive surviving order of flowering plants. They have several features that seem to be barely modified from a basal angiosperm structure.

structures with bisexual reproductive organs are common in the flowering plants, they are also present in the extinct bennettitaleans. Leaves from flowering plants have a fairly distinctive network of veins, but again such veins are also present in Jurassic seed ferns (**Gymnospermopsida**).

● BASAL ANGIOSPERMS?

Archaeofructus 154–5
nymphaealean sp. 158–9

● MAGNOLIALES (DICOTYLEDONS)

MAGNOLIACEAE
Magnolia 178–9

LAURACEAE
lauracean sp. 192–3

BERBERIDACEAE
Mahonia (barberry) 204–5

HAMAMELIDAE
Platanus (plane tree or sycamore) 180–1, 188–9

PROTEACEAE
Protea 218–19

STERCULIACEAE
Florissantia (chocolate family) 204–5

ROSACEAE
Rosa (rose) 204–5
Amelanchier (serviceberry) 204–5

CANNABACEAE
Celtis (hackberry) 188–9, 224–5

MORACEAE
Ficus sp. (fig) 192–3

ANACARDIACEAE
Rhus (sumac – cashew family) 190–1

SAPINDACEAE
Koelreuteria (goldenrain tree) 204–5

SIMARUBACEAE
Ailanthus (quassia family) 190–1

RHAMNACEAE
Rhamnus (buckthorn) 180–1

VITACEAE
Vitis (vine) 180–1

● LILIOPSIDA (MONOCOTYLEDONS)

ARACEAE
Zantedeschia (arum lily) 198–9

ARECIDAE
Sabalites (palm) 190–1
Ceriops (mangrove palm) 192–3
Epipremnum 202–3

TYPHACEAE
Typha (cattail) 190–1, 198–9

SPARGANIACEAE
Sparganium (bur-reed) 156–7

ANTHROPOIDEA (ANTHROPOID OR SIMIIFORM MAMMALS)

Eocene – Extant

WITHIN *Primates*
SEE ALSO *Hominoidea, Hominidae*

Monkeys and apes, the so-called "higher primates", are classified as platyrrhines, catarrhines, cercopithecoids, and **Hominoidea**, and grouped together as the anthropoids (also known as simiiforms). The New World monkeys are characterized by their broadly spaced "platyrrhine" nostrils and the presence of a prehensile tail in some forms, whereas the Old World monkeys and apes have narrow "catarrhine" snouts with downward-opening nostrils separated by a thin septum, and non-prehensile tails, if they have them at all. In general the anthropoids share characters such as rounded instead of slit-shaped nostrils, and large occluding pairs of canine teeth.

For a long time, the anthropoids were thought to have originated in Africa, but recent finds, including the oldest known anthropoid (*Eosimias*) in China and other fossils such as *Siamopithecus* from Thailand, have challenged this. However, proponents of the African origin claim that the Asian forms are not anthropoids, but belong to other more primitive groups, such as the omomyids and tarsiers.

● BASAL ANTHROPOIDS

EOSIMIIDAE
eosimiid sp. 200–1
Eosimias 200–1

● CATARRHINI

PROCONSULIDAE
Proconsul 210–11

PROPLIOPITHECIDAE
Aegyptopithecus 202–3

CERCOPITHECIDA
cercopithecoid (colobine) 214–15
Papio (baboon) 218–19
Macaca (macaque) 224–5

PLIOPITHECIDAE
Dendropithecus 210–11

ARCHAEOCYATHIDA

Cambrian

This extinct group of marine invertebrates is related to the sponges and includes some 250 genera. Most build a simple porous-walled and conical, calcareous skeleton, but some develop additional conelet buds. Growing from the seabed, the archaeocyathans were abundant enough in places to form some of the first patch reefs.

IRREGULARES
Cambrocyathellus 46–7
Okulitchicyathus 46–7

REGULARES
radiocyathid sp. 46–7

ARCHOSAURIA (ARCHOSAURIAN REPTILES)

Triassic – Extant

WITHIN *Diapsida*
COMPARE *Lepidosauria, Ichthyosauria, Plesiosauria, Placodontia*
SEE ALSO *Pterosauria, Crocodyliformes, Dinosauria (including Ornithischia and Saurischia)*

Crocodiles and birds are the extant representatives of the Archosauria, a group that also includes extinct lineages such as the pterosaurs and the non-avian dinosaurs, and whose ancestry can be traced back to Early Triassic times. The archosaurs are diapsid reptiles characterized by two openings in the skull behind the eye.

● ARCHOSAUROMORPHA

RHYNCHOSAURIDAE

● ARCHOSAURIA

Mandrill (*Mandrillus sphinx*). *These anthropoid primates are closely related to baboons. They live in large groups, and the bright face markings of the male are likely to be linked to sexual competition.*

ARTHROPODA (ARTHROPODS)

Early Cambrian – Extant

SEE ALSO *Chelicerata, Trilobita, Crustacea, Uniramia (including Hexapoda)*

This most diverse and successful group of all animals has well over a million living species (mostly insects). The arthropods range from minute insect mites to giant crustacean spider crabs and include several important extinct groups such as the marine trilobites and eurypterids, which occupied both marine and freshwater habitats. Their huge success stems from the adaptability of their basic body plan, which has allowed them to conquer most habitats from the deep oceans to high mountains and the air itself. The segmented body has a threefold subdivision into head, thorax, and abdomen, with many of the segments carrying paired and jointed appendages that range from antennae to jaws, gills, legs, and wings.

A tough organic "armour", the exoskeleton, protects arthropods from the elements and provides rigid attachments for the internal muscles that move the limbs. The downside is that in order to grow the exoskeleton must periodically be replaced (moulted), leaving the animal temporarily vulnerable to predators.

Arthropods originated in the seas, probably during late Precambrian times. Myriapods (millipedes and centipedes) were the first animals to colonize freshwaters in Ordovician times, and, along with trignotarbids, they went on to conquer the land in Silurian times. Arthropods were also the first animals to take wing, and relatives of dragonflies are known from the Carboniferous. Today, arthropods are vital members of the food chain through their pollination of essential food plants, as well as being major pests and carriers of life-threatening diseases such as malaria.

Recently, a major subdivision of the arthropods into schizoramians and atelocerates has been recognized.

The schizoramians are those arthropods with "biramous" limbs, and comprise arachnomorphs and crustaceomorphs.

The arachnomorphs incorporate the chelicerates (**Chelicerata**), a diverse group that includes the living horse-shoe crabs, scorpions, and spiders as well as the extinct eurypterids, and the trilobites (**Trilobita**), an entirely extinct marine group.

Canadaspis. This strange creature from Canada's Burgess Shale has many features that link it to the most primitive arthropods.

The crustaceomorphs are an expansion of the more familiar crustaceans, including the living crabs, shrimps, ostracodes, and barnacles, all of which have mineralized skeletons plus the terrestrial woodlice and the extinct phyllocarids.

The atelocerates comprise the **Uniramia**, whose limbs have a single branch (including myriapods such as the centipedes and millipedes), and the insects (**Hexapoda**), with nearly a million living species, mostly beetles but including cockroaches, dragonflies, ants, wasps, and flies.

LOBOPODS
Cambrian – Extant

Closely related groups include the onychophoran lobopods, also known as the velvet worms, which are today represented by some 80 terrestrial species mostly confined to the leaf litter of Gondwanan tropical forests. Each segment of their elongated bodies carries a pair of short unjointed lobopod appendages, ending in a small claw and covered in a thin, chitinous cuticle. These are sometimes grouped together as "lobopods". A number of fossils, especially from the Cambrian Chengjiang and Burgess Shale, may belong to this group.

"LOBOPODS"

ANOMALOCARIDS
Cambrian – Devonian

The earliest known carnivores are the marine anomalocaridids of Cambrian age, found mostly in the Burgess Shale of British Columbia and at Chengjiang,

China. Some of them grew to over a metre (40in) long, and their morphology shows features in common with some "worm" groups, lobopods, and arthropods. Their most striking features are the large pair of appendages for grasping their prey, the ring of hard plates that surrounds the mouth, and the lateral flaps on either side of the body for swimming in pursuit of prey.

ANOMALOCARIDIDAE
Anomalocaris 48–8, 56–7

UNKNOWN ARTHROPOD GROUPS
Extinct

The animals below are thought to belong to unnamed and extinct arthropod groups:

Canadaspis (crustacean?) 48–9, 56–7
Opabinia 56–7
Yohoia (stem group euarthropod?) 56–7
Sidneyia (stem group euarthropod?) 56–7
Mimetaster 72–3

Andean condor (*Vultur gryphus*). The largest flying land bird of the western hemisphere, the condor is also the only living member of the genus *Vultur*.

AVIALAE (BIRDS)
Late Jurassic – Extant

WITHIN *Theropoda*
SEE ALSO *Palaeognathae, Neognathae*

One of the most successful vertebrate groups, the birds number over 9000 living species, considerably more than the mammals. Classified into some 153 extant families, they have a fossil record that extends back into Late Jurassic times, with a further 77 extinct families including *Archaeopteryx*, the earliest known bird. The early birds not only lived alongside the dinosaurs, but are now known to be descended from a group of small, bipedal, feathered theropod dinosaurs **(Theropoda).**

Today, birds range from minute hummingbirds to the huge flightless ostrich, but in the past included even bigger flightless forms such as the 3.7m (12ft) *Aepyornis* from Madagascar. Being warm-blooded and insulated with feathers, the birds have a great advantage over other reptiles, and have been able to adapt to habitats ranging from icy polar seas to arid barren islands.

The fossil record of birds is patchy because of their light skeletons, lack of teeth, and mode of life. Until the discovery of Early Cretaceous bird fossils in China and Spain there was a great gap in the fossil record of birds between the Late Jurassic and the Late Cretaceous. Now it is clear that there were several unique extinct bird groups such as the Enantiornithes, Hesperornithformes, and Ichthyorniformes.

ENANTIORNITHES
Cretaceous

The limb structure of this group of extinct birds is different from that of living birds. To begin with most of the extinct "opposite birds" were sparrow-sized and strong fliers, such as *Sinornis*, but by Late Cretaceous times there were larger forms such as *Enantiornis*, with a 1m (40in) wingspan, and some flightless fast runners.

HESPERORNITHFORMES
Cretaceous

With reduced wings and toothed jaws, these extinct birds were probably flightless, diving fish-eaters that evolved from flying ancestors. Their long-toed feet may have been webbed for swimming.

ICHTHYORNIFORMES
Cretaceous

These gull-sized, fish-eating birds had well-developed wings, relatively large heads, toothed jaws, a deep-keeled sternum to which the flight muscles were attached, and a short, bony tail (pygostyle) like modern birds. They may be part of the radiation of more derived birds, called the Ornithurinae, that also includes the extinct Hesperornithformes and the surviving Palaeognathae and Neognathae.

NEORNITHES (MODERN BIRDS)
Paleogene – Extant

These are divided into the Palaeognathae and Neognathae, and have all evolved in an extraordinary radiation within the last 65 million years.

● PALAEOAVES

ARCHAEOPTERYGIDAE
Archaeopteryx 138–9, 140–1

CONFUCIUSORNITHIDAE
Confuciusornis 150–1

OMNIVOROPTERYGIFORMES
Sapeornis 154–5

BENNETTITALES (CYCADEOID PLANTS)
Middle Triassic – Late Cretaceous

WITHIN *Gymnospermopsida*
COMPARE *Pteridospermales, Cycadophyta, Ginkgoales, Coniferopsida, Angiospermae*

This extinct Mesozoic group of cycad-like plants produced clusters of seeds in cone-like structures, which were sometimes protected by scales (bracts). Their fossil leaves can be very difficult to distinguish from those of cycads. Some bennettitaleans had male and female reproductive parts arranged in a very similar way to those found in the flowering plants (angiosperms).

Nilssoniopteris 156–7
Pterophyllum 156–7

BOREOEUTHERIA
Late Cretaceous – Extant

WITHIN *Eutheria*
COMPARE *Afrotheria, Xenarthra*
SEE ALSO *Laurasiatheria, Euarchontoglires (including Primates)*

Since the 1970s, molecular analysis of the placental (eutherian) mammals has clarified many of the relationships between taxa and produced new groupings. There is a fundamental division between the **Afrotheria** (aardvarks, tenrecs, golden moles, and elephant relatives) on the one side, and the **Xenarthra** (armadillos, sloths, and anteaters) and Boreoeutheria on the other.

The Boreoeutheria is a very large group and is itself divided into two main branches – the **Laurasiatheria** (insectivores, bats, carnivores, and ungulates) and **Euarchontoglires** (primates, rodents, rabbits etc).

A number of extinct groups that evolved during the first phase of placental radiation in early Paleogene times are also placed in the boreoeutherians, such as the leptictids, taeniodonts, and pantodonts. The leptictids are small shrew-like insectivores found in the Paleocene–Oligocene age strata of Asia and North America, as are the plant-eating pantodonts. North America was also occupied by the taeniodonts, a small group of pig-sized animals that lasted from the Paleocene to the Eocene.

● LEPTICTIDA

LEPTICTIDAE
Prodiacodon 180–1

APATEMYIDA
Apatemys 192–3

PSEUDORHYNCOCYONIDAE
Leptictidium 198–9

GYPSONICTOPIDAE
Stilpnodon 180–1

TAENIODONTA
Conoryctes 180–1

● PANTODONTA

PANTOLAMBDIDAE
Pantolambda 180–1

CORYPHODONTIDAE
Coryphodon 192–3

● CONDYLARTHRA

PAROXYCLAENIDAE
Kopidodon 198–9

PANTOLESTIDAE
Palaeosinopa 192–3
Buxolestes 198–9

BRACHIOPODA

Early Cambrian – Extant

This large group of marine shellfish contains more than 4000 extinct genera, but only a few hundred survive today. With two calcareous shells, brachiopods look somewhat similar to bivalve clams (see **Mollusca**), but belong to a separate phylum. Most live attached to a substrate by a fleshy stalk called a pedicle, and filter organic particles from the seawater by means of a feathery structure called a lophophore. Brachiopods diversified in Paleozoic times, but declined through the Mesozoic as the clams became more successful.

◀ **Patagonian mara** (*Dolichotis patagonum*). The maras are cavy-like rodents, part of the Euarchontoglire branch of the great boreoeutherian mammal group.

LINGULIDA
Longtancunella 48–9
lingulid sp. 60–1
Lingula 68–9

ACROTRETIDA
orbiculoid sp. 62–3

RHYNCHONELLIDA
Sowerbyella 58–9
Protochonetes 66–7
Atrypa 66–7
Strophochonetes 68–9

BRYOZOA (MOSS ANIMALS)

Ordovician – Extant

These colonial aquatic animals can be mistaken for small colonial corals or graptolites. They are filter-feeders, more derived than corals, and are placed in their own phylum – Bryozoa. Today there are some 6000 species, mostly marine but with some freshwater forms. The colonies are arranged in fan, stick, or disc shapes with perhaps several thousand pinhead-sized individuals building a skeleton composed of calcium carbonate or organic material. The colonies grow attached to a firm substrate such as algae or shells. The mineralized skeleton ensures a good fossil record that extends back to Ordovician times when they were often important reef builders, but a number of groups became extinct at the end of Triassic times.

CHELICERATA (CHELICERATE ARTHROPODS)

Cambrian – Extant

WITHIN *Arthropoda*
COMPARE *Trilobita, Crustacea, Uniramia*

This huge group of arthropods includes animals such as spiders, mites, ticks, and scorpions (together known as arachnids), the horseshoe crabs (xiphosurans), and important extinct groups such as the eurypterids. There are some 74,450 living species of chelicerates, the vast majority of which are arachnids, and they are distinguished from other arthropods by a two-fold division of the body and a pair of pincers ("chelicerae")

Fat-tailed scorpion (*Androctonus* sp.). These chelicerates, found in the Middle East and Africa, are among the most dangerous of all scorpions.

developed from the first segment – most obvious in the scorpions. The eurypterids and scorpions were among the first animals to adapt to life in freshwater and on land – they included top predators and some of the largest known arthropods in Silurian and Devonian times.

Today the main groups are all arachnids (scorpions with some 1200 living species, spiders with some 35,000 species, and ticks and mites with another 30,000 species).

● BASAL CHELICERATES

Haikoucaris 48–9

● MEROSTOMATA

XIPHOSURA (HORSESHOE CRABS)
Limulitella 114–15
Mesolimulus 140–1

EURYPTERIDA
Megalograptus 58–9
Onychopterella 62–3
Pterygotus 64–5

● ARACHNIDA

SCORPIONIDA
Pulmonoscorpius 84–5
Protoischnurus 160–1
scorpionid sp. 90–1, 114–15

UROPYGIDA
Geralinura (whip scorpion) 88–9

AMBLYPYGIDA (WHIP SPIDERS)
Graeophonus 92–3
amblypygid sp. 94–5
Britopygus 160–1

ARANEAE (SPIDERS)
Nephila 204–5
Cretaraneus 158–9
Rosamygale 114–15

335

● DROMOPODA

OPILIONES
opilionid 84–5

● TRIGONOTARBIDA

ANTHRACOMARTIDAE
Coryphomartus 92–3

TRIGONOTARBIDAE
Palaeocharinus 70–1
Palaeotarbus 68–9

● ACARI (MITES)

PACHYGNATHIDA
Protacarus 70–1

PYCNOGONIDA ("SEA SPIDERS")
Silurian – Extant

This marine group of some 1300 living species is characterized by relatively small bodies and four pairs of stout or spindly walking legs. They are sometimes grouped within the chelicerates or as a sister group.

● PYCONGONIDA

Palaeoisopus 72–3

CHONDRICHTHYES (CARTILAGINOUS FISH)
Late Silurian – Extant

WITHIN *Gnathostomes*
COMPARE *Osteichthyes*

These primitive marine fish have a skeleton and scales made of an organic material called cartilage that is

Hammerhead shark (*Sphyrna* sp.). These bizarre-looking chondrichthyans have eyes and nostrils located on the tips of their head projections.

---

generally not mineralized, and so (except for the teeth) their remains do not fossilize well. In Paleozoic times the chondrichthyans were just as successful as the newly evolving bony fish (**Osteichthyes**), but both groups suffered a setback in the end-Permian extinction and the chondrichthyans never regained their earlier diversity. However, they are still a relatively successful group today, with some 840 living species.

● ELASMOBRANCHII

CLADOSELACHIDAE
Cladoselache 80–1

STETHACANTHIDAE
Falcatus 86–7
Stethacanthus 86–7

PETALODONTIDA
Belantsea 86–7

CTENACANTHIDAE
Ctenacanthus 80–1

HYBODONTIDAE
Hybodus 128–9
Palaeospinax 130–1

ANACORACIDAE
Squalicorax 176–7

BATOIDEA:
Heliobatis 190–1
Trygon 194–5

● HOLOCEPHALI

CHIMAERIFORMES
Echinochimaera 86–7

CHONDRENCHELYIFORMES
Harpagofututor 86–7

ACANTHODIA (SPINY "SHARKS")
Ordovician – Permian

This group includes some of the oldest known **Gnathostomes**, found in Late Ordovician strata, and survived into the early Permian. Mostly small fish, less than 20cm (8in) long, they are characterized by their arrays of long spines (up to six pairs on the belly) and their spiny pectoral and pelvic fins. The internal skeleton is rarely preserved, but the body is typically covered with small, close-fitting scales and the jaws generally lack teeth, indicating that they probably fed on small food items, swallowed whole.

● ACANTHODIFORMES

Acanthodes 84–5
acanthodian sp. 72–3

CHORDATA (CHORDATES)
Early Cambrian – Extant

SEE ALSO *"Agnatha", Gnathostomes*

With over 52,200 living species, this is a major group, mostly made up by the great diversity of familiar vertebrate animals from fish to humans. There are also some basal forms that show the fundamental characters of the group such as a dorsal axial stiffening rod, known as the notochord, which gives the body an anterior–posterior elongation and sideways flexibility for swimming when coupled with a series of paired muscles. The notochord provides the basis for the backbone of the vertebrates Above this lies the dorsal nerve cord, from which arise the paired nerves that activate the muscles, and the sensory apparatus, which tends to be concentrated at the front of the animal where it encounters the environment and its potential food. The anterior mouth leads through an elongate gut to a posterior anus, beyond which extends the tail. Main groups include myxinoids (hagfish) and petromyzontiforms (lampreys), **Chondrichthyes**, **Sarcopterygii**, and **Actinopterygii**.

● BASAL CHORDATES

Myllokunmingia 48–9
Pikaia 56–7

CNIDARIA (HYDROIDS, CORALS, SEA ANEMONES, JELLYFISH, ETC)
Late Precambrian? – Extant

COMPARE *Ctenophora*

The cnidarians include some 9000 living species that have radially or biradially symmetrical bodies constructed from just two germ cell layers (diploblastic), with a single body cavity and opening surrounded by tentacles and specialized stinging cells. They may be solitary or colonial and many are supported by calcareous organic skeletons. The hydroids are classified as hydrozoans, the jellyfish as scyphozoans, and the corals, sea anemones, gorgonians, and seapens as anthozoans. The latter include those cnidarians best represented in the fossil record – the zoantharian corals.

Corals are one of the most important marine fossil groups. They look like their sea anemone relatives,

---

but secrete a calcareous supporting "cup" on which the coral polyp lives. Asexual budding of the polyps produces colonial corals that can grow to several metres in size. Today, there are some 9000 coral species, and most are inhabitants of warm, shallow, tropical seas although there are also some deepwater solitary forms. The growth of tropical corals generates some of the biggest biological structures on Earth, such as Australia's Great Barrier Reef. As reef formers, they have contributed significantly to the rock record since Ordovician times, although Paleozoic corals became

Sea anemone (Hexacorallia sp.). Anemones have a central body surrounded by tentacles armed with stinging cells called nematocysts.

extinct during the end-Permian extinction event. These formed two groups – the colonial tabulates and the rugose corals, which had both solitary and colonial forms. Modern (scleractinian) corals are biologically different, having evolved from a group of soft corals in Triassic times.

● SCYPHOZOA (JELLYFISH)

RHIZOSTOMATIDA
Essexella 88–9

LIMNOMEDUSIDAE
Progonionemus 114–15

● ANTHOZOA

ACTINIARIA (SEA ANEMONES)
Archisaccophyllia 48–9

PENNATULACEA (SEAPENS)
Thaumaptilon 56–7

ZOANTHARIA (CORALOMORPHS AND CORALS)
coralomorph sp. 46–7

● TABULATA

HALYSITIDAE
Halysites 66–7

FAVOSITIDAE
Favosites 66–7

HELIOLITIDAE
Heliolites 66–7

● RUGOSA

CHONOPHYLLIDAE
Ketophyllum 66–7

CONIFEROPSIDA (CONIFERS)
Carboniferous – Extant

WITHIN *Gymnospermopsida*
COMPARE *Pteridospermales*, *Bennetitales*, *Cycadophyta*, *Ginkgoales*, *Angiospermae*

The coniferophytes or pinophytes are more commonly known as conifers and are the one major group of modern plants that evolved in Paleozoic times. They did not grow in the Carboniferous "coal measure" swamps but were better adapted for drier conditions. Most conifers are woody trees and, although abundant today, they were even more diverse in their Mesozoic heyday. They reproduce by seeds borne on the surface of cone scales, and normally both male and

Spruce cone (*Picea* sp.). Some species of this evergreen conifer can reach ages of up to 9500 years old.

female cones develop on the same tree. Some cones are modified to form fleshy berry-like structures (such as in the junipers), so they can attract animals to aid their seed dispersal, but more typical female cones have seeds borne on woody scales, while the smaller, male, pollen-bearing cones are less conspicuous.

● BASAL CONIFERS

VOLTZIACEAE
Voltziopsis 112–13
Voltzia 114–15

● PINOPSIDA

ARAUCARIACEAE
Araucarioxylon 126–7
Araucaria 156–7
Lindleycladus 158–9
cheirolepidacean sp. 136–7
Pagiophyllum 138–9

TAXODIACEAE
Taxodium 180–1

CEPHALOTAXACEAE
Cephalotaxus (plum yew) 198–9

CUPRESSACEAE
Protojuniperoxylon 120–1

UNCLASSIFIED
Podozamites 118–19

CONODONTA
Cambrian – Triassic

WITHIN "*Agnatha*"
COMPARE *Gnathostomes*

Small, tooth-shaped "conodont" fossils (1–4mm long), commonly found in Paleozoic seabed limestones, were a

biological puzzle for over a century as there was no other information about the creature they belonged to. The discovery of symmetrical pairs of conodont tooth elements at the front of an eel-shaped body with paired eyes, segmented pairs of muscles, and traces of a notochord showed that they were primitive chordates and perhaps even primitive jawless vertebrates.

Normally only the phosphate mineralized (apatite) tooth bars and plates are preserved, but show that these active and abundant predators evolved rapidly in early Palaeozoic times. They typically grew up to 4cm (1.6in) long, but rare giant species reached 40cm (16in). Their rapid diversification makes them very useful for "biostratigraphic" subdivision of rock deposits, especially in limestone strata where graptolite fossils are scarce.

● AGNATHA
Promissum 62–3

CROCODYLIFORMES (CROCODILES)
Triassic – Extant

WITHIN *Archosauria*
COMPARE *Pterosauria, Dinosauria*

Although there are only eight surviving genera of crocodiles (including the alligators and gharial), they are distributed around the world, thriving in both fresh and marine waters, and therefore represent the

remnants of an ancient and much larger group, which evolved in Late Triassic times and, unlike many of their archosaur relatives (except birds) survived the end-Cretaceous mass extinction event.

Although modern crocodilians might look superficially like a number of extinct Paleozoic amphibious tetrapods, with long flattened skulls, long bodies and tails, and short legs, early crocodilians were small, fast-running, insectivorous, and possibly bipedal. Possible relatives of crocodyliformes include the extinct ornithosuchids, phytosaurs, aetosaurs, and rauisuchians, with some very large animals all linked to the later crocodilians by their common ankle structure.

● CROCODYLIFORMES

THALATTOSUCHIA
Steneosaurus 130–1

SPHENOSUCHIDAE
Hesperosuchus 126–7

● NEOSUCHIA

BASAL NEOSUCHIAN
Susisuchus 160–1

GONIOPHOLIDAE
Goniopholis 144–5

CROCODYLIA
Asiatosuchus 198–9
Borealosuchus 190–1
Machimosaurus 138–9

Caiman (*Caiman* sp.). These alligators of Central and South America are distinguished by features of their skulls and chest armour.

CRUSTACEA (CRUSTACEAN ARTHROPODS)
Cambrian – Extant

WITHIN *Arthropoda*
COMPARE *Trilobita, Chelicerata, Uniramia*

The crustacean group consists of some 50,000 living arthropod species which, although predominantly marine, have adapted to niches in environments that range from deep seas to tropical forests. They include familiar animals as diverse as barnacles (cirripedes), woodlice, and crabs.

Less familiar are the shrimp-like phyllocarids, which were among the earliest crustaceans, and the tiny bivalved ostracods (mostly 1–2mm in size) – but both have played an important part in the history of life. The mineralized exoskeleton of most crustaceans means that they are well represented in the fossil record.

● PHYLLOCARIDA

ARCHAEOSTRACA
Ceratiocaris 64–5
Nahecaris (rhinocaridid) 72–3

LIPOSTRACA (BRANCHIOPODS)
Lepidocaris 70–1

● EUMALACOSTRACA

BELOTELSONIDEA
Belotelson 88–9

● EUCARIDA

DECAPODA (SHRIMPS, CRABS, LOBSTERS)
Crangopsis 86–7

Antrimpos 114–5
Aeger 134–5, 142–3
Eryma 134–5
Cycleryon 142–3

● EUTHYCARCINOIDEA
Leverhulmia 70–1

● ? CRUSTACEAN

THYLACOCEPHALA
Ainiktozoon (thylacocephalan arthropod?) 64–5
Dollocaris 134–5

CTENOPHORA (COMB JELLIES)
Cambrian – Extant

COMPARE *Cnidaria*

Allied to the cnidarians, and resembling medusae (jellyfish), the hundred living marine species of ctenophore are predominantly carnivores, feeding on zooplankton that they capture using tentacles laden with stinging cilia. Some occur in vast numbers, but their soft bodies are very delicate and rarely preserved as fossils.

Ctenorhabdotus 56–7

CYCADOPHYTA (CYCADS)
Permian – Extant

WITHIN *Gymnospermopsida*
COMPARE *Pteridospermales, Bennetitales, Ginkgoales, Coniferopsida, Angiospermae*

The cycads were one of the more advanced gymnosperm plants that largely displaced the seed ferns, becoming abundant during Jurassic and Cretaceous times only to decline in turn as the flowering plants came to dominance from the Late Cretaceous onwards. Today, they are represented by just a few living species. Like the bennettitaleans, they reproduce by means of cone-like clusters of seeds, mostly carried on modified fronds. The sexes are separate, with pollen produced from male cones. This innovation helped them to colonize relatively dry and cold environments from Permian times onwards – some Mesozoic vine-like forms even became deciduous and managed to grow in polar regions when there were no ice caps. By contrast, modern cycads are

Male cycad cone (Cycadels sp.). The pollen-bearing cone developed by this male plant is known as a strobilus.

restricted to frost-free regions, and typically have squat trunks with whorls of leathery evergreen fronds.

● CYCADALES

Nilssonia 144–5
cycadalean sp. 144–5
Taeniopteris 162–3

CYNODONTIA (CYNODONT "MAMMAL-LIKE REPTILES" AND MAMMALS)

Late Permian – Extant

WITHIN *Therapsida*
SEE ALSO *Mammalia (including Monotremata, Marsupialia, Eutheria)*

Cynodonts gave rise to and include the mammals. Non-mammalian cynodonts, however, thrived only from the Late Permian to the Early Jurassic, and were weasel- to dog-sized therapsid reptiles with mammal-like features in their jaw and palate. During the Triassic, successive cynodont groups evolved and became extinct, and the fossil record shows their development towards a more mammal-like form. This is especially seen in the posture, with the legs being brought in under the body, accompanied by significant changes in the limb joints. Meanwhile, the direction of flexure of the spine shifted from sideways to up and down. By

Late Triassic times, cynodont groups such as the tritylodonts had distinctly mammal-like form but still lacked the jaw articulation of the true mammals, which arose from cynodonts in the Late Triassic.

● CYNODONTIA

INDETERMINATE CYNODONT
Madygenia 118–19

PROCYNOSUCHIDAE
Procynosuchus 104–5

GALESAURIDAE
Cyanosaurus 104–5

THRINAXODONTIDAE
Thrinaxodon 112–13

DVINIIDAE
Dvinia 106–7

TRAVERSODONTIDAE
Exaeretodon 120–1

DIAPSIDA (DIAPSID AMNIOTES)

Late Carboniferous – Extant

WITHIN *Amniota*
COMPARE *Synapsida*
SEE ALSO *Lepidosauria, Archosauria, Ichthyosauria, Plesiosauria, Placodontia*

Diapsids are one of three main amniote groups, the others being the synapsids and anapsids. Together with the anapsids they also form a larger grouping known as the sauropsids. The most primitive and earliest diapsids are extinct forms such as *Youngina* and *Araeoscelis*, and the Late Permian diapsids form two groups that rose to considerable prominence with numerous genera still alive today – namely the **Archosauria** (including **Dinosauria**, **Crocodyliformes**, and **Avialiae**) and the **Lepidosauria** (lizards and snakes).

● LEPIDOSAUROMORPHA

THALLATOSAURIA
Askeptosaurus 116–17

YOUNGINIDAE
Youngina 104–5

ARAEOSCELIDAE
Araeoscelis 98–9

NOTHOSAURIA
Ceresiosaurus 116–17

UNKNOWN DIAPSID
Longisquama 118–19

DINOSAURIA

Triassic – Extant

WITHIN *Archosauria*
COMPARE *Crocodyliformes, Pterosauria*
SEE ALSO *Ornithischia, Saurischia (including Sauropodomorpha, Theropoda, and Avialae)*

This highly successful group of reptiles evolved in Middle Triassic times around 230 million years ago. Although there are not many genera of dinosaurs known (some 600), they became extraordinarily diverse in form and scale, ranging from fast-moving sparrow-sized runners to lumbering beasts some 30m (100ft) long, weighing up to 50 tonnes. The dinosaurs spread around the world and came to dominate terrestrial environments for more than 160 million years before largely becoming extinct at the end of Cretaceous times, 65 million years ago. However, before becoming extinct they gave rise to a highly successful group of small, feathered dinosaurs that are still alive today – the birds.

Two major groups of dinosaurs arose right at the beginning of dinosaur evolution – the **Ornithischia** and the **Saurischia**, distinguished by differences in the structure of their hip and pelvis. The dinosaurs as a whole belong to the larger reptile grouping called the **Archosauria** that includes, among others, the extinct flying **Pterosauria** and the surviving **Crocodyliformes**. The ancestors of the dinosaurs were Triassic archosaurs, and they probably emerged from

bipedal forms known as dinosauromorphs, such as *Marasuchus* from Argentina.

● DINOSAUROMORPH ARCHOSAURS

Dromomeron 124–5
Eucoelophysis 124–5

ECHINODERMATA

Early Cambrian – Extant

As their name implies, this is a varied group of "spiny skinned" animals. They are all marine, and include the familiar starfish and brittlestars (asterozoans and ophiuroideans), sea urchins (echinozoans), and less familiar groups such as the sea lilies (crinozoans) and sea cucumbers (also echinozoans). Altogether, there are some 6000 living species and many more fossil ones including entirely extinct groups such as the blastozoans (Early Cambrian–Permian) and homalozoans (Cambrian–Devonian). Today's most familiar groups all originated in the Ordovician.

Biologically, the echinoderms are interesting since their curious five-fold (pentaradial) body symmetry means they have no obvious head or tail and some (the sea lilies) even look like plants. Yet the echinoderms are

Furcaster. This beautiful fossil brittlestar from Germany's Hunsruck Slate highlights the typical five-fold radial symmetry seen in echinoderms.

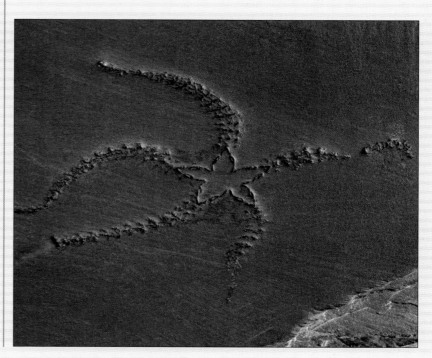

advanced invertebrates whose larvae have features that relate them to the chordates. They also use a unique hydraulic "water vascular" system of extensible tube feet to move around. The tube feet are also used for feeding, respiration, and chemoreception. The sophisticated echinoderm skeleton is made of porous calcareous plates or spines, and generally falls apart shortly after the death of the animal. The numerous individual plates enhance the fossilization potential of the group, and they are common components of "shelly" limestones.

● ASTEROZOA (STARFISH)

PALASTERISCIDAE
Salteraster 58–9
Palaeosolaster 72–3

● OPHIUROIDEA (BRITTLESTARS)

PROTASTERIDAE
Furcaster 72–3

OPHIACANTHIDAE
Ophiopinna 134–5

● ECHINOIDEA (SEA URCHINS)

CIDAROIDA
Temnocidaris 176–7

● CRINOIDEA (SEA LILIES)

CLADIDA
Gissocrinus 66–7
Imitatocrinus 72–3

ISOCRINIDA
Balanocrinus 58–9

SAGENOCRINIDA
Pentacrinites 128–9
Pentacrinus 130–1

UINTACRINIDA
Marsupites 176–7

"EDIACARA"
Late Neoproterozoic

There are now several hundred different taxa of soft-bodied fossil organisms that have been found within the late Precambrian (late Neoproterozoic), Ediacaran-age marine strata from around the world. Together they are known as the "ediacarans" or vendozoans. Most are a few centimetres in length, and variously frond, disc, sac, or ribbon-shaped. They have no fossilized hard parts, and are only preserved as casts and moulds in seabed sediments. When they were first discovered, attempts were made to shoehorn them into known living groups such as the cnidarian jellyfish, seapens, worms, molluscs, and even arthropods. Certainly, they show different degrees of body symmetry from bilateral or pseudobilateral to radial, which may elevate them to eumetazoan status, but there is little evidence of their tissues being organized into organs or organ systems.

With more detailed analysis, it became clear that very few of the Ediacaran fossils satisfied the diagnostic criteria for these various groups. Despite their size and the evident presence of relatively tough tissues that retained the body form through the processes of burial and fossilization, even their animal status has been questioned. An alternative argument is that they represent some failed evolutionary experiment and as such belong to a completely extinct group of organisms.

However, within their diversity of form, there seem to be a number of distinct groups that have been given some taxonomic recognition.

◀ **Ediacaran "spindles"**. These frond-like forms from Newfoundland's Mistaken Point are just one of many enigmatic species found in Precambrian strata around the world.

● ERNIETTAMORPHA

DICKINSONIDAE
Dickinsonia 42–3, 44–5
Phyllozoon (dickinsoniamorph?) 44–5

ERNIETTIDAE
Ernietta 44–5

PTERIDINIIDAE
Pteridinium 44–5

● RANGEOMORPHA

CHARNIIDAE
Charnia wardi 42–3
Charniodiscus 42–3, 44–5

RANGEIDAE
Bradgatia 42–3

● CYCLOZOA

CYCLOMEDUSIDAE
Cyclomedusa 42–3

UNKNOWN GROUPS
Spriggina 42–3
Thectardis 42–3
"spindles" 42–3
Tribrachidium 42–3, 44–5
Namalia 44–5

EUARCHONTOGLIRES (EUARCHONTOGLIRE MAMMALS)
Mid-Cretaceous – Extant

WITHIN *Boreoeutheria*
COMPARE *Laurasiatheria*
SEE ALSO *Primates (including Anthropoidea)*

As the name suggests, two major groups of placental mammals, the archontans and the glires, are cladistically combined into this single superorder. They include some of the most successful living animals – humans, rats and rabbits. The archontans include the **primates**, tree shrews (scandentians), and flying lemurs (dermopterans), while the glires include the rodents (rodentians) and rabbits etc. (lagomorphs). There are more than 2000 species of glires, some 40 per cent of all mammals, and their success has been largely due to rapid rates of reproduction and their continuously growing and self-sharpening incisor teeth. Despite small body size, their large and widely distributed populations have resulted in a good fossil record, especially of the teeth – a record that is much better than that of the primates.

Pika (*Ochotona* sp.). These small-eared mammals, with a somewhat hamster-like appearance, are in fact lagomorphs – Euarchontoglires related to the hares and rabbits.

● RODENTIA

ANOMALURIDAE
Paranomalurus (scaly-tailed squirrel) 210–11

SCIURIDAE
Marmota (marmot) 222–3

ISCHYROMYIDAE
Paramys 192–3

MYODONTA
Cricetops (muroid) 206–7
Eutamius 212–13
Neotoma 234–5
Papagomys armandvillei (giant rat) 238–9

● LAGOMORPHA

OCHOTONIDAE
Ochonta (pika) 206–7

LEPORIDAE
Oryctolagus 228–9
Lepus (hare) 226–7

EUTHERIA (PLACENTAL MAMMALS)

Early Cretaceous – Extant

WITHIN *Mammalia*
COMPARE *Monotremata, Marsupialia*
SEE ALSO *Xenarthra, Afrotheria,*
Boreoeutheria (including Laurasiatheria and
Euarchontoglires)

The living placental mammals (eutherians) range from insectivores such as shrews to the proboscidean elephants and cetacean whales. They have diversified into more than 5000 species that are placed in some 20 major living groups, and another 6 or more extinct groups. Despite a reasonable fossil record, our understanding of the relationships between these groups and their early evolution has been highly problematic. Recent molecular comparisons of

Eomaia. This small fossil mammal, from the Early Cretaceous of Liaoning Province in China, is the earliest known eutherian, and a possible ancestor of modern placental mammals.

living groups have clarified a number of the issues, but cannot help to assimilate the fossil groups.

Major eutherian branches so far identified include the **Afrotheria, Xenarthra, Laurasiatheria** and **Euarchontoglires**. Molecular clock estimates put the divergence of the eutherians from the marsupials between 185 and 130 million years ago, but the earliest fossil eutherian presently known is the 125-million-year-old *Eomaia* from the early Cretaceous of China.

● BASAL EUTHERIANS

Eomaia (marsupial-like placental) 154–5

OTLESTIDAE
Prokennalestes 156–7

● EPITHERIA

ANAGALIDA
Zalambdalestes 170–1

● CIMOLESTA

CIMOLESTIDAE
Cimolestes 178–9

FILICOPSIDA (FERNS)

Late Devonian – Extant

With their frond-shaped leaves and clusters of sporangia on the lower surfaces, the ferns are much more familiar today than other primitive plant groups such as the clubmosses (**Lycopsida**) and horsetails (**Sphenopsida**). Of all the primitive plants they have survived the best,

with around 12,000 living species, and they are by far the biggest living group of seedless vascular plants.

Flat, intricately divided fronds act as very efficient "solar panels" that can gather light even in the gloom of the forest floor, while a well-developed root system allows some ferns to survive harsh conditions by becoming deciduous, and thus grow even in seasonal climates. Some ferns have become vine-like climbers, and others have adapted to freshwater with floating fronds. One group of Late Devonian ferns evolved into so-called "tree-ferns", intertwining their roots and stems in a way that allowed them to grow into tree-sized plants such as the 18m (60ft) *Psaronius*, with a crown of divided fronds, each of which was up to 3m (10ft) long. They look superficially like modern tree-ferns but are not related.

OSMUNDALES
osmundacean sp. 144–5

FILICALES
gleicheniacean sp. (coral fern) 136–7

DICKINSONIACEAE
Coniopteris 146–7

SCHIZAEACEAE
Klukia 138–9
Ruffordia 158–9, 160–1

MATONIACEAE
Phlebopteris 138–9

INDETERMINATE FERNS
Anomopteris 114–15
sphenopterid spp. 114–15, 162–3

FORAMINIFERA

Cambrian – Extant

These tiny but remarkable aquatic organisms are single cells that secrete spiral-coiled shells and are related to the amoeba. Most foraminifers (often referred to as simply "forams") have millimetre-sized shells, but some extinct forms, such as the *Nummulites* that form the Eocene limestones used in building the Egyptian pyramids, grew into giant coin-shaped discs up to 6cm (2.4in) in diameter. Foraminiferan shells are either made from calcite, or "glued together" from sand grains and shell fragments. Most are marine, and they are often incredibly abundant in nutrient–rich ocean waters. Analysis of the balance of isotopes within shells recovered from the deep ocean has been of great importance as a proxy measure of climate change, especially over the last 5 million years or so of the recent Quaternary ice ages.

Fern (Filicopsida sp.) Although most modern ferns grow in the understorey of well-shaded woodlands, there are some species that thrive in more exposed landscapes, including alpine and even desert regions.

GINKGOALES (GINKGOS)

Late Triassic – Extant

WITHIN *Gymnospermopsida*
COMPARE *Pteridospermales, Bennetitales, Cycadophyta, Coniferopsida, Angiospermae*

Originally a highly diverse group of Mesozoic plants, the ginkgos are only survived by the Chinese maidenhair tree, *Ginkgo biloba*, which is often referred to as a "living fossil".

GINKGOACEAE
Baiera 138–9, 156–7
Ginkgoites 140–1
Gingkoites australis 162–3

GNATHOSTOMES (JAWED VERTEBRATES)

Early Silurian – Extant

WITHIN *Chordata*
COMPARE *"Agnatha"*
SEE ALSO *Chondrichthyes, Osteichthyes (including Sarcopterygii and Actinopterygii)*

The evolution of jaws was one of the most important events in the evolution of the vertebrates, and promoted a never-ending "arms race" between predators and prey. All vertebrates, from cartilaginous and bony fish (**Chondrichthyes** and **Osteichthyes**) such as sharks and salmon, to mammals such as humans, evolved from early gnathostomes. Their front bony gill supports became modified in stages to form the upper and lower jaws, reducing the number of gill slits. Early steps in the jaw-forming process are recorded by extinct groups such as the placoderms (Silurian–Devonian) and acanthodians (Silurian–Permian). The former were heavily armoured fish that occasionally grew to gigantic size, as seen in the 10m (33ft) predator *Dunkleosteus*. In contrast, the acanthodians were shark-like, and characterized by spines that supported the fins, acting as cutwaters and for protection.

The main groups include acanthodians, placoderms, **Chondrichthyes**, and **Osteichthyes** (which includes the major groups of **Actinopterygii** and **Sarcopterygii**).

● PLACODERMI

RHENANIDA
Gemuendina 72–3

Ginkgo (*Ginkgo biloba*). The extant ginkgo tree is unusual in many ways, with unique fan-shaped leaves, and a lifespan of centuries or more thanks to its resistance to weather damage, insects, and disease.

ANTIARCHI
Asterolepis 78–9
Bothriolepis 74–5, 76–7

PTYCHODONTIDA
Campbellodus 74–5

ARTHRODIRA
Eastmanosteus 74–5
Dunkleosteus 80–1
Groenlandaspis 82–3

GRAMINAE (GRASSES)

Eocene – Extant

WITHIN *Angiospermae*

Today there are some 10,000 species of grasses – they were one of the most important innovations in the evolution of the flowering plants in the Eocene or perhaps earlier. Their diversity has produced plants resistant to drought, frost, and waterlogging, and has made them important as food to many plant-eating mammals such as cattle and horses. Humans have also hybridized many domesticated varieties as staple food crops such as rice.

GRAPTOLITHINA (GRAPTOLITES)

Middle Cambrian – Late Carboniferous

As their name implies, the extinct graptolite fossils look a bit like "writing on the rock", fret saw blades, or primitive plants. They are in fact advanced marine invertebrates (hemichordates, closely related to the surviving pterobranchs) that lived in linear colonies (up to 2m/6.6ft long) of up to a hundred or more small asexually budded zooids. Each of these grew to only around 1mm in size, but they connected with each other by secreting an organic tubular skeleton up to 2m (6.6ft) long. Graptolites were filter feeders, with some living rooted to the seabed while others were free-floating. Their abundance, wide dispersal, and rapid evolution has made them very useful in identifying divisions of the early Paleozoic strata.

GRAPTOLOIDEA
Monograptus 66–7
Cyrtograptus 66–7

GYMNOSPERMOPSIDA (SEED-BEARING PLANTS)

Late Devonian (?) – Extant

COMPARE *Rhyniophyta, Sphenopsida, Lycopsida, Filicopsida*
SEE ALSO *Pteridospermales, Bennetitale, Cycadophyta, Ginkgoales, Coniferopsida, Angiospermae*

The seed-bearing or gymnosperm plants are a large group that includes the seed ferns (**Pteridospermales**), cycads (**Cycadophyta**), bennettitaleans and glossopterids, ginkgos (**Ginkgoales**), conifers (**Coniferopsida**) etc. They became increasingly common following the extinction of the large tree-like horsetails (**Sphenopsida**), clubmosses (**Lycopsida**), and ferns (**Filicopsida**) of Carboniferous times.

The name gymnosperm means

Glossopteris. The glossopterid trees, named from their distinctive tongue-shaped leaves, were a form of gymnosperm that spread across much of the southern hemspheere during the Permian Period.

"naked seed" because the seed does not have a completely enclosing outer protective covering (ovary). This represents an early stage in the evolution of the more advanced condition in which unfertilized seeds (known as ovules) are enclosed, as seen in the ovaries of angiosperms (flowering plants). In gymnosperms, the ovule has an egg and a large food store that is partly covered with a protective coating, but has an opening at one end for pollen to enter and fertilize it. Following fertilization, the developing embryo depends upon food stored within the seed and is protected by its coating.

● PROGYMNOSPERMOPSIDA

Archaeopteris 76–7

● BENNETTITALES (CYCADEOIDS)

Welwitschiostrobus 158–9

GLOSSOPTERIDALES

Glossopteris sp. 102–3
Glossopteris linearis 102–3
Plumsteadia 102–3
Australglossa 102–3
Dictyopterium 102–3

CZEKANOWSKIALES

Czekanowskia 144–5, 154–5

● GNETALES

EPHEDRACEAE

ephedroid sp. 160–1

● CORDAITALES

Cordaites 90–1

HEXAPODA (INSECTS)

Early Devonian – Extant

WITHIN *Uniramia*

By whatever measure – be it abundance, diversity, biomass, or almost anything else one can think of – the insects are some of the most remarkable animals ever to have evolved. This uniramian arthropod group (see **Uniramia**) includes cockroaches, dragonflies, ants, beetles, wasps, and flies – all creatures with six legs, as the name hexapod suggests. There are nearly a million species of living insect, including some 300,000 beetles alone, and altogether there have probably been more than 5 million species past and present. The insect fossil record stretches back to the Early Devonian, with primitive forms such as the springtail *Rhyniella* from the Rhynie Chert. However, the lack of a mineralized exoskeleton means that the fossilization potential for insects is not as good as for some other groups of **Arthropoda** such as the trilobites (see **Trilobita**). Nevertheless, certain quiet-water sediments, such as lake and lagoon muds, can preserve insect tissues very well (as can amber), and some 40,000 fossil species have been described so far.

The insect body has a three-fold division into head, thorax, and abdomen. The head has six segments, compound eyes, sensory antennae, and chewing mandibles, while the thorax typically has two pairs of wings and three segments, each carrying a pair of legs, and the abdomen has up to 11 segments. However, adaptation to many different environments, including most terrestrial ones and many aquatic ones, has brought many departures from this basic body plan.

Insects have played a vital role in the evolution of plants, and especially **Angiospermae** (flowering plants). With their abundance and diversity, many are adapted for specific niches and climatic conditions, and hence their remains are very useful in the reconstruction of ancient environments such as those of the recent ice ages.

● ENTOGNATHA

COLLEMBOLA

Rhyniella (isotomid springtail) 70–1

● ECTOGNATHAN PTERYGOTES

PALAEODICTYOPTERIDA

palaeodictyopteridan sp. 94–5
Dunbaria 98–9

EPHEMEROPTERA (MAYFLIES)

megasecopteran sp. 92–3
Voltziaphemera 114–15
Ephemeropsis 152–3
Ephemera 204–5

PRODONATA

Meganeuropsis 98–9

ODONATA (DRAGONFLIES)

Namurotypus 94–5
Arctotypus 106–7
Tarsophlebia 142–3
Araripeliupanshania 160–1
Miopodagrion (damsel fly) 204–5

POLYNEUROPTERAN "PROTORTHOPTERANS"

Eucenus 88–9
Gerarus 88–9

Golden ground beetle (*Carabus auratus*). These flightless European beetles show a typical hexapod body plan with three pairs of limbs on the thorax, a single set of antennae, and two pairs of mouthparts.

● POLYNEUROPTERAN ANARTIOPTERANS

DERMAPTERA

Labiduromma (earwig) 204–5

ORTHOPTERIDA

Gigatitan (titanopteran) 118–19

● DICTYOPTERAN "BLATTODEANS"

BLATTODEA (COCKROACHES)

blattodean spp. 88–9, 94–5, 114–15
Syscioblatta 100–1

ISOPTERA

"*Macrotermes*" (termite) 214–15

MANTODEA (MANTISES AND STICK INSECTS)

Santanmantis 158–9

● PARANEOPTERA

CICADOMORPHA

Quadraticossus (cicadoid) 132–3
cicadomorph sp. 152–3
Baeocossus (palaeontinid) 158–9
Tettagalma (palaeontinid) 158–9
cicada sp. 196–7

● HOLOMETABOLA (ENDOPTERYGOTA)

COLEOPTERA (BEETLES)

Notocupoides 118–19
Tetraphalerus 154–5
Hadeocoleus 118–19
buprestid sp. (jewel beetle) 198–9
"*Bledius*" (rove beetle) 204–5
Aphodius (scarab) 204–5

● NEUROPTERIDA

NEUROPTERA (LACEWINGS)

Leptolingia 132–3

Grammolingia 132–3
Marquettia 204–5
Baisopardus 160–1
myrmeliontid sp. 160–1
belostomatid sp. 158–9

● HYMENOPTERIDA

HYMENOPTERA (WASPS, BEES, AND ANTS)

Formicium 196–7
Ischnidium 150–1
Cretofedtschenkia (wasp) 158–9
Heriades (leaf-cutter bee) 204–5
Microstylum (robber fly) 204–5
Palaeovespa (hornet) 204–5

● ANTLIOPHORA

MECOPTERA (SCORPION FLIES)

Holcorpa 204–5
Mongolbittacus (hanging fly) 132–3

DIPTERA (TRUE FLIES)

Axioxyela 118–19
Syrphus (syrphid hoverfly) 204–5

● AMPHIESMENOPTERA

LEPIDOPTERA (MOTHS AND BUTTERFLIES)

Oligodonta (pierid) 204–5
Prodryas (nymphalid) 204–5
Vanessa (vanessid) 204–5

HOMINIDAE

Miocene – Extant

WITHIN *Hominoidea*

The hominids are a group of hominoid, catarrhine primates that can be divided into the southeast Asian orang-utans and their fossil relatives

343

(together known as the Ponginae), and the African apes – chimps and gorillas plus humans and their fossil relatives (together known as the Homininae). Their divergence resulted from the adoption of different modes of locomotion – the orang-utan swings by its arms ("brachiation") and climbs slowly, while the African great apes mostly use all four limbs to climb in a typical quadripedal manner, and humans are bipedal.

The earliest known hominids are fossil forms such as the Miocene age *Kenyapithecus* from eastern Africa and central Europe. There is a reasonable record of fossil Ponginae in Asia from Miocene times onwards, but there is a big gap in the record of African Homininae until around 6 million years ago. At present, the earliest fossil hominin is *Sahelanthropus*, a small ape-like creature from the late Miocene that lies close to the divergence of ape and human branches from their common ancestor.

Some 20 human-related species have diverged over the last 6 million years in Africa, but few of them have spread beyond this continent. One of these African species, *Homo sapiens*, arose some 200,000 years ago to become the most widespread and successful.

● PONGINAE

PONGIN SIVAPITHECINES
Sivapithecus 212–13

● HOMININAE

UNDETERMINED GROUP
Sahelanthropus 214–15

"AUSTRALOPITHECINES"
Australopithecus afarensis 216–17
Australopithecus africanus 218–19

"PARANTHROPINES"
Paranthropus boisei 220–1

HOMININI
Homo habilis 220–1
Homo antecessor 222–3
Homo erectus/ergaster 224–5
Homo sapiens 226–7, 230–1, 236–7, 240–1, 242–3, 244–5
Homo neanderthalensis 228–9
Homo floresiensis 238–9

HOMINOIDEA (APES)
Miocene – Extant

WITHIN *Anthropoidea*
SEE ALSO *Hominidae*

The apes, also known as hominoids, are a group that unites the gibbons (Family Hylobatidae) with the Hominidae (chimps, gorillas, orang-utans, humans, and their ancestors). Although there are few surviving ape species, their distribution in Africa and southeast Asia reflects a much wider diversity and distribution in the past. Unfortunately, the early ape fossil record is sparse, obscuring many details of their origins within the haplorhine primates around 25 million years ago. The best known early fossil representative of the group is the Miocene *Proconsul* from Africa. By late Miocene times, the apes were abundant and widely dispersed across Africa, Asia, and Europe. They were mostly fruit-eating, chimp-sized, and tailless, with short snouts, forward-facing eyes, and highly domed skulls.

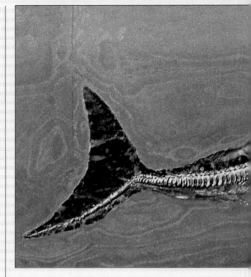

Stenopterygius. This relatively small, slender-skulled ichthyosaur from Holzmaden in Germany had a form strongly resembling a modern dolphin, and probably had a similar fish-eating lifestyle.

ICHTHYOSAURIA
Early Triassic – Late Cretaceous

WITHIN *Diapsida*
COMPARE *Lepidosauria, Archosauria, Plesiosauria, Placodontia*

The extinct ichthyosaurs or "fish lizards" were a very successful group of predatory marine diapsid reptiles throughout much of the Mesozoic. Their dolphin-shaped bodies were streamlined for the pursuit of fast-swimming prey, their tetrapod limbs were adapted as seal-like paddles for steering, and the main propulsion came from a muscular tail. Ichthyosaurs grew up to 15m (50ft) long, and had projecting beak-like jaws armed with sharp teeth. Their large eyes suggest that they depended upon sight for hunting.

● ICHTHYOSAURIA

MIXOSAURIDAE
Mixosaurus 116–17

SHASTASAURIDAE
Shastasaurus 116–17

ICHTHYOSAURIDAE
Ichthyosaurus 128–9

STENOPTERYGIIDAE
Stenopterygius 130–1
Platypterigius 164–5

OPTHALMOSAURIDAE
Aegirosaurus 140–1

◄ **Orang-utan** (*Pongo* sp.). Although these hominid apes are generally solitary, offspring stay with their mothers until the age of six or seven.

LAURASIATHERIA (LAURASIATHERIAN MAMMALS)

Early Cretaceous – Extant

WITHIN *Boreoeutheria*
COMPARE *Euarchontoglires*

Extending back to Early Cretaceous times, the first laurasiatherian mammals included the ancestors of living insectivores such as shrews, hedgehogs, moles (lipotyphlans), and bats (chiropterans). Among more advanced members is a large branch known as the ferungulates, whch includes the familiar cetartiodactyls (cattle, pigs, and whales), perissodactyls (horses, rhinoceroses, and tapirs), carnivorans (dogs, cats, weasels, seals, and bears), and the bizarre and less familiar pholidotans (pangolins).

The cetartiodactyls are further subdivided into the even-toed ungulates (artiodactyls) and, according to molecular analysis, the cetaceans. The former include the suiforms (pigs and hippos) and the selenodontids (cattle, deer, giraffes, antelopes, and camels) – their link with the cetatceans, which include whales, dolphins, and porpoises, might seem surprising but is now supported by fossil evidence. The artiodactyls seem to have originated in Eocene times from rabbit-sized plant-eaters such as *Diacodexis*. There is also fossil evidence that the cetaceans arose from a land-living artiodactyls such as *Pakicetus* from Pakistan.

Eocene times also saw the diversification of the perissodactyls, which have odd numbers of toes, to become the dominant browsing plant-eaters (and subsequently grazing plant-eaters once the grasses had evolved in Oligocene times). Two important perissodactyl groups that became extinct but left good fossil records are the chalicotheres and brontotheres. The latter were rhino-like browsers such as the North American *Brontops*, which stood more than 2m (6.6ft) high at the shoulder. The similar sized chalicotheres were strange animals with horse-like heads, long grasping forelimbs, and short back legs all ending in three-hooved digits.

Among the remaining laurasiatherians, the carnivorans are characterized by a pair of enlarged carnassial cheek teeth that work together as effective shears, cutting the flesh off their prey for rapid ingestion. Additionally, the canine "eye" or "dog" teeth have developed into dagger-like fangs, capable of puncturing tough hide to help hold and kill their prey. The pinnipeds (seals, sealions, and walruses) are an aquatic group of carnivorans that probably evolved from a bear-like ancestor in the Oligocene, perhaps similar to *Enaliarctos* from North America.

Finally, molecular analysis has revealed close links between the carnivorans and the strange scaly-skinned, ant-eating, and toothless pangolins. Although living pangolins are restricted to Africa and southeast Asia, they were once more widespread throughout North America and Europe. *Eomanis* is a wonderfully preserved pangolin found in the Eocene shales of Messel in Germany.

▶ **Flying fox** (*Pteropus* sp.). Large "fruit bats" and their smaller, insect-eating relatives show how the Laurasiatheria have evolved enormous variety.

Walrus (*Odobenus rosmarus*). It is hard to imagine two creatures much more different than a bat (previous page) and a walrus, but both have the same common ancestry in the Laurasiatheria.

LEPIDOSAURIA (LIZARDS, SNAKES ETC)

Late Triassic – Extant

WITHIN *Diapsida*
COMPARE *Archosauria, Ichthyosauria, Plesiosauria, Placodontia*

This large group of reptiles includes some 4470 living lizard species and 2920 living species of snake, plus a number of extinct groups and *Sphenodon*, the lizard-like tuatara from New Zealand whose ancestors date back to Late Triassic times. A number of lepidosaurs are also grouped as the squamates – these include various "lizard" groups such as the iguanas, geckos, skinks, ambisphaenids, anguimorphs, and snakes (Serpentes), all of which have flexible skulls. Their evolutionary inter-relationships are not yet clear, and the anguimorphs are the most diverse – in Late Cretaceous times some of them, such as the giant predatory mosasaurs, adapted to life in the sea.

● **SQUAMATA**

Green snake (*Opheodrys* sp.). Snakes represent a group of lepidosaurs that evolved to graduallly lose their limbs, most likely as an adaptation to a burrowing lifestyle. Some fossil snakes from the Cenomanian Stage of the Cretaceous, around 100 million years ago, show clear vestigial hindlimbs.

LEPOSPONDYLI

Carboniferous – Early Permian

WITHIN *Tetrapoda*
COMPARE *Temnospondyli, Amniota*

This group of extinct tetrapods includes the microsaurs, nectrideans, and aistopods. Most were land-living, but some became secondarily aquatic (eg the nectrideans), and some had either reduced limbs (some microsaurs) or lost their limbs altogether and became snake-like (eg aistopods).

● **MICROSAURIA**

● **NECTRIDEA**

● **AISTOPODA**

LISSAMPHIBIA (MODERN AMPHIBIANS)

Triassic – Extant

WITHIN *Temnospondyli*

There are some 4000 species of living amphibians (known as lissamphibians), with considerable diversity ranging from the familiar frogs and toads (anurans), through salamanders and newts (urodeles), to the strange snake-like caecilians (gymnophiones), which have become secondarily limbless. Despite the need to return to water for reproduction, they have occupied a remarkable diversity of aquatic and terrestrial habitats, from caves to marshes and mountainside forests. Most are fairly small, with delicate bony skeletons that do not generally fossilize well except in rare favourable environments such as lakebed muds. The earliest known members of the group are Triassic frogs, but these already had skeletons very similar to modern frogs, and the origins of the group must be more ancient than this.

Stag's-horn clubmoss (*Lycopodium clavatum*). This widespread lycoposid species grows mainly along the ground, with branches turning upright to produce spore-bearing cones.

● **ANURA (FROGS AND TOADS)**

● **URODELA (SALAMANDERS AND NEWTS)**

● **ALBANERPETONIDAE**

LYCOPSIDA (CLUBMOSSES)

Late Silurian – Extant

COMPARE *Rhyniophyta, Sphenopsida, Filicopsida, Gymnospermopsida*

The clubmosses or lycopsids are primitive plants that mostly consist of a simple upright stem growing from a horizontal, root-like, underground structure. Strap-shaped leaf-like "microphylls" grow in a helical spiral from the stem, and when detached leave a characteristic scar. Like most primitive plants they are dependent upon water for reproduction, with spores produced in sporangia on the upper surface of the microphylls where they attach to the stem.

The first clubmosses grew only 50cm (20in) or so tall, but the group diversified through Devonian times, and by the Carboniferous they dominated the world's low-lying tropical forests, producing giant tree-sized plants such as *Lepidodendron* with species growing to 30m (100ft) high. Decimated at the end-Permian extinction, the group survived but never really recovered – today there are only a few small genera such as the quillwort *Isoetes* and *Selaginella*.

DREPANOPHYCALES
Drepanophycus 82–3

LYCOPODIALES
Lycopodium 138–9

LEPIDODENDRALES
Stigmaria 94–5

SIGILLARIACEANS
Sigillaria 90–1, 92–3

● ZOSTEROPHYLLOPSIDA

ZOSTEROPHYLLALES
Asteroxylon 70–1
Serracaulis 82–3

MAMMALIA
Late Triassic – Extant

WITHIN *Cynodontia*
SEE ALSO *Monotremata, Marsupialia, Eutheria (including Xenarthra, Afrotheria, and Boreoeutheria)*

The 5400 and more species of living mammals, ranging from rats and bats to whales and humans, are characterized by a number of features including insulating hair, warm-bloodedness ("endothermy"), giving birth to live young, and suckling them with milk from the mother's mammary glands. Few of these characters (except occasionally hair) are preservable in fossil form, but there are also a number of skeletal characters particularly associated with the mammals – most notably a single jawbone, the dentary, which now has a single articulation with the skull. The cynodont ancestors of the mammals (see **Cynodontia**) had four bones in the jaw and during the evolution of the jaw, three of these transformed into the tiny bones of the middle ear. Development of the middle ear bones was particularly useful for hearing and balance in small

primitive mammals that were probably nocturnal tree dwellers.

Additionally, mammalian teeth are complex in form in order to help capture, kill, and "pre-process" food, helping to speed up the release of energy during digestion. This was particularly important for the first mammals, which were active insect-eating predators. Surprisingly, mammalian teeth, especially the cheek teeth, can often be identified down to the species level, which is lucky since teeth comprise much of the mammalian fossil record.

More than 20 major groups of mammals have evolved over the last 65 million years of Cenozoic time and survived until the present, with another eight or nine that evolved only to become extinct within the same period. Mammals can be divided into three groups that trace their origins to the Mesozoic – the egg-laying **Monotremata** (Jurassic – extant), the pouch-bearing **Marsupialia** (Middle Cretaceous – extant), and the "placental" mammals (**Eutheria**) (Early Cretaceous – extant), which are the dominant form today. Eutherian mammals have prolonged gestation of the young within the uterus, and are grouped with the marsupials (which give birth to very immature young) as "therians", distinct from the more primitive **monotremes** that produce small, leathery eggs.

In addition, there are several extinct groups of primitive mammals, mostly known from the Mesozoic. Until recently their fossil record, mostly of teeth, has been poorly known, but the discovery of entire skeletons, mostly in Mesozoic strata from China and Mongolia, is revolutionizing our understanding of the primitive mammals that coexisted with the dinosaurs.

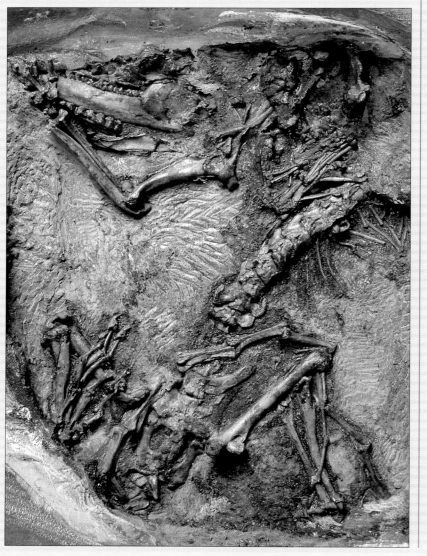

Henkelotherium. This fossil from the Jurassic deposits of Guimarota, Portugal, is among the earliest and most complete mammals known. A "basal therian", it is related to both marsupials and placentals.

EXTINCT MAMMAL GROUPS
Late Triassic – Oligocene

The most primitive true mammals are incompletely known, thanks to the vicissitudes of the fossil record when it comes to preserving any small, land-living vertebrates. Fossils such as the *Adelobasileus* from the Late Triassic of North America show mammalian features in the braincase, but also retain some cynodont features. Extinct Jurassic morganucodonts such as *Megazostrodon* are the best known among the first true mammal groups. They were small (around 10cm/4in long), shrew-like active insectivores that still had a somewhat sprawling posture to their forelimbs.

Other short-lived early mammals include the kuenotheriids (Early Jurassic), docodonts (Middle–Late Jurassic), triconodonts (Early Jurassic–Cretaceous), multituberculates (Late Jurassic–Oligocene), and symmetrodonts (Late Jurassic–Cretaceous). These last were closest to the Theria – it is not known whether they laid eggs or bore live young, but the very small size of the pelvic opening suggests that they were most likely to have been viviparous with very immature babies, like the present-day marsupials. In general, these primitive mammals were mostly small, insect-eating, shrew to hedgehog-sized mammals and are mainly known from their teeth.

However, recent additions to the fossil record show that the early mammals underwent an extensive radiation and thrived to a much greater extent during the Mesozoic than previously thought. Discoveries of more complete mammal fossils in China and Mongolia are now beginning to show that they were also more diverse in form and habit than previously thought. For instance, the opossum-sized triconodont *Repenomamus* weighed around 14kg (31lb), and was one of the largest early mammals. A superbly preserved fossil has been found with the remains of a baby psittacosaur dinosaur in its stomach, suggesting that it was an active predator or scavenger.

● BASAL MAMMALIAFORMS

SHUOTHERIIDAE
Pseudotribos 132–3

VOLATICOTHERIIDAE
Volaticotherium 132–3

DOCODONTA
Castorocauda 132–3
Haldanodon 138–9

UNKNOWN AFFINITY
Fruitafossor 144–5, 146–7

● EUTRICONODONTA

Jeholodens 154–5

GOBICONODONTIDAE
eutriconodontid sp. 156–7

REPENOMAMIDAE
Repenomamus 152–3

● ALLOTHERIAN MULTITUBERCULATES

paulchoffatiid sp. 138–9

CIMOLODONTA
Kryptobataar (djadochtatheriidan) 170–1
Nemegtbataar 172–3

PTILODONTIDAE
Ptilodus 180–1

● CLADOTHERIA

DRYOLESTIDAE
Henkelotherium 138–9

PERAMURIDAE
Tendaguratherium 136–7

● TRIBOTHERIA

AEGIALODONTIA
Kielantherium 156–7

● METATHERIA

DELTATHERIDIIDAE
Deltatheridium 174–5

MARSUPIALIA (MARSUPIAL MAMMALS)

Cretaceous – Extant

WITHIN *Mammalia*
COMPARE *Monotremata, Eutheria*

Living marsupials, such as the kangaroos, wombats, koalas, Tasmanian devils, bandicoots, and marsupial moles are mostly found in Australia, but there are also American marsupials such as opossums. This distribution suggests that they were much more widely spread in the past, and the fossil evidence seems to show a first diversification in the Early

▶ **Eastern grey kangaroo** (*Macropus giganteus*). Of all marsupials, kangaroos display their distinctive pouch or marsupium most obviously.

Cretaceous from the Americas into Antarctica and Australia, followed by an early Cenozoic extension from North America into Eurasia and Africa. However, they were later displaced across most of their range by the rise of the placental eutherian mammals, except within Australia, which by then had become isolated from the rest of Gondwana.

The Australian marsupials, including 143 living species, provide us with a wonderful example of convergent evolution with the placentals. As they evolved to fill all the available ecological niches, they developed everything from wolf-like predators, such as the recently extinct *Thylacinus*, to Pleistocene hippo-like plant-eaters such as *Diprotodon*. Convergent evolution in South America also produced marsupial "cats" (eg *Thylacosmilus*), "shrews", and 'bears', alongside some strange placental mammals, but all of these were largely replaced by new placentals during late Cenozoic times (around 3 million years ago), when he continent reconnected to North America.

● BASAL MARSUPIAL

Sinodelphys 154–5

● DIDELPHIMORPHA

DIDELPHIDAE
Herpetotherium 204–5

● AUSTRALIDELPHIA

YALKAPARIDONTIDAE
Yalkaparidon ("thingodontan") 208–9

THYLACINIDAE
Nimbacinus (Dickson's thylacine) 208–9

Nautilus (*Nautilus* sp.). Despite appearances, the various nautilus species are probably not closely related to the extinct ammonites – the soft-bodied cephalopod molluscs (octopuses, squids, and cuttlefish) are probably closer. Nevertheless, nautiloids are true living fossilss, largely unchanged for 500 million years.

● AUSTRALIDELPHIAN DIPROTODONTIANS

VOMBATIFORMES
Litokoala (rainforest koala) 208–9

WYNARDIIDAE
Namilamadeta 208–9

DIPROTODONTIDAE
Neohelos (zygomaturid) 208–9

● MACROPODIFORMES

HYPSIPRYMNODONTIDAE
Hypsiprymnodon (musky rat-kangaroo) 208–9
Ekaltadeta (carnivorous kangaroo) 208–9

PSEUDOCHEIRIDAE (POSSUMS)
Pseudochirops (woolly rainforest ringtail possum) 208–9
Paljara (primitive ringtail possum) 208–9

THYLACOEONIDAE
Priscileo (marsupial "lion") 208–9

PHALANGERIDAE
Strigocuscus (ground cuscus) 208–9

ACROBATIDAE
Distioechurus (feather-tailed possum) 208–9

BURRAMYIDAE
Burramys (rainforest mountain pygmy-possum) 208–9

● MACROPODIFORMES KANGAROOS AND WALLABIES

MACROPODIDAE
Macropus sp. 230–1

MOLLUSCA

Early Cambrian – Extant

One of the most diverse and important groups of shelly invertebrates, molluscs, includes some 50,000 living species and another 60,000 known extinct species – mostly snails (gastropods) and clams (bivalves). Evolving at first in the sea, they have since invaded freshwaters, and snails have also successfully made the move to land.

The group also includes the marine cephalopods – some 650 living species of squid, cuttlefish, and octopus, as well as several thousand species of extinct nautiloids and ammonoids with distinctive chambered shells, which mostly became extinct at the end of the Cretaceous.

● BASAL MOLLUSCS

Odontogriphus 56–7
Wiwaxia 56–7

● GASTROPODA (SNAILS)

SINUITIDAE
Sinuites 58–9

ARCHAEOPULMONATA
Protodiscus 90–1

BUCCINIDAE
Nassarius kraussianus 226–7

• BIVALVIA (CLAMS)

AVICULOPECTINIDAE
Aviculopecten 88–9

POSIDONIDAE
Bositra 134–5

OSTREIDAE
Hyotissa 176–7

LIMIDAE
Plagiostoma 176–7

MYTILOIDAE
Mytilus (mussel) 228–9

• HETEROCONCHIA

UNIONIDAE
Velesunio (freshwater mussel) 230–1

CEPHALOPODA (OCTOPUS AND SQUID)
Cambrian – Extant

The Class Cephalopoda includes a number of important fossil and extant subclasses that are listed below.

• NAUTILOIDEA

ORTHOCERIDA
Orthoceras 58–9

NAUTILIDA
nautiloid sp. 62–3

• AMMONOIDEA

CERATITIDAE
Eoprotrachyceras 116–17

ANCYCLOCERATINA
Scaphites 176–7
Microderoceras 128–9
Harpoceras 130–1

• BELEMNITIDA

Passaloteuthis 128–9, 130–1
Acanthoteuthis 142–3
Belemnitella 176–7

• COLEOIDEA

OCTOPODA
Proteroctopus 134–5

TEUTHIDA
Rhomboteuthis 134–5

• INDETERMINATE MOLLUSC?

Tullimonstrum 88–9

▶ **Short-beaked echidna** (*Tachyglossus aculeatus*). Despite being one of only two surviving monotreme genera, the Australian "spiny anteater" is remarkably widespread and successful. Its relatives, confined to New Guinea, are far rarer.

◀ **Ruby-throated hummingbird** (*Archilocus ccolubris* – female). Endemic to the Americas, these birds are specialist nectar eaters, and have co-evolved with the flowers on which they feed.

MONOTREMATA
Jurassic (?) – Extant

WITHIN *Mammalia*
COMPARE *Marsupialia, Eutheria*

The most primitive of living mammals are the Australian egg-laying monotremes – the duck-billed platypus and spiny echidna. Unfortunately their fossil record is very sparse, partly because they are very specialized and (except in the young platypus) do not have teeth. The discovery of similar teeth from the Paleocene of Argentina suggests that monotremes were originally a much more widely distributed Gondwanan group, and their ancestry has now been linked to Cretaceous and Jurassic fossils from Madagascar and Australia. The addition of these Gondwanan fossil representatives forms a wider grouping called the Australosphenidans.

AUSKTRIBOSPHENIDA
Bishops 162–3

NEOGNATHAE
Eocene – Extant

WITHIN *Avialae*
COMPARE *Palaeognathae*

From songbirds (Passeriformes) to penguins (Sphenisciformes) and condors (Falconiformes), the wonderful diversity of these "new-jawed birds" has evolved within the last 55 million years since Eocene times, and is displayed across some 140 living families. One fossil relative of the living condors (*Teratornis*), *Argentavis*, had a 55cm (22in) skull and a wingspan that may have reached 7.5m (25ft). At the height of neognath diversity in Pleistocene times, there may have been as many as 20,000 species, more than half of which have become extinct within the last few hundred thousand years.

• BASAL NEOGNATHS (ANSERIFORMES)

PRESBYORNITHIDAE
Presbyornis (basal neognath) 190–1

DROMORNITHIDAE
Bullockornis 208–9

• GASTORNITHIFORMES

GASTORNITHIDAE
Gastornis 198–9

• GALLIFORMES (CHICKENS ETC)

GALLINULOIDIDAE
Gallinuloides (basal neognath) 190–1

PHASIANIDAE
Numida (Guinea fowl) 216–17

The following taxa are grouped under the Neoaves.

• GRUIFORMES

Sarothrura (rail) 202–3

• PELECANIFORMES

FREGATIDAE
Limnofregata 190–1

• CICIONIFORMES (LONG-LEGGED WADERS)

ARDEIDAE
Ardea (purple heron) 202–3
Nyctiocorax (night heron) 202–3

THRESKIORNITHIDAE
Rhynchaeites (ibis) 198–9

TERATORNITHIDAE
Teratornis merriami 234–5

• CHARADRIIFORMES

JACANIDAE
Actophilornis 202–3
charadriid sp. (plover) 204–5

STERNIDAE
Sterna paradisaea (Arctic tern) 236–7

ALCIDAE
Pinguinus impennis (great auk) 236–7

• FALCONIFORMES (HAWKS ETC)

PANDIONIDAE
Pandion (osprey-like form) 202–3, 228–9

ACCIPITRIDIDAE
Haliaeetus (fish eagle) 202–3, 228–9
Stephanoaeatus (crowned eagle) 218–19

VULTURIDAE
Vultur 240–1

• STRIGIFORMES

PALAEOGLAUCIDAE
Palaeoglaux (owl) 196–7

• APODIFORMES (SWIFTS)

AEGIALORNITHIDAE
Aegialornis 198–9

• PROCELLARIFORMES

"*Marinavis*" 192–3

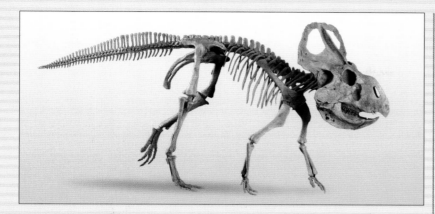

● BALAENICIPITIFORMES

Balaeniceps (shoebilled stork) 202–3

● PICIFORMES

Primozygodactylus
(woodpecker-like bird) 196–7

● CORACIIFORMES

Eocoracias (roller-like bird) 198–9

● PASSERIFORMES

meliphagid sp. (honey-eater) 208–9
paradisaeid sp. (bower bird) 208–9
Petroica (South Island robin) 244–5

ORNITHISCHIA (ORNITHISCHIAN DINOSAURS)

Late Triassic – End Cretaceous

WITHIN *Dinosauria*
COMPARE *Saurischia*

As their name suggests, the ornithischian dinosaurs have a bird-like hip structure. They were all plant-eaters, and include well-known groups such as the ceratopsians, stegosaurs, ankylosaurs, and ornithopods, which had very bird-like feet. They were relatively rare until Jurassic times, and are further divided into the cerapods and thyreophorans. The former include the horned ceratopians, thick-skulled pachycephalosaurs, and bipedal ornithopods, while the latter include the stegosaurs and ankylosaurs, both armoured against predators with bony plates.

● THYREOPHORA

Scelidosaurus 128–9

STEGOSAURIDAE

Stegosaurus 146–7
Kentrosaurus 136–7

Protoceratops. With its distinctive beak and neck frill, this ornithischian dinosaur shows a clear relationship to later, larger ceratopsians.

ANKYLOSAURIDAE

Minmi 164–5
Shamosaurus 156–7
Pinacosaurus 170–1

NODOSAURIDAE

Edmontonia 178–9

● ORNITHOPODA

EUORNITHOPODA

Anabisetia 166–7
Leaellynasaura 162–3

IGUANODONTIA

Dryosaurus 136–7
Camptosaurus 146–7
Iguanodon atherfieldensis 148–9
Iguanodon bernissartensis 148–9
Iguanodon 156–7
iguanodontid sp. 150–1
Jinzhousaurus 152–3
Muttaburrasaurus 164–5

● IGUANODONTIAN HADROSAURIFORMES

HADROSAURINAE

Edmontosaurus 178–9

LAMBEOSAURINAE

Lambeosaurus 174–5

● ORNITHISCHIAN CERAPODS

PACHYCEPHALOSAURIDAE

Stegoceras 174–5

● CERAPOD CERATOPSIANS

PSITTACOSAURIDAE

Psittacosaurus 152–3, 156–7

PROTOCERATOPSIDAE

Protoceratops 170–1

CERATOPSIDAE

Chasmosaurus 174–5
Triceratops 178–9

OSTEICHTHYES (BONY FISH)

Silurian – Extant

WITHIN *Gnathostomes*
COMPARE *Chondrichthyes*
SEE ALSO *Actinopterygii, Sarcopterygii
(including Tetrapoda)*

The bony fish are by far the most successful and diverse of all vertebrate groups, with some 21,000 living species in 260 families. They split into two main groups, the ray-finned fish (**Actinopterygii**) and lobe-finned fish (**Sarcopterygii**) in late Silurian times, and diversified in the following Devonian Period.

The two branches (clades) were thought to be distinct until the recent discovery of an ancestral form, *Psarolepis*, from Siluro-Devonian strata in China. This intriguing fossil has a skull that shows a mixture of actinopterygian and sarcopterygian features, and in addition it has bony pectoral and dorsal spines characteristic of the extinct acanthodians and placoderms.

PALAEOGNATHAE

Paleocene – Extant

WITHIN *Avialae*
COMPARE *Neognathae*

The most primitive of the modern birds, the palaeognaths retain some features relating them to theropod dinosaurs, especially in the palate (roof of the mouth). They may have originated in the northern hemisphere, but today they are essentially Gondwanan, known mainly from South America, Africa and Australia, where they include flightless ratites such as the rhea and ostrich.

Apteryx (kiwi) 244–5
Dinornis giganteus (moa) 244–5
Euryapteryx (moa) 244–5
Emeus (moa) 244–5

Ostrich (*Stuthio camelus*). The largest extant species of bird, the African ostrich, and its similar relatives on other continents are thought to have evolved independently from flying ancestors.

PLACODONTIA

Triassic

WITHIN *Diapsida*
COMPARE *Lepidosauria, Archoasuria, Ichthyosauria, Plesiosauria*

This short-lived group of marine reptiles had relatively unspecialized legs, and some of them looked more like land animals. However, loose articulation of the limbs shows that they were being used for swimming rather than walking, and the teeth are also modified for eating shellfish.

PLACODONTIDAE
Paraplacodus 116–17

PLESIOSAURIA

Late Triassic – Late Cretaceous

WITHIN *Diapsida*
COMPARE *Lepidosauria, Archosauria, Ichthyosauria, Placodontia*

This extinct group of large marine reptiles includes the short-necked pliosaurs and the long-necked plesiosaurs. The pliosaurs had streamlined bodies and two pairs of paddle-like limbs – they were fast-swimming active predators equipped

with long heavy skulls, powerful jaws and sharp teeth. By contrast, the plesiosaurs were slower swimmers with long necks, relatively small skulls and large, heavy paddles.

● **PLESIOSAUROIDEA**

Plesiosaurus 128–9

● **PLIOSAUROIDEA**

RHOMALEOSAURIDAE
Rhomaleosaurus 130–1

PLIOSAURIDAE
Kronosaurus 164–5

POLYCHAETA (POLYCHAETE WORMS)

Cambrian – Extant

This group of marine annelid worms are commonly known as bristle worms because of the numerous lateral bunches of hair-like bristles they carry on the fleshy-lobed, paired appendages projecting from each body segment. The appendages are used used for swimming, burrowing and protection.

Canadia 56–7
Bundenbachochaeta 72–3

Bearded fireworm (*Hermodice carunculata*). These reef-dwelling polychaete worms are found in the Atlantic and Mediterranean. Their bristle-like hairs can inject a powerful neurotoxin on contact.

Giant barrel sponge (*Xestospongia muta*). These hard sponges, found in the Caribbean at depths of more than 10m (33ft), have internal hollows to increase their efficiency at filtering food particles from water.

PORIFERA (SPONGES)

Late Precambrian – Extant

They might not look like it, but sponges are clusters of animal cells that combine to construct a multicellular colony growing from an inorganic substrate. Sponges are simple filter-feeding organisms, with highly porous walls through which water is sieved for food and oxygen before being expelled through a single large exhalant opening. The walls are supported by spicules made of silica (glass sponges – Hexactinellida), calcium carbonate (calcisponges – Calcarea), organic protein (bath sponges), or a combination of all three materials (Demospongiae).

Of the 10,000 or so living sponge species, most are marine and occupy all depths of water, but there are also freshwater forms. They are highly variable in structure, but most are vase or globe-shaped. Sponges are important contributors to reefs and have at times been plentiful enough to form entire reefs on their own, such as the Capitan Reef of Permian Texas, which extended for some 500km (300 miles).

● **DEMOSPONGIAE**

Quadrolaminiella 48–9

POECILOSCLERIDA
Paraleptomitella 48–9

VERONGIIDA
Vauxia 56–7

HADROMERIDA
Pirania 56–7

STROMATOPOROIDEA
Ordovician

These extinct marine relatives of sponges built low, dome-shaped, and layered calcareous mounds that were important components of Paleozoic reefs. Originally placed in a group of their own, they are now distributed between the calcisponges and demosponges

stromatoporoid sp. 74–5

PRIAPULIDA (PRIAPULID 'WORMS')

Early Cambrian – Extant

These worm-like marine predators have a low diversity today with only 16 species, but they have an evolutionary history stretching back to the early

Ottoia. This priapulid from Canada's Burgess Shale may have been an active burrower. Fossil gut contents show it fed on small invertebrates.

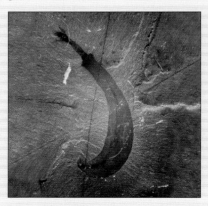

Cambrian. Their elongated cylindrical bodies are divided into three parts and covered in a chitinous cuticle that groups them in the "Ecdysozoa", alongside the arthropods and nematodes. Priapulids burrow into the sediment using a protactile proboscis, then lie in wait to snare their prey using curved hooks that cover this proboscis.

Paraselkirkia 48–9
Ottoia 56–7

PRIMATES

Mid-Cretaceous (?) – Extant

WITHIN *Euarchontoglires*
SEE ALSO *Anthropoidea (including Hominoidea)*

The name primate originates from the original classifications of animals and plants produced by eminent 18th-century Swedish botanist Carl Linnaeus. Today, this small group is taken to include 376 species of living humans, apes, gibbons, monkeys, tarsiers, lorises, galagos, and lemurs, and their numerous but sometimes obscure fossil ancestors. Throughout their history, most primates have been woodland dwellers, and as a result their fossil record is very poor, consisting largely of teeth and tough bony remnants left by predators and scavengers.

Typically, these animals are small to medium-sized omnivores, ranging from the 30g (1oz) pygmy mouse lemur to the 180kg (400lb) gorilla. Critical evolutionary advances from their insectivore ancestors include the acquisition of an opposable thumb, flat nails instead of claws, a shortened snout, and closely spaced, forward-pointing eyes. The limbs and acute stereoscopic eyesight are all adapted for tree climbing, and there has also been enlargement of the brain, delayed sexual maturity, reduced reproduction, and increased maternal care and social behaviour.

Molecular analysis separates two major groups within the primates, the strepsirrhines (lemurs and lorises) and the haplorhines (tarsiers and anthropoids). Their divergence dates back some 50 million years to Eocene times and the appearance of lemur-like adapid primates, such as *Cantius*, and tarsier-like omomyid primates such as *Omomys* and *Shoshonius* from North America.

Verreaux's sifaka (*Propithecus verreauxi*). Lemurs colonized the island of Madagascar some time after it separated from Africa. They are the lone survivors of a once-widespread primate group.

● PLESIADAPIFORMES

PLESIADAPIDAE
Plesiadapis 180–1

● STREPSIRRHINI

ADAPIDAE
Cantius 188–9, 192–3
Adapoides 200–1

● HAPLORHINI

? Darwinius masillae 196–7

ANTHROPOIDEA
Apidium (parapithecid) 202–3

PTERIDOSPERMALES (SEED FERNS)

Carboniferous – Cretaceous/Extant?

WITHIN *Gymnospermopsida*
COMPARE *Bennetitales, Cycadophyta, Ginkgoales, Coniferopsida, Angiospermae*

The seed ferns or pteridosperms were an important group of ancient plants that had fern-like foliage and conifer-like woody tissue. Their fossil fronds are often confused with those of true ferns. However, the seed ferns retained enlarged "female" spores within a sporangium and partly enclosed within a structure to generate a functional seed with its own food supply.

This so-called "naked seed" was a major innovation in plant evolution and gave rise to a larger grouping called the **Gymnospermopsida** that includes the cycads (**Cycadophyta**), **Bennettitales**, conifers (**Coniferopsida**) etc. Their evolutionary relationships are still disputed, and it is possible that cycads, for example, may be living seed ferns. During Carboniferous times, the seed ferns diversified to produce trees, shrubs, climbing and ground creepers, dominating the more primitive free-sporing plants. However, they in turn were increasingly eclipsed and dominated from Jurassic times by more advanced gymnosperms and flowering plants.

NEUROPTERIDACEAE
Neuropteris 88–9, 90–1

SPHENOPTERIDACEAE
Sphenopteridium 84–5

PTEROSAURIA

Late Triassic – End Cretaceous

WITHIN *Archosauria*
COMPARE *Crocodyliformes, Dinosauria*

This extraordinary group of more than 100 species of flying archosaur dominated Mesozoic skies until the birds began to replace them in the Late Cretaceous. They included giants such as *Quetzalcoatlus*, which had a 12m (39ft) wingspan and was the largest flying animal known. They typically had short bodies, long necks, and large heads, while the fourth finger on each arm was enormously elongated to support a wing membrane. The earliest forms such as *Eudimorphodon* had long, bony tails, but this was greatly reduced in many later forms. They seem to have developed a variety of feeding habits, ranging from eating insects to fish and sieving plankton.

● BASAL PTEROSAURS

DIMORPHODONTIDAE
Dimorphodon 128–9

ANUROGNATHIDAE
Jeholopterus 152–3

RHAMPHORHYNCHIDAE
Rhamphorhynchus 136–7, 138–9, 140–1
Dorygnathus 130–1

'REPTILIA'

Carboniferous – Extant

WITHIN *Tetrapoda*
COMPARE *Lepospondyli, Temnospondyli*
SEE ALSO *Synapsida (including Therapsida),
Diapdia (including Lepidosauria, Archosauria,
Ichthyosauria, Plesiosauria, Placodontia)*

Living reptiles were traditionally considered as cold-blooded, egg-laying tetrapods with scaly skins. However, many squamates give birth to live young and birds, which are now known to be dinosaurs, are warm, blooded. There are some 16,000 living species – mostly lizards and snakes (**Lepidosauria**), **Crocodyliformes**, tortoises and turtles (chelonians),

Hawksbill turtle (*Eretmochelys embricata*). The sea turtles are reptiles that have adapted to an almost entirely aquatic life – in the case of hawksbills, feeding on sponges around shallow lagoons and reefs.

amphisbaenids (limbless lizards), and the tuatara, sole survivor of the Order Spenodontia. There are also a number of extinct groups – notably the **Ichthyosauria, Plesiosauria, Pterosauria** and **Dinosauria**. The birds (**Avialae**) are also an integral part of the evolution of the group, and account for a further 9000 species.

There is a broad division among reptiles based on skull type. The primitive anapsid skull type, with no skull openings apart from the eyes, is seen in late Carboniferous fossils, such as *Hylonomus* and *Paleothyris* from Joggins, and in the present-day chelonians.

Amniotes whose skulls have an additional pair of lower temporal openings are known as synapsids, and include a number of extinct groups from which the mammals, also synapsids, emerged. The presence of two pairs of temporal openings is characteristic of the large and varied grouping of diapsid amniotes, which includes the extinct dinosaurs and pterosaurs, crocodylinas, avialians (birds), lizards and snakes (the lepidosaurs).

Finally, with the loss of the lower openings, the euryapsid condition with a single pair of upper temporal openings is seen, particularly in the extinct Mesozoic marine 'nothosaurs', plesiosaurs, and ichthyosaurs.

***Hylonomus*.** This well-preserved fossil from Joggins, Nova Scotia, is the oldest confirmed reptile species known (*Westlothiana* from East Kirkton in Scotland may be older, but its affinities are still not fully understood). *Hylonomus*' name is a hybrid of Greek and Latin words, meaning "forest mouse".

RHYNIOPHYTA (RYHYNIOPHYTOID PLANTS)

Devonian

COMPARE *Sphenopsida, Lycopsida,
Filicopsida, Gymnospermopsida*

Land plants, known as embryophytes, like land animals, have to be able to cope with the harsh environment and life exposed to the atmosphere instead of the more supportive and protective environment of water. The most primitive land plants are the bryophytes (mosses, liverworts, and hornworts, Ordovician – Recent), none of which have supported or fluid conducting (vascular) tissues and so do not grow much above the ground surface. However, they do have their reproductive tissues enclosed and protected and reproduce by means of spores.

The evolution of fluid-conducting (vascular) tissues allowed some vertical growth against gravity and towards sunlight, and the addition of internal strengthening tracheids provided support for the development of upright stems. Simple division of these stems into dichotomous branches results in more numerous terminal reproductive sporangia.

Cooksonia. These early land plants from the late Wenlock Stage of the Silurian Period are thought to lie close to the evolutionary branch between the Rhyniophyta and the clubmosses.

These advances characterize this group of vascular plants as the polysporangiophytes and are represented by fossils such as *Cooksonia*.

Reproduction by means of spores requires the plants to live in damp environments with a reproductive cycle that includes both sporophyte and gametophyte generations in the same species. Such an alternation of generations has been recognized in some fossil plants from the Early Devonian deposits of Rhynie, and these are categorized as rhyniopsids, while the term "rhyniophytes" has been applied to other primitive land plants found at Rhynie that might not be true rhyniopsids.

● POLYSPORANGIOPHYTES

Cooksonia 68–9

● TRACHEOPHYTES

"RHYNIOPHYTES"

Steganotheca 68–9
Rhynia 70–1
Aglaophyton 70–1
Horneophyton 70–1

● UNKNOWN AFFINITIES

Nothia 70–1

▶ *Eusthenopteron*. This lobe-finned fish from Miguasha, Quebec, shows unmistakable signs of evolution towards the later tetrapod species.

SARCOPTERYGII (LOBE-FINNED FISH)

Devonian (?) – Extant

WITHIN *Osteichthyes*
COMPARE *Actinopterygii*
SEE ALSO *Tetrapoda (including Lepospondyli, Temnospondyli, and Amniota)*

Although represented today by just a few lungfish and the coelacanth, the lobefins and their evolution are pivotal in the story of vertebrate life. During Devonian times, when all of the group were still confined to the water, their muscular paired fins became adapted to form the tetrapod limb. Subsequently this limb was seconded for movement out of water and on to land. Another important preadaption was the development of lungs able to obtain oxygen directly from air when oxygen levels were low in the water.

A succession of fossils record the transitional stages from the purely fish-like lobefins (eg *Eusthenopteron*), through *Tiktaalik* with its strengthened fore-fins and shoulder or pectoral girdle, to the fully tetrapod *Ichthyostega*, which was able to emerge fully from the water.

ONYCHODONTIFORMES
Onychodus 74–5

POROLEPIFORMES
Laccognathus 78–9

ACTINISTIA (COELACANTHS)
Chinlea 126–7
Allenypterus 86–7
Caridosuctor 86–7

● DIPNOI (LUNGFISH)

DIPNORHYNCHIDAE
Holodipterus 74–5
Griphognathus 74–5
Scaumenacia 76–7

CERATODONTIFORMES:
Ceratodus 146–7
dipnoan sp. 98-9

● TETRAPODOMORPHA

"OSTEOLEPIFORMES"
Gogonasus 74–5
Eusthenopteron 76–7

ELPISTOSTEGALIA
Tiktaalik 78–9

SAURISCHIA (SAURISCHIAN DINOSAURS)

Late Triassic – Extant

WITHIN *Dinosauria*
COMPARE *Ornithischia*
SEE ALSO *Sauropodomorpha, Theropoda (including Avialae)*

As their name suggests, the saurischian dinosaurs had a more primitive and lizard-like pelvic structure than the ornithischian

dinosaurs. A major subdivision separates the saurischians into two groups – the four-legged, plant-eating **Sauropodomorpha** and the two-legged, meat-eating **Theropoda**.

HERRERASAURIDAE (BASAL THEROPOD-LIKE SAURISCHIANS)
Herrerasaurus 120–1
Eoraptor 120–1
Chindesaurus 124–5

SAUROPODOMORPHA (SAUROPOD DINOSAURS)

Late Triassic – End Cretaceous

WITHIN *Saurischia*
COMPARE *Theropoda*

This well-known group of saurischian dinosaurs were large, sometimes truly gigantic, four-legged terrestrial plant-eaters with a relatively simple body plan. The basal prosauropods that arose during the Triassic Period included the plateosaurs, which may have been able to stand on their hind legs, but true sauropods were firmly quadrupedal. They first appeared in the Late Triassic, and include such well-known Jurassic species as *Diplodocus* and *Brachiosaurus*. Characterized by massive bodies supported on elephant-like legs, they had long necks and small-brained skulls counterbalanced with long,

Diplodocus. The skull of this giant herbivore clearly displays its peg-like teeth, used for stripping foliage. The hole at the top of the skull is formed by the nostrils.

muscular tails. In some forms, the nostrils were positioned high on the head, and close to the eyes. The teeth, meanwhile, were often reduced to pencil or peg-shapes that were used to rake foliage from branches.

The sauropods had an almost global distribution, and survived for almost 100 million years until the end of the Cretaceous Period. Although their heyday of diversity and abundance was in the Late Jurassic, some of the Cretaceous titanosaurs grew to great size and even developed armour.

● DIPLODOCOIDEA

DICRAEOSAURIDAE
Dicraeosaurus 136–7

DIPLODOCIDAE
Diplodocus 144–5, 146–7
Apatosaurus 146–7
Barosaurus 136–7

● MACRONARIA

CAMARASAURIDAE
Camarasaurus 144–5

BRACHIOSAURIDAE
Brachiosaurus 136–7

TITANOSAURIA
Andesaurus 166–7
Argentinosaurus 168–9

SPHENOPSIDA (HORSETAILS)
Late Devonian – Extant

COMPARE *Rhyniophyta, Lycopsida, Filicopsida, Gymnospermopsida*

The horsetails or sphenopsids are a group of primitive plants that once formed the dominant vegetation of the low-lying, wet Carboniferous tropical forests, but are reduced today to just 18 species, all belonging to the genus *Equisetum*.

They grow from both true roots and horizontal root-like "rhizomes", and have distinctively segmented upright stems, out of which whorls of stiff bristles grow. Like the clubmosses (lycopsids), they reached gigantic proportions in Carboniferous times, with tree-sized plants such as *Calamites* growing up to 40m (130ft) tall.

SPHENOPHYLLALEAE
Sphenophyllum 88–9, 92–3, 102–3

EQUISETALEAE
Archaeocalamites 84–5

EQUISETACEAE
Calamites 88–9
Sphenopteris 102–3

SPHENOPHYLLALEAE
Phyllotheca 102–3

▶ **Stromatolites.** Modern stromatolites still build up around salty lakes and lagoons, perhaps over thousands of years. These examples line the shores of Lake Thetis in Western Australia.

STROMATOLITES
Archean – Extant

The oldest readily visible fossil-related structures are layered, mound-shaped stromatolites from the 3.5-billion-year-old Apex Chert of Western Australia. Although no organic remains are normally found within stromatolites, their growth and structure is the result of interaction between microbial mats of algae and cyanobacteria, and sedimentary processes. Microbial mats growing over the seabed trap fine sediment, and the light-seeking microbes grow upwards to re-establish a new surface mat. Repetition of the process produces a finely laminated and organically bonded pile of sediment. Local wave and current activity interacts with seabed topography to produce a variety of shapes and sizes, but commonly they are mounds, tens of centimetres in diameter and up to a metre (40in) or so tall, depending upon the depth of the water. Stromatolites still form in shallow, clear, and warm tropical seas.

"egg carton" form 36–7
"cuspate swale" form 36–7
"encrusting/domical" form 36–7
thrombolite 46–7

CYANOBACTERIA?
Primaevifilum 38–9
Archaeoscillatoriopsis 38–9

SYNAPSIDA (SYNAPSID AMNIOTES)
Late Carboniferous – Extant

WITHIN *Amniota*
COMPARE *"Reptilia", Diapsida*
SEE ALSO *Therapsida (including Cynodontia)*

Synapsids are those amniotes with only one pair of lower temporal openings behind the eye in the skull. As a group, they diverged from the anapsids and **Diapsida** (together known as the **"Reptilia"**) in the middle Carboniferous Period, and expanded enormously in the Permian to become the dominant land animals. Six basal synapsid groups, such as the edaphosaurids and sphenacodontids, were particularly important in early Permian times and are collectively known as the "pelycosaurs". These were succeeded in late Permian times by newly evolving plant and meat-eating synapsids known as the **Therapsida**.

● PELYCOSAURIAN SYNAPSIDS

OPHIACODONTIDAE
Archaeothyris 94–5
ophiacodont sp. 96–7

EDAPHOSAURIDAE
Edaphosaurus 98–9

SPHENACODONTIDAE
Dimetrodon 98–9, 100–1

TEMNOSPONDYLI

Carboniferous – ?Extant

WITHIN *Tetrapoda*
COMPARE *Lepospondyli, Amniota*
SEE ALSO *Lissamphibia*

This major group of small Carboniferous tetrapods, with over 140 genera, diversified and thrived until well into Triassic times, and still survived with reduced diversity into the Early Cretaceous. They are distinguished by a palate structure similar to that seen in frogs and salamanders today, which allows them to swallow food and air. This may indicate that the modern **Lissamphibia** arose from the temnospondyls.

● CAPITOSAUROIDEA

CAPITOSAURIDAE
Wetlugasaurus 110–11

BENTHOSUCHIDAE
Benthosuchus 110–11

● BRACHYOPOIDEA

BRACHYOPIDAE
brachiopoid sp. 112–13
Koolasuchus 162–3

DENDRERPETONTIDAE
Dendrerpeton 92–3
Balanerpeton 84–5

▶ ***Balanerpeton.*** This early temnospondyl from East Kirkton in Scotland probably breathed by gulping in air with its mouth, rather than by expanding its ribcage and drawing air inwards.

ERYOPIDAE
Eryops 98–9

● "BRANCHIOSAURS"

branchiosaur tadpole 94–5

TETRAPODA (FOUR-LIMBED VERTEBRATES)

Devonian – Extant

WITHIN *Sarcopterygii*
SEE ALSO *Lepospondyli, Temnospondyli (including Lissamphibia), Amniota (including Synapsida and Diapsida)*

All backboned animals with a basic four-limbed appendicular skeleton, at least in the early stages of their development, are tetrapods. These include all the living amphibians, reptiles, birds, and mammals, as well as numerous extinct groups such as the Devonian aquatic forms *Tiktaalik* and *Acanthostega*, the Paleozoic **Temnospondyli,** and reptile-like tetrapods etc. The most basal living tetrapods are the **Lissamphibia,** animals such as frogs, toads, newts, and salamanders, but these did not evolve until Triassic times whereas the earliest reptiles evolved in Carboniferous times. Consequently, the ancestry of reptiles and mammals lies within extinct groups of tetrapods whose relationship to the living amphibians is unclear, so it is preferable to use the name "tetrapod" rather than "amphibian" as a general group name for the extinct Paleozoic forms.

● BASAL TETRAPODS

ICHTHYOSTEGIDAE
Ichthyostega 82–3

ACANTHOSTEGIDAE
Acanthostega 82–3

● BAPHETIDS

Baphetes 90–1

THERAPSIDA

Permian – Extant

WITHIN *Synapsida*
SEE ALSO *Cynodontia (including Mammalia)*

This group of synapsid tetrapods replaced the dominant pelycosaur synapsids of earlier Permian times. They included both plant and meat-eaters with heavily built and muscular limbs, and some of them grew to considerable size, such as the 5m-long (16.5ft) plant-eating *Moschops* from South Africa. In contrast, the top predatory carnivores, known as cynodonts, were somewhat smaller at not much more than a metre (40in) long, but were armed with powerful jaws and specialized teeth including large, dagger-shaped canines. All extant mammals are synapsid "cynodonts".

● DICYNODONTIA

DICYNODONTIDAE
Dicynodon 106–7

LYSTROSAURIDAE
Lystrosaurus 104–5, 110–11, 112–13

ROBERTIIDAE
Diictodon 104–5

● GORGONOPSIA

Inostrancevia 106–7

● THEROCEPHALIA

Moschorhinus 104–5

THEROPODA

Late Triassic – Extant

WITHIN *Saurischia*
COMPARE *Sauropodomorpha*
SEE ALSO *Avialae (including Palaeognathae and Neognathae)*

A major subdivision among the **Saurischia** separates the two-legged meat-eating theropod dinosaurs from the four-legged, plant-eating

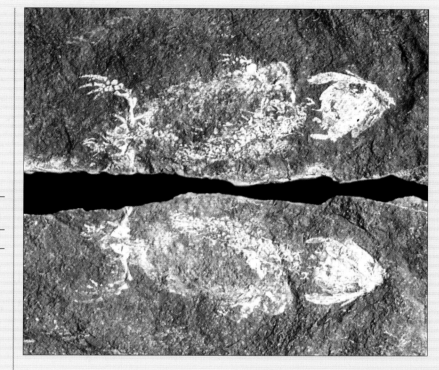

sauropods. Recent discoveries of well-preserved theropods from China have radically transformed understanding of the group, and confirmed that the birds (**Avialae**) originated within a group of small theropods known as the maniraptorans.

Theropods ranged in size from the gigantic tyrannosaurs and allosaurs that reached lengths of 13m (43ft) and weights of up to 6 tonnes, down to tiny dinosaurs such as the feathered 50cm (20in) *Microraptor*. Feathered dinosaurs all belonged to a theropod subdivision called the eumaniraptorans, from which avialian dinosaurs and the birds arose.

Overall, the diversity of theropods showed a wide range of size and form, from small bird-like animals to the largest predatory meat-eating animals that have ever lived.

● NEOTHEROPODA

COELOPHYSOIDEA
Coelophysis 126–7

● CERATOSAURIA

Elaphrosaurus 136–7

ABELISAURIDAE
Aucasaurus 168–9

● TETANURAE

SPINOSAURIDAE
Irritator 158–9

ALLOSAURIDAE
Allosaurus 146–7

CARCHARODONTOSAURIDAE
Giganotosaurus 166–7

The following species are grouped as tetanuran coelurosaurs:

COMPSOGNATHIDAE
Compsognathus 138–9, 140–1
Sinosauropteryx 150–1, 154–5

THERIZINOSAURIDAE
Beipiaosaurus 152–3

● MANIRAPTORIFORMES

ALVAREZSAURIDAE
Mononykus 170–1

EUMANIRAPTORA
Pedopenna 132–3

TYRANNOSAUROIDEA
Dilong 152–3

◀ **Heron** (*Ardea* sp.). These fish-eating neognath birds are direct descendants of flesh-eating theropod dinosaurs.

TYRANNOSAURIDAE
Tyrannosaurus 178–9
Gorgosaurus 174–5

ORNITHOMIMIDAE
Struthiomimus 174–5

● EUMANIRAPTORAN TROODONTIFORMES

TROODONTIDAE
Saurornithoides 170–1
Troodon 174–5

DROMAEOSAURIDAE
Microraptor 152–3
Sinornithosaurus 154–5
Buitreraptor 166–7

VELOCIRAPTORIDAE
Velociraptor 172–3

OVIRAPTORIDAE
Protoarchaeopteryx 150–1, 154–5
Caudipteryx 154–5
Oviraptor 172–3

TRILOBITA

Early Cambrian – Late Permian

WITHIN *Arthropoda*
COMPARE *Chelicerata, Crustacea, Uniramia (including Hexapoda)*

Some 15,000 known species belong to this famous group of extinct marine arachnomorph arthropods, which bear a passing resemblance to the living horseshoe or king crabs. Mineralization of the exoskeleton and its repeated moulting during growth have enhanced their fossil potential, but complete specimens are rare.

Trilobites had flattened bodies with the head and tail shields separated by a flexible segmented body that allowed the animal to curl up for protection. Unlike the chelicerate arthropods (see **Chelicerata**), the trilobites lacked a

Dalmanites. Just one of countless trilobite genera, *Dalmanites* spread through all the world's oceans during Ordovician and Silurian times.

Yellow-spotted millipede (*Harpaphe haydeniana*). This distinctive modern uniramian is also known as the cyanide millipede, on account of its ability to secrete toxic hydrogen cyanide when threatened.

telson, and the pygidium was a fused single dorsal tergite (body segment) that covered several pairs of legs. Most trilobites were seabed-dwelling scavengers that used their numerous jointed legs for walking, but some could swim. The first trilobites appeared in the evolutionary explosion of the early Cambrian, and their heyday was in Ordovician times, but they did not finally die out until the end Permian extinction.

● **BASAL ARACHNOMORPHS**

HELMITIIDAE
Kuamaia 48–9

MARELLOMORPHA
Marella 56–7

● **TRILOBITA**

REDLICHIIDAE
Olenoides 56–7

PHACOPIDAE
Calymene 66–7
Dalmanites 66–7

ASAPHIDAE
Homotelus 58–9

NARAOIIDAE
Soomaspis 62–3
Retifacies 48–9

UNIRAMIA (UNIRAMIAN ARTHROPODS)

Cambrian (?) – Extant

WITHIN *Arthropoda*
COMPARE *Trilobita, Chelicerata, Crustacea*
SEE ALSO *Arthropoda*

This major group of arthropods includes the familiar **hexapod** insects, along with some 13,000 living species of millipedes and centipedes (together known as myriapods on account of multisegmented bodies and numerous pairs of legs). Uniramians are distinguished by their possession of a single pair of antennae and two pairs of mouthparts – a pair of mandibles and a pair of maxillae.

With a marine origin, the myriapods were among the earliest animals to colonize freshwaters (during Ordovician times) and then move onto land (in Silurian times). Most of the land-living myriapods eat rotting vegetation, and during Carboniferous times the high proportion of oxygen in Earth's atmosphere allowed giant forms to evolve, such as *Arthropleura*, which grew to mor e than 2m (6.6ft) in length and is the largest known land invertebrate of all time.

● **DIPLOPODA**

AMYNILYSPEDIDAE
Amynilyspes 92–3

SPIROBOLIDA
Xyloiulus 88–9

SYMPHYLA
Cratoraricrus 160–1

● **CHILOPODA**

SCUTIGEROMORPHA
Latzelia 88–9

● **ARTHROPLEURIDA**

ARTHROPLEURIDAE
Arthropleura 90–1

EOARTHROPLEURIDAE
Eoarthropleura 68–9
millipede sp. 84–5
myriapod sp. 116–17

XENARTHRA

Early Paleogene – Extant

WITHIN *Eutheria*
COMPARE *Afrotheria, Boreoeutheria*

This strange group of placental mammals (**eutherians**) includes ant-eaters, sloths, and armadillos, most of which have reduced teeth because of specialized feeding. Until the relatively recent extinctions of the Pleistocene Ice Age "megafauna" just over 10,000 years ago, the group included some remarkable giants in the Americas, including *Megatherium*, the 6m-long (20ft) giant ground sloth, and *Glyptodon*, the 2-tonne giant armoured armadillo.

● **POSSIBLE STEM XENARTHRAN**

Eurotamandua 198–9

● **PILOSA**

MYLODONTIDAE
Paramylodon 234–5

UNKNOWN

The stratigraphic record has revealed many fossils that cannot easily be placed in any known group of living or extinct organisms. Often this is becaause so little information is preserved, especially about the original soft tissues. One of the most important outstanding groups of unknown fossils is generally referred to as the "Ediacara", and they are discussed under this heading, but their biological affinities are still very much a matter of debate. There are a number of other fossils from Precambrian/Cambrian boundary strata that are important for our understanding of the evolution of invertebrates, but whose affinities are not clear.

● **UNKNOWN AFFINITY**

Namacalathus 46–7
Cloudina 46–7
chancellorid 46–7
Eldonia 48–9, 56–7
Dinomischus 56–7

Nine-banded armadillo (*Dasypus novemcinctus*). The most widespread modern xenarthran evolved in South America, moving north when the Americas joined together around 3 million years ago.

GLOSSARY

Actinopterygian
A member of a large group of **osteichthyans** that includes most of the bony fish, (ie those with paired "rayfins") and many extinct groups with bony scales. They have evolved from late **Silurian** times until the present.

Adaptation
A biological modification or process that makes an organism better suited for a particular way of life and environment.

Advanced
A state or feature of an organism that is transformed in comparison to a **primitive** state or feature.

Afrotherian
An African group of mammals that were the first main branch of placental mammals to evolve, possibly in the Late **Cretaceous**. Many are now extinct, but the survivors include elephants, aardvarks, and sea cows.

Agnathan
A group of jawless fish-like vertebrates that includes many extinct fossil marine and freshwater groups, from which only the lampreys and hagfish survive (**Cambrian** to Recent).

Algae
A collective name give to several unrelated groups of relatively simple **photosynthetic** organisms that live in moist or wet environments. They include both unicellular and multicellular forms and, unlike plants, they generally lack stems, roots and leaves.

Amniote
A vertebrate, either living or fossil, whose embryo develops within a protective membrane (the amnion). The amniotes include reptiles, birds, mammals, and some extinct **tetrapods**, but not the **lissamphibians**.

Amphibian
A term still commonly used to describe **tetrapods** that develop from unprotected eggs laid mostly in water, into a juvenile tadpole stage with gills, and then into an adult with lungs. However, the group is **polyphyletic**, and has been replaced by the **Lissamphibia** and various extinct groups, including the **temnospondyls**.

Anapsid
A group of reptiles, including turtles, tortoises, and various extinct forms, characterized by having no openings in the skull behind the eye.

Angiosperm
Commonly known as the "flowering plants", this major division of plants have their seeds protected by an enclosed ovary. Their early evolution is still somewhat obscure but occurred in Early **Cretaceous** times.

Archaea
A diverse and primitive group of microscopic **prokaryotes** that occur as single cells, filaments, or aggregates, and often colonize extreme environments.

Archean Eon
The earliest division of **Precambrian** time. The Archean does not yet have a formally defined lower limit, but ranges from around 4 billion years ago until 2500 million years ago.

Archosaur
This large group of **diapsid** reptiles arose in Early **Triassic** times and includes the living crocodiles, birds, and extinct groups such as the non-bird dinosaurs and pterosaurs.

Arthropod
The largest and most diverse grouping of animals, both living (eg insects) and extinct (eg trilobites), which originated in **Cambrian** times and are characterized by the possession of jointed limbs and a tough outer skin (exoskeleton).

Bacteria
This large, diverse, and primitive group of **prokaryote** microorganisms originated in **Archean** times and uses a great variety of metabolic pathways including **photosynthesis** with and without oxygen.

Basal
Referring to the first group or clade to branch off within a larger clade. Basal clades (sometimes also referred to as "stem" groups), sometimes contain similarities to the last common ancestor of the larger clade, when compared to later-branching clades with more **derived** features.

Biota
The total fauna and flora of a particular time or region.

Bipedalism
A mode of standing, walking, or running (locomotion) on two legs rather than four (compare **quadrupedal**).

Blastopore
In embryological development, the initial mouth or opening of the bowl-shaped early embryo.

Boreoeutheria
A large group of **placental** mammals identified by molecular analysis, which originated in Late Cretaceous times and is subdivided into the **Laurasiatheria** and **Euarchontoglires**.

Bryophyte
A group of plants including the mosses and liverworts that lack true roots and leaves and cannot grow far above ground level.

Similar plants first colonized land in **Ordovician** times.

Calcium carbonate
A non-toxic calcium mineral that is readily soluble in slightly acidic water, crystallizes at low pressures and temperatures, and is used by many organisms for skeletal structures and shells. It also forms rocks.

Cambrian Period
The first period of the **Phanerozoic** Eon and **Palaeozoic** Era, extending from the end of **Precambrian** times (542 million years ago) until the beginning of the **Ordovician** (488 million years ago), when most of the main animal groups first evolved in the sea.

Carapace
The hard protective body covering of animals such as some arthropods, and reptiles such as turtles.

Carboniferous Period
The period of geological time from the end of the **Devonian** (359 million years ago) to the beginning of **Permian** times (299 million years ago). In North America it is divided into the older Mississippian (359–318 million years ago) and younger Pennsylvanian (318–299 million years ago). Widespread tropical forests developed at this time, leaving extensive coal deposits and the remains of the first land-living **tetrapod** communities.

Carnivoran
A group of **laurasiatherian** mammals that evolved in **Paleogene** times, and is characterized by well-developed dagger-shaped canine teeth for killing prey and specialized slicing molars for eating flesh.

Carnivore
An animal that eats the flesh of other animals, but is not necessarily a carnivoran mammal.

Cenozoic Era
An interval of geological time that includes the **Paleogene** (65–23 million years ago) and **Neogene** (23 million years ago–present) periods, during which the flowering plants, insects, bony fish, birds, and placental mammals have all greatly diversified.

Character
An attribute that can be used to determine relationships between organisms.

Chert
A rock formed in seabed sediments by silica-rich solutions, which often encloses and fossilizes the remains of a variety of organisms from microbes to sponges.

Chondrichthyes
A group of primitive fish, originating in Late **Silurian** times, whose skeletal

material and dermal scales are made of a hard but non-mineralized and flexible organic material called cartilage – sometimes called "cartilagenous" fish.

Chordate
A major group of animals, originating in **Cambrian** times and including vertebrates, whose bodies are stiffened by a flexible dorsal rod called the **notochord**, which runs from head to tail and forms the basis for the **vertebrate** backbone.

Chromosome
A coiled strand of **DNA** that carries genetic information ("genes") for the manufacture of specific proteins in an organism's cells, and a variety of other functions.

Clade
An organism or group of organisms forming a **monophyletic** branch or group that contains an ancestor and all of its descendents.

Cladistics
A method and system of **phylogenetic** classification, based on **monophyletic** taxa or clades that are identified by the analysis of **characters** in an attempt to establish the evolutionary sequence of change.

Clubmoss (lycopsid)
A type of primitive plant that originated in Late **Silurian** times and grows upright from a horizontal root-like underground **rhizome**, with strap-shaped leaf-like structures in helical spirals along the stem and leaves with characteristic diamond-shaped scars. They reproduce by means of spores grouped in cone-like structures. Giant 30m (100ft) tree-sized forms were important members of the **Carboniferous** coal forests.

Conifer
A major group of living plants that originated in **Carboniferous** times and is characterized by woody tissues and scale-like leaves. Their seeds and pollen are borne in single-sex cones for reproduction, which typically develop on the same tree.

Convergent evolution
The phenomenon whereby different groups of organisms evolve similar structures in response to the same environmental pressures – for example the development of wings in pterosaurs, birds, and bats.

Cretaceous Period
The most recent period of the **Mesozoic Era** of geological time, which lasted between 145 and 65 million years ago and ended with a major extinction event.

Cryogenian period
The middle of three periods of geological time within the **Neoproterozoic Era**

(850–630 million years ago), marked by periods of widespread global glaciation.

Cycad
A kind of palm-like seedplant, originating in **Permian** times, with a massive stem terminated by a crown of fern-like leaves. They became abundant in **Mesozoic** times and occasionally grew to 20m (66ft) high.

Cynodont
An advanced mammal-like **tetrapod** with some mammal-like features, belonging to a group which originated in Late **Permian** times and includes the ancestors of the mammals.

Derived character
A novel **character** that first evolves in the common ancestor of a set of species and is unique to that particular group or **clade**.

Deuterostome
An organism in which the embryonic **blastopore** develops into the anus, with the mouth developing secondarily (compare **protostome**).

Devonian period
A period of geological time within the **Palaeozoic Era** from 416–359 million years ago. It is marked by the development of early land plants and the formation of the first forests, with tree-sized plants and the first land-going vertebrates.

Diapsid
A member of a major group of **amniote tetrapods** originating in Late **Carboniferous** times and characterized by a pair of openings in the skull immediately behind the eye socket. Diapsids include the **dinosaurs** and other extinct marine reptiles, along with the crocodiles, lizards, snakes, and birds.

Dinosaur
A member of an extinct and diverse group of land-living **diapsid** reptiles that flourished from Late **Triassic** times and died out, apart from the birds, in the End-**Cretaceous mass extinction**. They are characterized by the structure of the hip and pelvis and subdivided into the **ornithischians** and **saurischians**.

DNA
Deoxyribonucleic acid, the genetic material that determines the inherited characteristics of most living organisms. During cell division, DNA is duplicated so that each new daughter cell receives the same enetic DNA code.

Echinoderm
A group of marine **deuterostomes** that originated in Early **Cambrian** times, and which includes the starfish, sea urchins, sea lilies, sea cucumbers, and some extinct forms. They are characterized by five-fold body symmetry and a calcium carbonate skeleton.

Ecosystem
A natural unit made up of a community or communities of organisms and the environment in which they live and interact.

Ediacara fauna
A group of extinct soft-bodied marine organisms found globally within strata from the **Ediacaran Period**. Their nature is unclear, since few seem to have characteristics typical of the main groups of living **metazoans**. They are named after the Ediacaran Hills of the Flinders Range in South Australia, from where they were first described in the late 1940s.

Ediacaran Period
The latest geological period (630–542 million years ago) of the **Proterozoic Eon** of the **Precambrian**. Characterized by the presence of the soft-bodied **Ediacara fauna**.

Embryo
An animal in its earliest stage of development between fertilization and hatching or birth.

Embryology
The study of the formation and development of the **embryo**.

Eocene Epoch
One of three epochs (56–34 million years ago) in the **Paleogene Period** of the **Cenozoic Era**, characterized by a significant **radiation** of early **placental** mammals, **teleost** fish etc.

Erosion
The transport processes, typically involving wind, water, gravity, or ice, that remove rocky material that has already been broken down by **weathering**.

Euarchontoglires
A major group of **placental mammals**, identified from molecular data, which originated in Middle **Cretaceous** times and includes the primates, rabbits, and rodents.

Eukaryote
A cell or organism whose **chromosomes** are surrounded by a membrane, forming a discrete cellular nucleus.

Exoskeleton
The tough protective covering of an arthropod body. It is periodically shed (moulted) to allow the animals to grow. It is mineralized in some groups, such as the extinct trilobites and the extant crabs.

Extinction event
See **Mass extinction**

Fern
A member of a large group of primitive terrestrial plants that originated in Late Devonian times and is characterized by frond-shaped leaves with clusters of reproductive spore-bearing structures (sporangia) on their lower surfaces.

Gamete
A "haploid" cell, containing just one set of chromosomes from one sex, which fuses with another haploid cell from the opposite sex to form a "diploid zygote" with two sets of chromosomes – the key to sexual reproduction.

Gametophyte
The **gamete**-forming "haploid" phase, with just one set of chromosomes, in plant reproduction.

Genetics
The biology of heredity and variation.

Genus
A group of species with a close genetic relationship that is descended from a common ancestor. However, a genus may contain only one known species.

Ginkgo
A member of a group of plants with woody stems, fan-shaped leaves, and parallel veins. The group originated in Late **Triassic** times and diversified in later **Mesozoic** times, but has just one surviving species, the maidenhair tree *Gingko biloba*.

Glacial
A cold phase within an **ice age**, during which glaciers advance and sea levels fall.

Glossopterid
An extinct group of plants whose strap-shaped leaves are commonly found in the **Permian** and **Triassic** strata of the **Gondwanan** continents of India, Australia, South Africa, South America, and Antarctica. They are an important part of the fossil evidence for the early amalgamation of these continents in a single southern hemisphere **supercontinent**.

Gondwana
A **supercontinent** formed in the southern hemisphere by the amalgamation of many large crustal plates (India, Australia, South Africa, South America, and Antarctica) and numerous smaller crustal fragments, such as Florida. The pieces were first assembled in **Cambrian** times, and joined with **Laurasia** during the **Permian** to form the even bigger supercontinent of **Pangea**. It began to break up in **Jurassic** times, having lasted for some 300 million years.

Grass
A member of a large group of flowering plants (Graminae or Poaceae), whose **Cenozoic** evolution and **radiation** was closely linked to the evolution of grass-eating (grazing) mammals and ruminants (cud-chewers) such as the deer, cattle, and horses.

Greenhouse
A period of warm global climates, during which increased quantities of carbon dioxide and other atmospheric gases trap a greater proportion of the energy in sunlight than usual, causing global warming.

Gymnosperm
A member of a large group of seed-bearing plants whose "naked" seeds are not enclosed in ovaries. They originated in Late **Devonian** times, and include ferns, **conifers**, and extinct groups such as the bennettitaleans.

Hadean Eon
A geological period that lasted for around 700 million years from the catastrophe that formed the Moon about 4510 million years ago, to the end of the heavy asteroid bombardment (around 3850 million years ago).

Herbivore
An animal that feeds primarily on plant material and generally has specially adapted teeth and stomach for chewing and digesting plant tissues.

Holocene
The current epoch of geological time, which began 11,500 years ago at the end of the last **Pleistocene glacial**. It has been marked by global climate warming and rising sea levels.

Hominid
A member of a group of **primates** that originated some 13 million years ago and which includes gorillas, chimps, humans and their fossil ancestors. They are characterized by the bone structure of the wrist and hand.

Hominin
A member of a group of **primates** that originated some 7 million years ago, which includes humans, our fossil ancestors, and our nearest living relatives the chimps. They are characterized by molecular data and the bony structure of the palate.

Horsetail
A member of the sphenopsids, a group of primitive plants that originated in Late **Devonian** times and were significant members of the **Carboniferous** tropical coal forests, with tree-sized forms growing to 40m (130ft). They are characterized by segmented stems, with whorls of stiff bristle-like leaves that grow from horizontal root-like **rhizomes** and true roots. Their few surviving species are much smaller, but retain the primitive form and require damp conditions for growth and reproduction.

Ice age
A period of several million years of alternating warm and cool global climate, leading to the development of polar ice caps, ice-sheets, glaciers, changing sea levels, and shifting climate zones.

Icehouse
A phase of Earth history with cool climates and ice ages connected to periodic cycles of Earth's orbit, low levels of atmospheric **greenhouse** gases, and perhaps changes in ocean circulation patterns.

Igneous
A type of rock formed by the cooling and crystallization of molten rock material (magma).

Insectivore
In older classifications, a member of a group of small and mainly nocturnal mammals with a primary diet of insects. They included the living hedgehogs, shrews and their extinct ancestors dating from late

Cretaceous times and were characterized by their teeth, but molecular data shows that the group is **polyphyletic**. The name is still used to describe any insect-eating animal.

Interglacial
A warm climate phase between cold **glacials**, within an **ice age**.

Invertebrate
A traditional and convenient collective term used to lump together non-vertebrate **metazoan** animals. It has no **phylogenetic** meaning.

Jurassic Period
A **Mesozoic** period of geological time from 200–145 million years ago. It is named after the European Jura mountains, where rock strata of this age are well exposed.

Laurasia
A supercontinent formed in the northern hemisphere during **Ordovician** times by the joining of Laurentia (North America) and Eurasia (parts of Europe and Asia, excluding India).

Laurasiatheria
A large group of **placental** mammals with origins in Middle **Cretaceous** times, identified through molecular data.

Limestone
A **sedimentary** rock formed from carbonate minerals (especially calcium carbonate) in both marine and fresh waters. It often contains a high proportion of shell remains.

Mammal
A member of a group of warm-blooded and hairy **amniote** vertebrates that feed their young with milk from the mother's mammary glands. They originated in **Triassic** times, and are also characterized by features of the jaw and teeth.

Marsupial
A member of a group of primitive **mammals**, which give birth to immature live young that are nurtured in a skin pouch until they can fend for themselves. The group arose in Late **Jurassic** or Early **Cretaceous** times, and were once far more widespread than their present Australian and South American distribution.

Mass extinction
An event in which a significant percentage of the biota dies out over a relatively short time. Some five mass extinction events have been identified throughout the **Phanerozoic**.

Mesozoic
An era of geological time including the **Triassic** (251–199 million years ago), **Jurassic** (199–145 million years ago), and **Cretaceous** (145–65 million years ago) periods. It was characterized by the great evolutionary success of various reptile groups.

Metazoan
An animal whose body is made from many cells organised into discrete cell layers and tissues, coordinated by a nervous system.

Miocene Epoch
The first epoch of the **Neogene Period** of geological time, lasting from 23 million to 5 million years ago.

Molecular clock
A means of estimating the timing of evolutionary divergence, based on measures of the "genetic distance" between organisms and the rates of genetic change.

Mollusc
A large group of metazoans that is characterized by a mantle of tissue covering the gills and other body organs and whose members commonly secrete a protective carbonate shell. The molluscs originated in **Cambrian** times, and include the living snails, bivalves, squid, and octopuses, and extinct groups such as the ammonoids and belemnoids.

Monophyletic
A descriptive term for a group of species with a single common ancestor they do not share with any other group. Identification of monophyletic groups forms the basis of **cladistics**.

Monotreme
A group of primitive mammals, including the platypus and echidna, that originated in the **Cretaceous** and retained some reptile-like features such as laying eggs.

Multituberculate
A member of an extinct group of rodent-like primitive mammals, distinguished by their teeth, that originated in Middle **Jurassic** times and became extinct in the **Oligocene**.

Natural selection
A key evolutionary process whereby organisms best adapted to their environment are more likely to survive and raise offspring, and therefore increase in number over successive generations relative to less well adapted forms.

Neogene Period
The most recent period of geological time, beginning with the **Miocene Epoch** 23 million years ago, and lasting until the present **Holocene Epoch**.

Neoproterozoic Era
The last era of the **Proterozoic Eon**, which lasted from 1000–542 million years ago.

Niche
A space in the **ecosystem** occupied by a particular organism, defined by characteristics such as temperature tolerance, feeding habits and interaction with predators.

Notochord
A flexible dorsal stiffening rod within the **chordate** body, which runs from head to tail and forms the basis of the backbone in **vertebrates**.

Oligocene Epoch
The last epoch, extending from 40–23 million years ago, of the **Paleogene** Period.

Omnivore
An animal with a diet that includes both vegetable matter and meat.

Ordovician Period
A period of the **Palaeozoic Era**, lasting from 488–444 million years ago, with an **ice age** in its later stages.

Ornithischian
One of two subdivisions of the dinosaurs (the other being the **saurischians**) these plant eaters are characterized by the bird-like structure of the hip. They originated in Late **Triassic** times, and died out at the end of the **Cretaceous**.

Osteichthyes
A large, diverse, and successful group of bony fish that originated in Silurian times and is characterized by features such as the structure of the skull, jaws, and teeth.

Paleocene Epoch
The earliest epoch (65–56 million years ago) of the **Paleogene Period** and the **Cenozoic Era**, characterized by significant changes in terrestrial **biotas** following the end-**Cretaceous extinction event**.

Paleogene Period
The first period of the **Cenozoic** Era, lasting from 65–23 million years ago.

Paleozoic Era
An era of time that began 542 million years ago with the **Cambrian Period** and ended with the **Permian Period**, 251 million years ago. It is characterized by many extinct groups and the colonization of land by plants and animals.

Pangea
A supercontinent that extended from pole to pole and was formed by the joining of **Gondwana**, **Laurentia**, Baltica, and Siberia at the end of Triassic times.

Pelycosaur
A basal member of the group of **synapsid tetrapods** that originated in Late **Carboniferous** times and evolved some mammalian features, especially in the teeth and jaws. Some pelycosaurs had webbed sail structures rising from the backbone, which may have helped to regulate their temperature.

Permian Period
The last period of the **Palaeozoic Era**, from 299–251 million years ago, which ended in the greatest **mass extinction** in the history of life.

Phanerozoic Eon
An eon that began 542 million years ago with the **Cambrian Period** and continues to the present. It represents the main phase in the history and evolution of visible **metazoan** life.

Photosynthesis

A process that utilizes atmospheric carbon dioxide, water, and light energy from the sun trapped by chlorophyll, to form organic compounds and build living cellular matter. It is most familiar from plants, and produces oxygen as a by-product.

Phylogeny
The history of development of lineages of organisms through evolution. By extension, also a branching diagram showing the evolutionary relationships and descent of species sharing common ancestors.

Placental
A **mammal** possessing a placenta in the uterus – a spongy tissue that allows nutritional, respiratory, and waste materials to be exchanged between foetus and mother.

Plate tectonics
The geological process by which Earth's crustal plates are moved by convection currents, generated in the mantle. The plates may diverge to form new oceans, slide past one another along major faults, or converge to form island arcs and continental mountain ranges.

Pleistocene Epoch
An epoch of geological time within the **Neogene Period**, lasting from 1.8 million years ago to 11,500 years ago, and characterized by **ice ages** and the extinction of many large mammals.

Pliocene Epoch
An epoch, from 5.3–1.8 million years ago, within the **Neogene Period**.

Pollen
Microscopic spores and grains produced in huge numbers for reproduction by the male reproductive organs of flowering plants (**angiosperms**) and other seed-bearing (**gymnosperm**) plants.

Polyphyletic
A descriptive term for a group of species that cannot be united by a single and unique common ancestor they do not share with any other group. Removal of polyphyletic groups is a key technique in **cladistics**, since only **monophyletic** groups can be used to build an accurate **phylogeny**.

Precambrian
An immensely long period of Earth's history, from its formation around 4.56 billion years ago until the beginning of the **Phanerozoic Eon** 542 million years ago. It is formally divided into the **Archean** and **Proterozoic Eons**, with the lower limit of the former not defined. The **Hadean** is sometimes used to refer the earliest phase of Earth history, but is not a formally recognized term. The earliest indications of life date from around 3.6 billion years ago, but life remained microscopic until around 580 million years ago.

Primate
A group of **placental mammals**, originating in early **Paleogene** times, that includes monkeys, apes, humans, and their fossil

ancestors. For the most part they are arboreal (tree-dwelling), but some have become **secondarily** terrestrial. They are characterized by an opposable thumb and forward-facing eyes that give true binocular vision.

Primitive
A state or feature of an organism that is relatively close to the condition of the organism's ancestors compared with an **advanced** state or feature.

Prokaryote
A cell or organism whose DNA is not organized into **chromosomes** or enclosed by a nuclear membrane.

Proterozoic Eon
The younger of two long eons in the **Precambrian**, extending from 2500–542 million years ago.

Protostome
A **metazoan** animal in which the initial **blastopore** opening of the embryo forms the mouth or mouth and anus (compare **deuterostome**).

Quadruped
An animal that walks on all four legs.

Quaternary
An interval of geological time that overlaps with the definition of the **Pleistocene** and dates from around 2.6 million years ago to the present.

Radiation
In evolutionary terms, a period of rapid increase in the number and diversity of organisms within a certain group, often seen in groups that survive **mass extinctions**.

Reptile
In traditional classifications, a member of a class of **amniotes** characterized by possession of a tough horny skin with scales, scutes or plates, functional lungs, and a four-chambered heart. However, these characters are now taken to define a "grade of organization" that originated in **Carboniferous** times, rather than a true clade or group, since their descendants include the birds.

Rhizome
A horizontal plant stem usually found underground, which produces roots and shoots from its nodes.

Rodent
A member of a large and diverse group of small **herbivorous** or **omnivorous placental mammals** that originated in **Paleocene** times and is characterized by chisel-shaped front teeth that grow continuously.

Sarcopterygian
A group of lobe-finned fish that originated in **Devonian** time. Fish-like sarcopterygians are today only represented by lungfish and coelacanths. However, the **tetrapods** are also sarcopterygians: their paired muscular lobe-fins were adapted for supporting the body and evolved into the tetrapod limb.

Saurischian
A member of one of two groups of dinosaur, that originated in Late **Triassic** times and died out at the end of the **Cretaceous**, the other group being the **ornithischians**. Characterized by the possession of a lizard-like pelvis, this group diverged to produce plant-eating sauropodomorphs and meat-eating **theropods**.

Sauropod
A major group of herbivorous sauropodomorph saurischian dinosaurs that originated in Late **Triassic** times and were characterized by small heads, long necks and tails, massive bodies, and pillar-like legs. They became extinct at the end of the **Cretaceous**.

Secondary adaptation
An adaptation in which a primary adaptive feature, such as the use of all four **tetrapod** legs for walking, is further adapted, as in humans and other animals animals that have become **bipedal**. Secondary adaptations can often produce misleading features that lead to the mistaken classification of organisms.

Sedimentary
A rock or feature formed by sedimentation – typically the accumulation of small particles in surface layers that are later compressed by gravity, wind, water, or ice.

Seed fern
An extinct and important group of fossil **gymnosperms**, also known as pteridosperms, characterized by fern-like leaves on which seeds developed. They originated in Late **Devonian** times and died out in the Late **Cretaceous**.

Seed plant
A seed-bearing or **gymnosperm** plant and member of a large group that originated in Late **Devonian** times. They include the **seed ferns**, living **cycads**, **ginkgos** and **conifers**, and other extinct groups, such as the **glossopterids**.

Silurian
A period of the **Paleozoic** Era from 444–416 million years ago, characterized by the early development of land plants.

Snowball Earth
An extensive **ice age** in which glacial ice seems to have persisted in low latitudes and may have been global in extent. Two or three such events have been identified within **Precambrian** times.

Species
A closely related group of organisms whose genetic similarity allows them to interbreed and produce fertile young. Since this fundamental reproductive test cannot be applied to fossil species, they are generally defined on the basis of anatomical features.

Spore
A very small reproductive body produced by some plants, bacteria, and other unicellular organisms.

Strata
Layers of rock formed by sediment deposited at Earth's surface through the action of wind, water, or gravity. Successive layers deposited over time become buried and compressed, changing from loose sediment into hard **sedimentary** rock strata. Subsequent earth movements may dislocate and deform the originally horizontal layering to produce folds and faults.

Stratigraphy
The study of rock strata, their formation, history of deposition, relative age and correlation between different places that may have become separated by tectonic movements.

Supercontinent
A group of continental crust plates brought together by tectonic movements to form a bigger mass such as **Gondwana** or **Pangea**.

Synapsid
A member of a group of **amniotes** that originated in Late **Carboniferous** times and are characterized by a lower opening in the skull behind the eye. They included the extinct **pelycosaurs** and therapsids, and ultimately gave rise to the **mammals**.

Taxonomy
The study of the classification of life, its rules, principles, and practice.

Teleost
A member of a very large group of **actinopterygians** that originated in Late **Triassic** times and dominate the freshwater and marine fish faunas today. They are characterized by thin bony scales and the structure of the jaw and tail.

Tertiary
An old but still-used geological time period covering most of the **Cenozoic Era** from 65 million years ago until the beginning of the **Quaternary** around 2.6 million years ago.

Tethys Ocean
A largely tropical ocean that extended from east to west and separated Eurasia and **Gondwana** between **Permian** and **Miocene** times. It was largely closed by the northwards movement of the African and Indian continental plates.

Tetrapod
A member of a major group of **vertebrates** that originated in Late **Devonian** times and is characterized by the possession of paired muscular limbs with digits (the "tetrapod limb") derived from the paired fins of **sarcopterygians**. The tetrapod limb first appears in extinct water dwelling forms and was subsequently adapted for walking on land.

Theropod
A member of a major subdivision of the **saurischian dinosaurs**, which were bipedal, mostly meat-eating, and included the largest land-living predators such as tyrannosaurs. They originated in Late **Triassic** times and became extinct at the end of the **Cretaceous**.

Tillite
A rock formed from **glacial** deposits and left behind when the ice melts. Tillites are composed of an unsorted mixture of mud, silt and rock debris, and their surfaces are often scratched or "striated" as a result of transport by ice.

Trace fossil
A kind of fossil comprised of marks left by some behaviour of an organism, ranging from burrows in sediment to footprints and toothmarks.

Tree fern
A member of an ancient group of "pteridopsid" plants that originated in Early **Carboniferous** times. Some of them grew to the size of trees but few representatives survive.

Triassic
The first period of the **Mesozoic** Era, from 251–200 million years ago. Following the **mass extinction** at the end of the **Permian**, the Triassic saw a gradual recovery of life on land and in the seas, with the rise of important **reptile** groups such as the **dinosaurs**.

Trilobite
A member of an extinct but large group of marine **arthropods** that originated in Early **Cambrian** times and is characterized by a threefold lengthways division of the head, body, and tail into a central and two flanking lateral lobes. They became extinct in Late **Permian** times.

Ungulate
A hoofed, grazing **placental mammal**. Ungulates were traditionally thought to be part of a natural monophyletic group of similar animals, but are now known to have separate origins.

Vascular plant
Plants, including most land plants, with special cells for carrying water and nutrients from the soil through the root system and up the stem against the force of gravity. The cells in the stem are strengthened with lignin to allow it to grow upwards away from the ground and towards the sunlight. Vascular plants most probably originated in Late **Ordovician** times, but certainly by the Early **Silurian**.

Vendian
A term that has been used for a period of Late **Precambrian** time, now superseded by the formally recognized **Ediacaran Period** (630–542 million years ago).

Vertebrate
A member of a large group of **chordates** that originated in Cambrian times and is characterised by the replacement of the notochord with an internal jointed and articulated skeleton.

Weathering
A collective term for a range of chemical, thermal and mechanical processes that break down rocks into smaller fragments.

INDEX

ACKNOWLEDGEMENTS

Peter Barrett

In grateful acknowledgement and fond memory of my agent Virgil Pomfret whose unfailing support and encouragement played a vital part in the conception of this book and who sadly died in January 2009 before its completion.

Douglas Palmer

I would like to thank Peter Taylor, Giles Sparrow, Phil Gilderdale, Jenny Faithfull, and Dr Roger Benson of the University of Cambridge, whose substantial help and forbearance have allowed this book to appear. I also wish to thank my wife Tamsin, who has put up with it all.

The publishers would like to thank

Dr Charles Crumly, Dr Kevin Padian, Suzanne Arnold, Carole Ash, Yasia Williams, Colin Ziegler, Patrick Mulrey, Trudy Brannan, Jenny Lawson, Gary Almond, Geoff Borin, Jaspreet Bahra, Georgina Atsiaris, Jennifer Vladmirsky, George Philip Ltd, and Kenny Grant for their help with this project.

Picture credits

1 Getty Images / Kim Taylor
2 Getty Images / David Muir
8 Science Photo Library
10 (top left) Alamy / Kathy deWitt
10 (centre left) Alamy / Phil Degginger
10 (bottom) Frank Lane Picture Agency Ltd / Mitsuaki Iwago / Minden Pictures
11 (top left) Alamy / Phil Degginger
11 (top right Science Photo Library / Volker Steger
12 (top left) Science Photo Library / Sheila Terry
12 (centre left) Natural History Museum Picture Library
12 (bottom) Natural History Museum Picture Library
13 (top left) Alamy / Colin Underhill
13 (top right) Alamy / North Wind Picture Archives
16 (left) Science Photo Library / George Bernard
16 (right) Getty Images
18 (centre left) Getty Images
18 (top left) Science Photo Library / Science Source
19 (top right) Natural History Museum Picture Library
19 (centre right) Natural History Museum Picture Library
20 (centre) AKG
20 (bottom) Getty Images / Jim Ballard
20 (centre left) Getty Images / Time & Life Pictures
20 (top left) Science Photo Library / George Bernard
20 (bottom right) Science Photo Library / Jean-Loup Charmet
21 (top centre) Science Photo Library
21 (top right) Natural History Museum Picture Library
21 (top left) Natural History Museum Picture Library
22 (top left) Douglas Palmer
23 (centre right) Science Photo Library / James King-Holmes
23 (top left) Natural History Museum Picture Library
23 (top right) Science Photo Library James Holmes / Oxford centre for Molecular Sciences
24 (top right) Science Photo Library / Dr Mark J Winter
24 (centre left) Science Photo Library / Pasieka
24 (bottom Natural) History Museum Picture Library
25 (top left) H J Hofmann, McGill University
25 (left) Michael S Engel Image reproduced with permission from "Evolution of the Insects" (David A Grimaldi and Michael S Engel, Cambridge University Press, 2005)
26 (bottom) Alamy / Visual & Written SL
26 (top left) Science Photo Library / Sinclair Stammers
27 (top left) Alamy / blickwinkel
27 (bottom left) Science Photo Library David Parker
27 (centre right) Natural History Museum Picture Library
27 (top right) Tony Waltham Geophotos
28 (bottom) Getty Images The Bridgeman Art Library / Copper Age
28 (top left) Santiago Ramirez
29 photolibrary.com/OSF
29 (top left) Shutterstock / Ismael Monteru verdu
30 (bottom left) Science Photo Library / Sheila Terry
30 (bottom right) Douglas Palmer
30 (top left) Professor Eelco J Rohling & Jens Kallmayer
31 (top left) Alamy / Ace Stock Limited
31 (bottom) right Corbis UK Ltd / Charles Jean Marc/Corbis Sygma
31 (top right) University of California Museum of Palaeontology Florissant Fossil Beds National Monument, photo by Herb Meyer
32 (bottom) Science Photo Library / Philippe Plailly / Eurelios
32 (top left) Shutterstock / suravid
33 (top left) Science Photo Library / Eye of Science
33 (top right) Science Photo Library / Herge Congé, ISM
33 (bottom) Shutterstock
37 (left) Lochman Transparencies
37 (centre) Abigail Allwood
37 (right) Reg Morrison
39 (right) Martin Brasier
39 (left) J William Schopf
41 (left) Professor Michael J Hambrey
41 (right) Dr J G Gehling
43 (left) Douglas Palmer
43 (right) Reg Morrison
43 (centre) Guy Narbonne
45 (right) Douglas Palmer
45 (left) Ken McNamara
46 Wes Watters
47 (left) Wes Watters
45 (right) Stephen Bengtson
49 (centre) Derek Siveter
49 (right) Derek Siveter
49 (top left) Derek Siveter
49 (bottom left) Professor Degan Shu
50 Getty Images / Jeff Rotman
51 (left) photolibrary.com / Don Farrall
51 (top right) Dr J G Gehling
51 (bottom right) Science Photo Library / Martin Bond
54 (far top) Fotolia / Eric Gevaert
54 (top) Fotolia / javarman
54 (centre top) photolibrary.com / OSF / Richard Manuel
54 (centre) Reg Morrison
54 (centre bottom) Science Photo Library / Andrew J Martinez
54 (bottom) Fotolia / Bradford Lumley
54 (far bottom) Fotolia / BERA
55 Getty Images / Jeff Rotman
57 (centre right) Professor Simon Conway Morris
57 (left) Professor Simon Conway Morris
57 (centre left) Douglas Palmer
57 (right) Douglas Palmer / DEG Briggs
59 (centre) Humboldt State University Natural History Museum
59 (right) Museum of Comparative Zoology, Harvard University
61 (centre) Philippe Janvier
63 (centre) Professor Richard Aldridge
65 (top centre) left Hunterian Museum and Art Gallery
65 (top right) Hunterian Museum and Art Gallery
65 (bottom right) Hunterian Museum and Art Gallery
67 (top left) Derek Siveter
67 (right) Dr Alan Thomas
69 (left) Dianne Edwards
69 (centre right) Dianne Edwards
69 (centre left) Dianne Edwards
71 (left) Paul Selden / Paul Selden & John Nudds: Fossil Ecosystems published by Manson Publishing Ltd 2004
71 (right) Paul Selden / Paul Selden & John Nudds: Fossil Ecosystems published by Manson Publishing Ltd 2004
73 (left) Hunsruck Museum
73 (bottom right) Hunsruck Museum
73 (top right) SPM/GDKE-Erdgeschichte RLP
75 (right) Ken McNamara
75 (left) John Long
77 (left) Dr Brian Chatterton
77 (right) Dr Brian Chatterton
79 VIREO / Ted Daeschler
81 (top) American Museum of Natural History
81 (bottom) American Museum of Natural History
83 (right) Dr Jennifer A Clack
83 (left) Musee National de l'Histoire Naturelle, Paris
85 (left) ©The trustees of the National Museums of Scotland
85 (centre right) ©The trustees of the National Museums of Scotland
85 (right) ©The trustees of the National Museums of Scotland
87 (left) Dr Richard Lund
87 (top centre) Dr Richard Lund
87 (bottom centre) Dr Richard Lund
89 (left) Rich Paselk
89 (centre) Rich Paselk
91 (centre) Natural History Museum Picture Library
91 (left) Dr John Calder
91 (right) Dr John Calder
93 (right) Corbis UK Ltd
93 (left) Dr John Calder
95 (left) Dr Boris Ekrt
95 (centre left) Dr Boris Ekrt
95 (centre right) Dr Boris Ekrt
97 (right) Professor Michael J Hambrey
97 (bottom left) Roger Smith
99 (right) Alamy / Natural Visions
99 (left) Michael S Engel Courtesy of Michael S Engel
101 (bottom right) Thomas Martens
101 (top left) David Berman
103 (left) Science Photo Library
103 (right) Mary E White / Jim Frazier
103 (centre) Mary E White / Jim Frazier
105 (left) Roger Smith
105 (top right) Roger Smith
105 (bottom right) Roger Smith
107 (right) Giles Sparrow
109 (bottom left) Shutterstock / Serge Zastavkin
111 (left) Roger Smith
113 (centre) Juan Carlos Cisneros
113 (right) Juan Carlos Cisneros
115 (left) Jean-Claude Gall
115 (centre) Jean-Claude Gall
115 (right Jean-Claude Gall
117 (left) Heinz Furrer
117 (right) Heinz Furrer
119 (left) Terry Jones
119 (centre) Terry Jones
121 (top left) Paul Sereno/Project Exploration
121 (right) Paul Sereno/Project Exploration
123 (top right) Natural History Museum Picture Library
123 (bottom left) Elgin Museum
123 (centre) Elgin Museum
125 (left) Professor Randall Irmis
125 (right) Professor Randall Irmis
126 Ardea / Francois Gohier
127 (left) Natural History Museum Picture Library
129 (top left) Natural History Museum Picture Library
129 (bottom left) Natural History Museum PictureLibrary
129 (right) Natural History Museum Picture Library
131 (centre left) Urwelt-Museum-Hauff

TIMELINE

THE GEOLOGICAL AND EVOLUTIONARY TIMELINE SHOWN ACROSS THESE FOUR PAGES
TELLS THE STORY OF LIFE ON OUR PLANET FROM ITS ORIGINS SOME 4.6 BILLION YEARS AGO,
THROUGH SUCCESSIVE PERIODS OF COMPLEX LIFE, TO THE RECENT ARRIVAL OF HUMANITY.

EDIACARANS p.42–5
The earliest complex animals
preserved in the fossil record are
the mysterious Ediacarans, which
may include ancestral jellyfish
alongside entirely extinct forms
of life.

SNOWBALL EARTH p.40–1
Around 640 million years ago,
dramatic climate change saw the
Earth go through a series of worldwide
glaciations that left their traces
in ancient rocks.

CAMBRIAN EXPLOSION p.48–9
A sudden increase in the variety and
complexity of life around 520MA is
recorded at Chengjiang, China.

EARLIEST LIFE p.36–9
Fossilized stromatolites from Western
Australia preserve the first signs of life
on Earth, revealing the presence of
cyanobacteria some 3.4 billion
years ago.

600   590   580   570   560   550   540   530   520   510

CAMBRIAN

NEOPROTEROZOIC ◄ PRECAMBRIAN

PALEOPROTEROZOIC

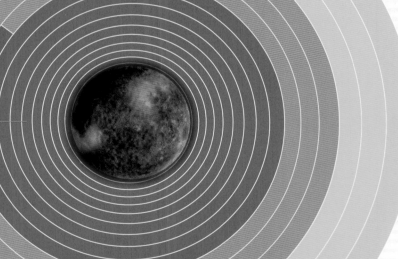

FIRST PREDATORS p.46–7
The first traces of predatory
animals appear in the very early
Cambrian "small shelly fauna" at
Nama Group, Namibia.

CHANGING GLOBE

Earth's crust is split into a number
of tectonic plates that are driven
around the planet with geological
slowness by currents from within.
Fresh crust is created in regions
where plates move apart, opening
rifts that grow into new seas and
oceans, while older crust is
destroyed where plates collide,
often driven down into the Earth's

ARCHEAN

MESOPROTEROZOIC

TEMPERATURE AND SEA LEVEL
The scales below show variations in Earth's average
temperature over geological time, and changes in sea
level compared to the present day.

─ 565MA   ─ 515MA

Warmer than present   Cooler than present

200m
150m
100m
50m
0m
-50m

BURGESS SHALE p.56–7
Canada's famous fossil deposits reveal a later stage in the Cambrian Explosion, and include the earliest known ancestors of many modern animal groups.

FLOURISHING FISH p.64–5
Silurian deposits such as those deposited at Lesmahagow in Scotland, some 430 million years ago, reveal that the jawless agnathan fish survived and diversified after the End-Ordovician extinction, along with a great many invertebrates.

RHYNIE CHERT p.70–1
This Scottish site, dating to 408 MA, is one of the best preserved early land ecosystems – a peat bog where a variety of vascular plants and small land arthropods flourished.

LOBE-FINNED FISH p.76–7
The rocks of Gogo, Western Australian preserve a wide range of both ray-finned and lobe-finned fish.

| 500 | 490 | 480 | 470 | 460 | 450 | 440 | 430 | 420 | 410 | 400MA | 390 | 380 | 370 | 360 | 350 | 340 |

ORDOVICIAN   SILURIAN   DEVONIAN   CARB

PALEOZOIC

ORDOVICIAN INVERTEBRATES p.58–9
The Ordovician deposits of Trenton, USA show the abundance and diversity of invertebrate life in ancient seas some 460 million years ago.

FIRST LAND PLANTS
The first plants began to move onto land towards the end of the Ordovician, around 450 MA or earlier. Initially, they were restricted to small mossy bryophytes that did not spread far from the water.

EARLY JAWED FISH p.72–3
Spiny acanthodias from the Hunsruck Slate in Germany were some of the earliest fish to develop a functioning jaw, around 407 MA.

LAND TETRAP
Ichthyostega (pi
Acanthostega, t
known tetrapo
developed limb
in 366-million-y
from East Gree

interior where the heat from its melting triggers volcanoes at the surface. When thick landmasses collide head on, the land itself buckles and folds to form new mountain ranges. These geological movements can be reconstructed with some certainty back into Precambrian times, as chronicled by the sequence of maps along the bottom of these pages.

FIRST FISH p.60–1
The first jawless "agnathan" fish emerged in the mid-to-late Ordovician Period. Some of the best preserved are from 455 million-year-old deposits at Cochabamba, Bolivia

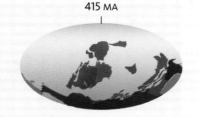

VASCULAR PLANTS p.68–9
Deposits from Ludford Lane on the Welsh Borders, around 419 MA, record the first known vascular plant, *Cooksonia*, with upright stems but no leaves.

THE 'FISHAPOD' p.78–9
The remarkable *Tiktaalik*, discoverd in Canadian Arctic rocks, is a lobe-finned fish that shows arm-like modifications to its fins – a vital step on the road to the later tetrapods.

STETHACANTH
Rare fossils of ca
preserved at Be
USA give a glim
early shark with
dorsal structure

455 MA   415 MA   375 MA

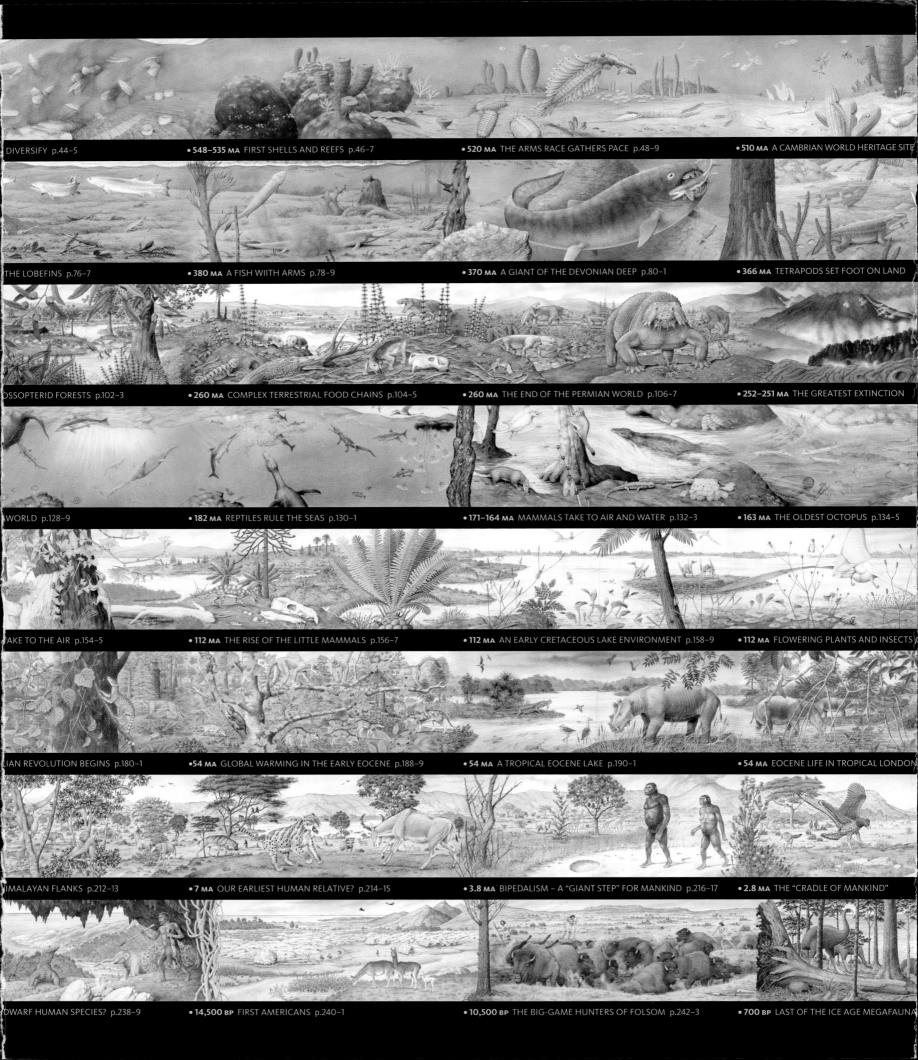

EAST KIRKTON p.84–5
Dating to around 328 MA, the fossils of this Scotttish site preserve a diversity of early tetrapods that lived in and around a freshwater lake.

GRES A VOLTZIA p.114–5
Unique preservational conditions in this ancient river delta, dating to 242 MA, have fossilized a variety of Triassic arthropods, including the first funnel spider, *Rosamygale*.

SAILBACKED SYNAPSIDS p.98–9
The Permian saw the synapsid amniotes develop from reptile-like forms such as the sail-backed *Dimetrodon* towards more mammal-like forms.

FIRST REPTILES p.92–5
The earliest known reptiles, including *Hylonomus*, are found in the fossilized forests of Joggins, Nova Scotia.

DIVERSIFYING MAMMALS p.132–3
The Daohuguo fossils from Inner Mongolia, China, date to around 168 MA and show surprising diversity among the small early mammals that lived alongside the dinosaurs.

FIRST DINOSAURS p.120–1
The earliest and least specialized dinosaurs are found in fossils from Argentina's Valley of the Moon.

330   320   310   300MA   290   280   270   260   250   240   230   220   210   200MA   190   180   170

ONIFEROUS   PERMIAN   TRIASSIC   JURASSIC

ME

...ODS p.82–3
...tured) and ...e earliest ...s with fully ..., are both found ...ear-old rocks ...land.

LATE CARBONIFEROUS ICE AGE p.96–7
The end of the Carboniferous Period around 299 MA was marked by an extensive ice age, the evidence of which is found is displayed in South Africa's Karoo Basin.

END-PERMIAN EXTINCTION
Around 250 MA, widespread volcanism in Siberia triggered a climate crisis and the largest mass extinction in the history of the world.

FIRST FLIERS p.118–9
The gliding diapsid reptiles *Sharovipteryx* and *Longisquama*, from the Middle Triassic Fergana Valley of Kyrgyzstan, show early steps towards the tetrapod conquest of the air.

GIANT DINOSA...
Fossils from Tendaguru... formed in coastal swa... 152 MA, include som... giant sauropo...

...US p.86–7
...rtilaginous fish ...r Gulch, Montana, ...se of this strange ...its brushlike

SWAMP FORESTS p.88–9
Throughout the Carboniferous, dense forests of giant clubmosses, seed ferns and horsetails spread around the world, as seen in the 314-million-year-old fossils of Mazon Creek, Illinois, USA.

MARINE REPTILES p.116–7
By the Middle Triassic, a number of reptile groups had returned to the seas where they occupied specialized niches as marine predators. An entire ecosystem from around 233 MA is preserved at Monte San Giorgio on the Italian-Swiss border.

LYME REGIS p.128–9
The famous fossils of England's Jurassic Coast, dating to around 195 MA, preserve a flourishing marine environment filled with diversifying life.

320 MA

255 MA

220 MA

185 MA

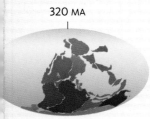

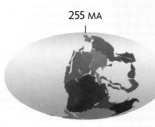

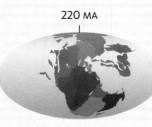

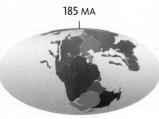

BISHOPS p.162–3
This small mammal from Australia's Dinosaur Cove site is though to be an early relative of the present-day monotremes.

MAMMAL EXPLOSION p.180–1
In the aftermath of the Cretaceous-Tertiary Extinction, mammals rapidly diversified and expanded to fill vacant ecological niches, as seen in the 62-million-year-old fossils from Crazy Mountain in Montana, USA.

FEATHERED DINOSAURS p.150–5
Fossils from Liaoning, China reveal that several groups of theropod dinosaurs had developed feathers by the Cretaceous.

CRETACEOUS-TERTIARY EXTINCTION
65.5 million years ago the entire dinosaur lineage (except for the birds), was driven to extinction, along with many other species, by a climate crisis linked to the impact of a 10km (6-mile) asteroid in what is now the Gulf of Mexico.

EARLY BIRD p.140–1
Archaeopteryx, the first known bird, lived near a lagoon at Solnhofen, Germany, around 151 MA.

CRETACEOUS PARK p.174–5
Canada's Judith River deposits, dating to around 74 MA, preserve the remains of many well known dinosaurs.

DAWN OF MAN p.220–1
Stone tools from Olduvai, Tanzania, coincide with fossils of *Homo habilis*, the first member of the genus *Homo*.

50   150   140   130   120   110   100MA   90   80   70   60   50   40   30   20   10   PRESENT

CRETACEOUS   PALEOGENE   NEOGENE

SOZOIC   CENOZOIC

FLOWERING PLANTS p.158–61
Brazil's Crato Formation, dating to 112 MA, preserves some of the earliest evidence for the evolutionary interrelationship between flowering plants and insects.

MESSEL p.196–7
The oil shales from Messel in Germany reveal a spectacular fossilized environment including diversifying mammals, birds and insects.

FIRST HOMINID p.214–5
The 7-million-year-old *Sahelanthropus tchadensis*, discovered in Chad's Djurab desert, lies close to the evolutionary divergence between humans and apes.

URS p.136–7
n Tanzania,
nps around
of the first
dinosaurs.

PATAGONIAN GIANTS p.166–9
South American fossils from around 98 MA reveal plant-eating sauropods and carnivorous theropods that grew to enormous size after the continent became isolated.

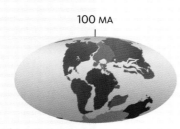

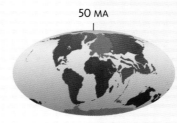

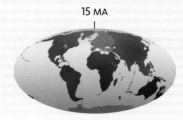

MORRISON FORMATION p.144–7
Mid-Jurassic fossils from the midwestern USA include some of the best known dinosaurs, including *Apatosaurus*, *Allosaurus*, and *Stegosaurus*.

NESTING DINOSAURS p.170–3
At Ukhaa Tolgod in Mongolia, 80-million-year-old fossils include the nesting *Oviraptor* (pictured), herbivorous *Protoceratops*, and the predatory *Velociraptor*.

WALKING ON TWO LEGS p.216–7
The famous "Laetoli footprints" from Tanzania vividly capture the development of upright walking hominids around 3.8 million years ago.

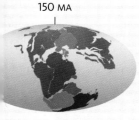

150 MA   100 MA   50 MA   15 MA

PANORAMIC VIEW

FLOWED TOGETHER, THE ARTWORKS FEATURED THROUGHOUT THIS BOOK CREATE A
SWEEPING PANORAMA OF LIFE'S HISTORY, FROM THE FIRST SINGLE-CELLED ORGANISMS
TO THE PRESENT DAY.

• **3430 MA** IN THE BEGINNING p.36–7

• **430 MA** SEA SCORPIONS AND JAWLESS FISH p.64–5   • **425 MA** REEF LIFE RECOVERS p.66–7   • **419 MA** GREENING THE LAND p.68–9

• **314–313 MA** REPTILE BEGINNINGS p.90–1   • **314–313 MA** WILDFIRE IN THE CARBONIFEROUS p.92–3   • **310 MA** TETRAPODS DIVERSIFY p.94–5

• **233 MA** MARINE REPTILES DIVERSIFY p.116–17   • **228 MA** INSECTS DIVERSIFY WITH NEW PLANTS p.118–19   • **227 MA** DINOSAUR BEGINNINGS p.120–1

• **151 MA** LIFE AND DEATH IN A JURASSIC LAGOON p.142–3   • **155–148 MA** THE REAL JURASSIC PARK p.144–5   • **155–148 MA** THE MORRISON ECOSYSTEM p.146–7

• **84–80 MA** BRINGING UP BABY – A DINOSAUR NURSERY p.168–9   • **80 MA** LATE CRETACEOUS LIFE IN MONGOLIA p.170–1   • **80 MA** DINOSAUR PARENTING SKILLS p.172–3

• **45 MA** OUR EARLIEST ANTHROPOID RELATIVES p.200–1   • **34 MA** PRIMATES PROLIFERATE IN EOCENE EGYPT p.202–3   • **34 MA** FLOWERS AND INSECTS EVOLVE TOGETHER p.204–5

• **75,000 BP** THE FIRST HUMAN ART p.226–7   • **50,000 BP** NEANDERTHAL LIFE p.228–9   • **40,000 BP** FIRST AUSTRALIANS p.230–1